Ecological Engineering:
Concepts and Applications

Ecological Engineering: Concepts and Applications

Edited by Colby Whedon

SYRAWOOD
PUBLISHING HOUSE

New York

Published by Syrawood Publishing House,
750 Third Avenue, 9th Floor,
New York, NY 10017, USA
www.syrawoodpublishinghouse.com

Ecological Engineering: Concepts and Applications
Edited by Colby Whedon

© 2022 Syrawood Publishing House

International Standard Book Number: 978-1-64740-132-0 (Hardback)

Cataloging-in-Publication Data

Ecological engineering : concepts and applications / edited by Colby Whedon.
    p. cm.
Includes bibliographical references and index.
ISBN 978-1-64740-132-0
1. Ecological engineering. 2. Environmental management. 3. Ecology. I. Whedon, Colby.
GE350 .E26 2022
628--dc23

TABLE OF CONTENTS

PREFACE

This book has been a concerted effort by a group of academicians, researchers and scientists, who have contributed their research works for the realization of the book. This book has materialized in the wake of emerging advancements and innovations in this field. Therefore, the need of the hour was to compile all the required researches and disseminate the knowledge to a broad spectrum of people comprising of students, researchers and specialists of the field.

The use of ecology and engineering to predict, design, construct or restore, and manage ecosystems is known as ecological engineering. It is aimed at integrating human society with its natural environment. The applications in ecological engineering can be categorized into 3 spatial scales: mesocosms, ecosystems and regional systems. Mesocosms range from a single centimeter to hundreds of meters, ecosystems range from a single kilometer to ten kilometers, and regional systems are those systems which span over ten kilometers. There is an increase in the complexity of the design usually observed with an increase in the spatial scale. Applications of ecological engineering are focused on the creation or restoration of ecosystems such as wetlands and greenhouses. From theories to research to practical applications, case studies related to all contemporary topics of relevance to the field of ecological engineering have been included in this book. The detailed analyses and data will prove immensely beneficial to professionals and students involved in this area at various levels.

At the end of the preface, I would like to thank the authors for their brilliant chapters and the publisher for guiding us all-through the making of the book till its final stage. Also, I would like to thank my family for providing the support and encouragement throughout my academic career and research projects.

Editor

Impacts of Climate Change on Hydrological Regime and Water Resources Management of the Narew River in Poland

Łukasz Malinowski[1*], Iwona Skoczko[1]

[1] Bialystok University of Technology, Faculty of Civil and Environmental Engineering, Department of Technology and Environmental Engineering Systems, Wiejska 45A, 15-351 Białystok, Poland

[*] Corresponding author's e-mail: lukasz.malinowski66@wp.pl

ABSTRACT

The amount of water required to support a river ecosystem in proper condition are of particular importance in the areas of high natural value. The hydrological threats for the protected areas are region-specific and vary from region to region. The local hydrological conditions depend largely on the temporal and spatial variations of the hydrologic cycle, of the main components and physiographic conditions on site. Future climate change is projected to have a significant impact on the hydrological regime, water resources and their quality in many parts of the world. The water-dependent ecosystems are exposed to the risk of climate change through altered precipitation and evaporation. Investigating the current climate changes and their hydrological consequences are very important for hydrological issues. This analysis may be a very important foundation for determining the causes observed in the recent period of anomalous growth – both hydrological and climatic. The aim of the research is to assess the effect of projected climate change on water resources in lowland catchment the Narew River in Poland. The hydrological reaction to climate warming and wetter conditions includes changes in flow and water level. This paper describes the directions of changes climatic and hydrological conditions and the impact of climate change on the Narew River. The data such as: daily air temperature, precipitation obtained from the Bialystok climate station located within the Narew river and hydrological data such as water flows and water states observed in water gauges were used for the analysis of climate variability and their hydrological consequences. The results show a significant decrease in winter outflows in river, as well as a delayed increase in the spring melt flow. It has also been observed that this is the initial phase of changes in maximum water levels and maximum flows.

Keywords: Narew River, water resources, ecosystem, hydrological conditions, global warming

INTRODUCTION

A climate change exerts both direct and indirect impacts on various sectors of the economy and society by affecting the biological and physical components of ecosystems, such as air, water, soil, and biodiversity. On a global scale, the climate change is becoming more and more visible, and the process is constantly aggravating. It should be assumed that in the future they will be more visible and severely noticeable for society and the economy [Piniewski, 2014]. Global warming is unquestionable, primarily due to the observation of an increase in the average global air temperature, ocean temperature, disintegration of the floating ice cover (ice melting) and the rise of the global mean sea level. Increasing the air temperature near the Earth's surface is a common phenomenon on a global scale, although its intensity indicates spatial diversity. Increasing the global air temperature is conducive to the increase in the frequency and intensity of occurrence of many climate-related phenomena, which include extreme weather events, such as heat waves, rainstorms, storms, hail, tornadoes and sandstorms [Sadowski et al., 2013].

TEMPERATURE INCREASE

Temperature is a key indicator characterizing the climate change. The last century was characterized by the variability of air temperature [Oschlies et al., 2017]. It was observed that the temperature at the Earth's surface increased by 0.74°C and continues to rise. Long-term observations and measurements indicate a faster increase in warming in the land areas of both hemispheres than the oceans. According to the data from the Intergovernmental Panel on Climate Change (IPCC), the temperature rise in the last two decades was twice as high above the lands than over the ocean and amounted to 0.27°C and 0.13°C, respectively [IPCC, 2007]. The second half of the 20th century and the first decade of the 21st century turned out to be a particularly warm period. The most visible changes were observed at high altitudes in spring and winter [Sadowski et al., 2013, Majewski, Walczykiewicz, 2012].

The general trend of temperature changes is conducive to the increase in the frequency of extreme precipitation, as well as expansion of the areas where drought or desertification occurs. Long-term changes in the amount of atmospheric precipitation, despite their spatial and temporal diversity, have been noted in many large areas. In the period from 1900 until 2005, the amount of precipitation increased in the northern regions of Europe, in the eastern parts of North and South America and also in the northern and central Asia. In turn, a significant decrease in the occurrence of rainfall is observed in the Mediterranean basin, in the Sahel, in the southern part of Asia and in southern Africa [Sadowski et al., 2013].

The incidence and intensity of certain extreme weather events has changed over the past 50 years, and the main changes that can be distinguished are primarily: occasional occurrence of cold nights and days as well as frosts on most land areas in medium latitudes, hot days and hot nights are more frequent. The occurrence of heat on most land areas, volatile rainfall and increase of the share of volatile rainfall is observed as well. Since 1975, a noticeable occurrence of extremely high sea level has been observed in many areas [Sadowski et al., 2013, Arcipowska, Kassenberg, 2007].

The most noticeable consequence of climate change is the global warming, resulting mainly from the human activity (combustion of fossil fuels to obtain energy, continuously increasing emission of exhaust fumes from cars, aircraft and ships, deforestation). Greenhouse gases have a total transmission of solar shortwave radiation in the range of 0.15–4.0 nm, which heats the Earth, and

also retains long-wave thermal radiation emitted from its surface. By partially reflecting the heat towards the surface of our planet, it warms it up and the rest of the radiation escapes into space. The increased content of greenhouse gases causes that the majority of long-wave radiation is directed back to the Earth's surface, and in this way are retained in the "trap" while acting in the same way as a greenhouse [James et al., 2017].

Agriculture also contributes to and experiences the effects of climate change. All agricultural crops require appropriate soils, water, heat and sunlight so that they can grow optimally. The increase in air temperature has already affected the length of the growing season in a large area in Europe. Both flowering and harvesting cereal crops occur a few days earlier and it is expected that these changes will progress in many regions. The land designated for agricultural crops is at risk. It may be dislodged or disappear completely, leading to economic problems in countries, including lack of food. The existing extreme weather phenomena have a direct or indirect impact on a significant increase in the risk of failed harvests, also on the soil, causing a decrease in the organic matter content, which is the main factor ensuring its fertility. The direct impact is mainly a change in the atmospheric conditions for the productivity of crops, including sums of atmospheric precipitation, changes in thermal conditions, frequency and intensity of extreme phenomena. The 2004 European Environment Agency report states that compared to the 1990s, a two-fold increase in the number of climate disasters occurred, and the value of losses caused by them exceeded in 2005 the amount of up to 200 billion dollars, while at the turn of the 20th and 21st century alone, they reached several billion dollars a year [Trzpil, 2008, Tubiello et al., 2007, Olsen et al., 2011].

PRECIPITATION

Precipitation is an inseparable element of the weather and climate, which is extremely important for the economy and everyday life of people. Understanding the current state and changes in the size and structure of precipitation in the scale of the whole globe, the country as well as in the regional aspect is particularly important. Precipitation, due to the lack of continuity in time and space, is very difficult to estimate [Wibig, Jakusik, 2012]. The air pollution also affects the composition of precipitation, which reacts with compounds introduced directly into the atmosphere. This type of rainfall is called acid rain. Very often, in some areas, the precipitation changes into sulfuric or nitric acid. The most severe effects of their impact are the contamination of water reservoirs (especially drinking water), the negative impact of vegetation. [Majewski, Walczykiewicz, 2012, Ziemniański, Ośródka, 2012, Szczygieł, 2008]. Precipitation increases the river flows and contributing to the occurrence of floods, which bring about many economic, economic and social losses. By draining agricultural areas with salt water, these areas become useless. The island of Tegua, which is part of the Vanuatu state, located in the South Pacific, experienced the effects of global warming to a great degree. In 2007, the inhabitants of this island were forced to leave their homes as a result of the flood and flooding of the island and were the first to be recognized by the United Nations as victims of global warming [IPCC, 2014].

In addition to flooding, there are other problems related to the operation and use of water. Heavy atmospheric precipitation is a consequence of the epidemic of diseases transported by water as a result of pathogens becoming active or by strong water pollution as a result of flooding sewers. In turn, reduced summer water flows can cause bacteriological and chemical risks. Higher water temperatures cause more frequent blooms of harmful algae [COM, 2009].

WATER CONDITION

Water is the basic resource of the environment; therefore, managing it according to specific rules is particularly important. The physical parameters of surface water, such as temperature, smell, taste, turbidity, transparency and color, mainly depend on the origin of water, season, climatic zone, and sewage discharged to it. The right temperature of water has a huge impact on the life of biocenosis and the course of biological, physical and chemical processes [Okruszko, Kijańska, 2009]. The water temperature is thus the predominant parameter combining climatic changes with the course of phenomena and reactions taking place in the water environment. All changes in the meteorological conditions have a direct impact on the formation of thermal conditions in all water bodies, and these in turn cause

changes in the course and processes occurring in them [Miller, Frydel, 2014]. Climate changes may also affect the quality status of surface waters designated for providing people with drinking water, and even prevent the supply of healthy water, of the appropriate quality required by the regulations of the European Union and the Ministry of Health. The global warming and desertification have a major impact on the water cycle in nature. These changes relate to the amount of precipitation, water levels in rivers and lakes, snow cover supplying processes and soil moisture [Stagl et al., 2014]. They may necessitate the modification of the processes applied in water treatment technology. Limited access to water is one of the most dangerous consequences of climate change [Majewski et al., 2009].

Climate changes directly affect the intensification of anthropogenic pressures. The development of civilization caused that river valleys, lakes and wetlands are subjected to considerable abiotic and biotic loads. The most common factors that cause pressure and deterioration of water ecosystems include: water retention, river regulation, navigation, water abstraction, water metering, drainage, hydropower, tourism and recreation, point and area pollution, management of the catchment including flood plains and other activities related to the progress of civilization [Sadowski et al., 2013].

Increasing the temperature on the globe is common, especially higher in the high latitudes of the northern hemisphere. Over the last century, the average temperature in the Arctic has increased at least twice as much as on a global scale. The land areas become warm faster than the oceans. The IPCC Report presents the results of observations conducted since 1961, which show that at least to the depth of 3000 m, the average temperature of the all-weather has risen, and that the ocean has stored over 80% of the heat that has been transferred to the climate system [Ziemiański, Ośródka, 2012]. New data analysis and satellite measurements of the central and lower temperature of the troposphere indicate warming of the same order as on the Earth's surface. In detailed hydrological systems, a greater outflow and earlier maximum spring flow of rivers with glacial and snow supply is observed, as well as an increase in rivers and lakes temperature in many areas, which has a significant impact on the thermal structure and water quality [IPCC, 2007, Okruszko, Kijańska, 2009].

RESEARCH AREA

The research area is the Narew River flowing through the north-eastern part of Poland and its selected tributaries. The river is one of the largest right-bank tributaries of the Vistula. It originates from the vast area of Dziki Swamp at the watershed of Narewka, Narew and Jesiołdy (tributary of Prypiat) located on the Belarusian part of the Białowieża Forest. Below the boggy area, it flows in a slightly meandering, almost natural riverbed, but on a stretch of several kilometers before the reservoir, Siemianówka takes on the character of a multi-crystal river (anastomosing). The length of the river is 484 km, the majority of which (448 km) is in Poland, and the rest in Belarus [Malinowski, 2016]. After entering the territory of Poland, the river spreads into Lake Siemianowskie; then it flows from the dam in Bondary to Lake Zegrzyński, where it connects with the river Bug. After another 22 km, in the town of Nowy Dwór Mazowiecki, Narew flows into the Vistula. As a typical lowland river, it forms vast areas of swamps, marshes and peat bogs. In the section from Suraż to Rzędziany, it belongs to the Narew National Park where it is protected. Throughout the stretch, the river flows from the north-east towards the south-west. The topography of the valley is not uniform, and the outlines of individual terraces are not clear and continuous, the edges have been blurred and distorted in the processes of erosion of side waters of tributaries. The riverbed is delicately indented, regular, with only highly meandering sections, and the longitudinal slope of the river is 0.125 ‰ [Orłow-Gozdowska, 2005].

The Narew River, along its entire length up to the Zegrze Reservoir, is unregulated, with a natural flow, forming many branches, bends and oxbow lakes. This makes the banks of the river very difficult to access. Active branches and oxbow lakes are very often reached from the main Narew riverbed. The river is characterized by a snow-supply type of regime with one clear maximum and one minimum during the year [Marcinkowski, 2017]. The largest water resources of the Narew River correspond to the early spring period, as a result of snow supply, giving in the final result the maximum values of the average monthly water levels up to the retention period, i.e. late summer and autumn. Narew is a typical lowland river, characterized by the highest flows in the period from March to April, while the lowest ones from June to October [Dembek, Okruszko, 1996]. The

climatic conditions within the Narew River and its tributaries were characterized on the basis of average annual air temperatures and mean annual precipitation heights measured at the synoptic station Białystok. The hydrological conditions and their variability were determined on the basis of long-term daily measurements of the flow rate and water status in water gauge profiles: Bondary, Narew and Strękowa Góra (Figure 1).

METHODS

In this paper, the daily data of air temperature, atmospheric precipitation for the Białystok climate station located within the Narew river and hydrological data such as flows and water levels observed in the Narew, Bondary and Strękowa Góra water level profiles were used to analyse the climate variability and their hydrological consequences. The climate data, i.e. the sum of rainfall, average air temperature and hydrological data: water levels and flows from the same research period 1981–2014 were subjected to the statistical analysis. The trend method was used to determine the tendency of variability of climate parameters over many years and to assess the shaping of flows and water levels in the catchment area under consideration. The non-parametric Mann-Kendall statistical test was used for the study, which consists in verification of the hypothesis about the lack of trend based on the nonparametric correlation coefficient. The Mann-Kendall test analysis shows the difference between successively measured values. The newly measured value is compared to all previously measured values. The verification of this hypothesis is carried out according to the Mann-Kendall statistic defined by the formula (1):

$$S = \sum_{k=1}^{n-1} \sum_{j=k+1}^{n} \text{sgn}(x_j - x_k) \qquad (1)$$

where:

$$\text{sgn}(x) = \begin{cases} +1 \text{ for } x>0 \\ 0 \text{ for } x=0 \\ -1 \text{ for } x<0 \end{cases}$$

$\{x1, x2, ..., xn,\}$ – set in form of a time series

In order to determine the correlation between two variables, a correlation coefficient was determined. The nonparametric equivalent of the cor-

Fig. 1. The location measurement and observation stations in the study area

relation coefficient used in the Mann-Kendall test is the rank correlation coefficient in the data string and the sequence of corresponding time moments, also referred to as the Kendall tau coefficient. The values of the coefficient around zero testify to the hypothesis that there is no trend, negative values support the occurrence of a declining trend, while large positive values of the coefficient indicate the occurrence of an increasing trend. The level of significance (p-value) has been determined for each test performed. In this study, a change in the significance level above 94% (above 0.06) was assumed to be statistically significant (increasing or decreasing). The level of significance in the range of 90–94% (0.06–0.10) indicates a near statistical significance, while only the tendency of changes was assumed to be the level of significance in the range of 75–90% (0.10–0.25). The significance level below 75% (below 0.25) was considered irrelevant and without a specific direction of change.

Analysis of time variability of climate characteristics

The analysis of temporal variability of climate characteristics, such as annual precipitation (P) and mean air temperature (T) in the year was made on the basis of statistical analysis of chronological sequences from the Białystok climate station from the measurement period 1981–2014. Demonstration of trends in changes in climate characteristics will determine the hydrological consequences of the current climate change.

As far as the statistical analysis is concerned, it can be concluded that the analysed precipitation sum from the 1981–2014 measurement period shows a growing significantly statistical trend of this characteristic on the significance level of 0.045 (p-value). The highest values were recorded in 2009 and 2010, where the average annual precipitation fluctuated in the order of 700 and 850 mm, respectively. Currently, the growing trend of changes in precipitation sums from year to year is observed. Changes in atmospheric precipitation in the period 1981–2014 are shown in Figure 2.

In the average air temperature measured at the Bialystok climate station in the period 1981–2014, a growing, statistically significant trend – at the significance level of 0.188 (p-value) – was observed. The course of the average annual temperature in the years 1981–2014 indicates its large fluctuations from year to year (Figure 3). The year 2010 was the warmest, where the average temperature was 8.65°C, and 1987 was the coldest, with the average temperature of 5.14°C.

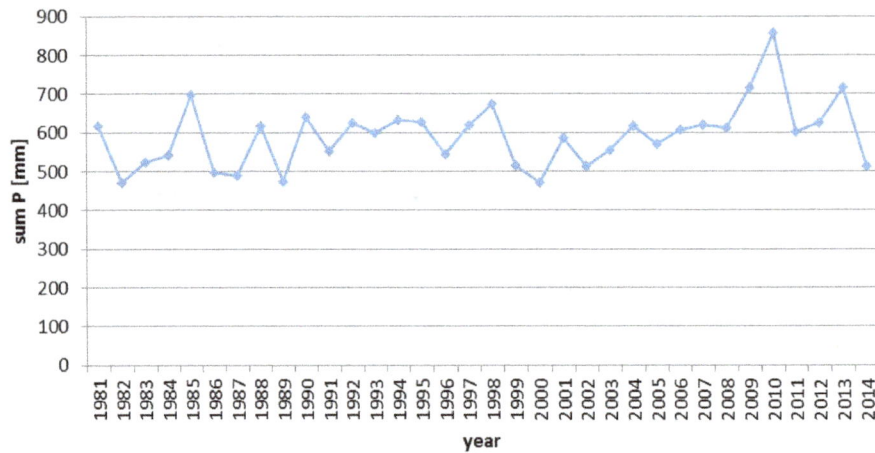

Fig. 2. The average total annual precipitation recorded at the climate station in Bialystok in the period 1981–2014

Analysis of time variability of hydrological characteristics

The Narew River is unregulated, having a natural flow character, forming many branches, bends and oxbow lakes. Its water resources and their variability along the river characterize its hydrological conditions. Therefore, the variability of maximum flows and water levels along the course of the river was analyzed on the example of three water gauge profiles – Bondary, Narew and Strękowa Góra.

The analysis of time variability of hydrological characteristics was performed on the basis of the statistical analysis of chronological sequences in individual water gauges. The variability of surface waters in the above-mentioned water level profiles was analyzed: maximum water levels (H) and variability of maximum annual flows (Q) in the period 1981–2014.

The statistical analysis of maximum water levels in the Bondary water level profile showed a decreasing trend of statistically significant changes (Fig. 3) at the significance level of 0.046 (p-value). The lowest water level was recorded in 1992, 2003 and 2004 and reached a level of about 197 cm. The highest water levels were observed in 1988 and 2010, where the water level reached about 270 cm. The reason for this was the high level of precipitation in these years. In 1988, the amount of precipitation reached an average of over 600 mm/year, while in 2010 – about 850 mm/year. Similar changes were observed in the Narew and Strękowa Góra water level profiles. The declining trend of changes in water levels is statistically significant at the significance level of 0.046 and 0.031 (p-value) (Fig. 4).

The statistical analysis of maximum flows from the 1981–2014 measurement period measured in the Bondary, Narew and Strękowa Góra

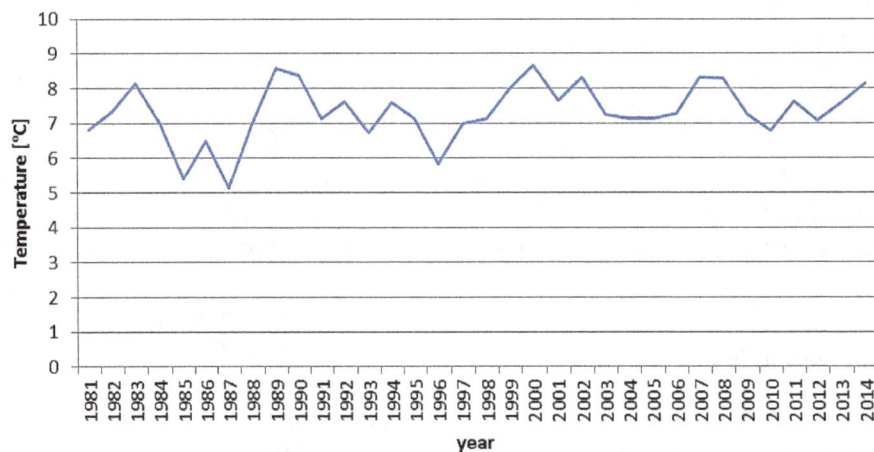

Fig. 3. The course of the average air temperature measured at the Białystok climate station in the period 1981–2014

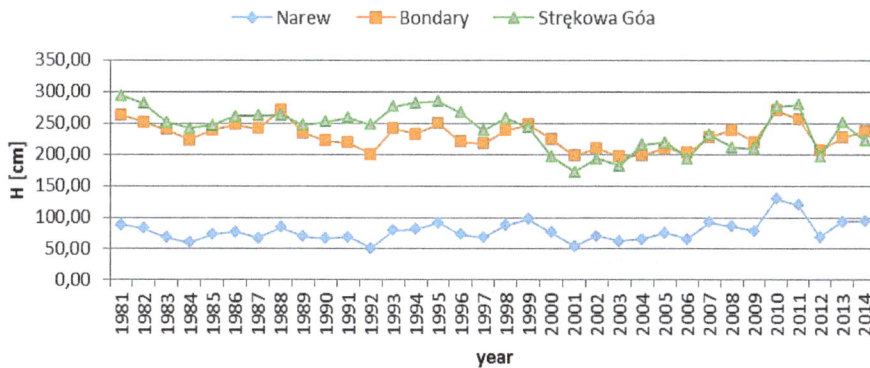

Fig. 4. Maximum annual water levels recorded in the water gauge profile: Bondary, Narew and Strękowa Góra, Narew river in the period 1981–2014

water gauge showed only descending trends statistically insignificant (Figure 5). The lowest flows were observed in the years in which low average annual amounts of precipitation were observed, such as 1992, 2003–2004. The highest flows in each water level profile were observed in 2010–2011, which were dependent primarily on the amount of atmospheric precipitation recorded in these years and the temperature course of the air during this period.

CONCLUSIONS

In recent years, the climate of Poland has changed in comparison with the climate information established on the basis of data from previous research periods –a warming has clearly occurred. The importance of climatic conditions in the modification of hydrological conditions has spatial variation due to the regional variability and en-

vironmental conditions of individual catchments. Certain confirmation of climatic conditions, according to Wrzesinski [2014], can be observed in many rivers, a significant decrease in winter outflows and a delayed increase in the spring thaw in the 1950s and 1960s. Similar works were carried out by Pociask-Karteczka [2003], thus confirming the tendency of changes in water level and flows in rivers, obtained as a result of statistical analysis. In turn, Romanowicz et al. [2014] stated that the frequency changes related to the climate are very complex and also depend on the generating mechanisms.

The statistical analysis of data (especially climatic data) allows to conclude that climate warming is noticeable. The analysis of hydrological parameters such as maximum water states and flows illustrates these changes to a lesser extent. However, it is still worth paying attention to. It can be said that this constitutes the initial phase of changes in the reduction of maximum water

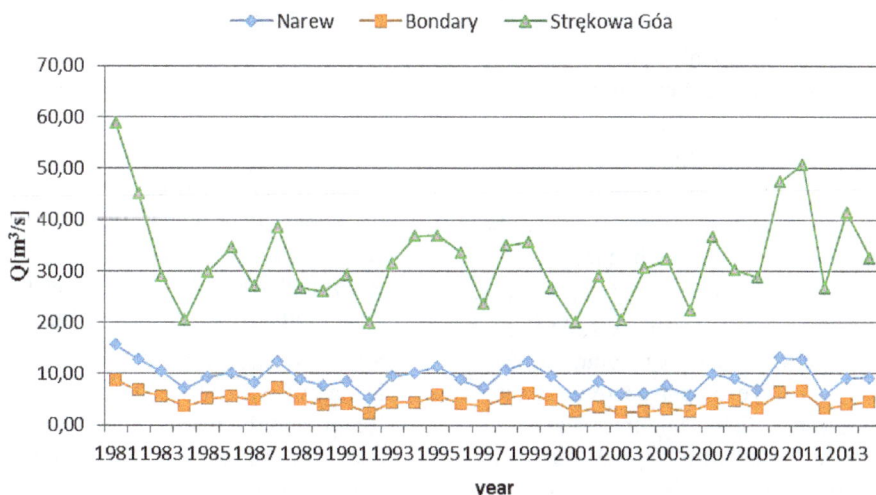

Fig. 5. Maximum annual water flows recorded in the water gauge profile: Bondary, Narew and Strękowa Góra, Narew river in the period 1981–2014

levels and maximum flows. The study of changes in climatic conditions, i.e. observed and forecast increase in air temperature, may affect the physical, chemical and biological processes in watercourses and water reservoirs, and may lead to deterioration of water quality. The negative impact of climate change mainly includes an increase in the frequency and intensity of extreme events such as heavy rainfall, strong wind, storms, glaze, and mists.

REFERENCES

1. Arcipowska A., Kassenberg A. 2007. Small ABC ... climate protection. Institute for Sustainable Development (in Polish). Warsaw.

2. COM. 2009. Commission of the European Communities (2009)147.White paper. Adapting to climate change: Towards a European framework for action (in Polish).

3. Dembek W., Okruszko H., 1996. Economic and social issues concerning the Upper Narew Valley (in Polish). Advances of Agricultural Sciences Problem Issues. 428, 7-13.

4. IPCC 2007. Pachauri R.K. et al.. 2007. IPCC 2007: Climate change 2007. Synthetic report. Contribution of Working Groups I, II and III to the Fourth Assessment Report of the Intergovernmental Panel on Climate Change. Publisher Institute of Environmental Protection, Warsaw.

5. IPCC 2014: Climate Change 2014: Mitigation of Climate Change. Contribution of Working Group III to the Fifth Assessment Report of the Intergovernmental Panel on Climate Change [Edenhofer, O., R. Pichs-Madruga, Y. Sokona, E. Farahani, S. Kadner, K.Seyboth, A. Adler, I. Baum, S. Brunner, P. Eickemeier, B. Kriemann, J. Savolainen, S. Schlömer, C. von Stechow, T. Zwickel and J.C. Minx (Eds.)]. Cambridge University Press, Cambridge, United Kingdom and New York, NY, USA.

6. James, R., Washington, R., Schleussner, C.-F., Rogelj, J. and Conway, D. 2017. Characterizing half-a-degree difference: a review of methods for identifying regional climate responses to global warming targets. WIREs Clim Change, 8.

7. Majewski W., Stepnowski R., Wita A. 2009. Determination of the current state of the Vistula and its basin and expected changes in the medium (5 years) and long term (15–20 years) perspective – stage 1 (9.1), subtasking area: Vistula bottom, project CLIMATE, task. 9. (in Polish). IMGW-PIB, Warszawa.

8. Majewski W., Walczykiewicz T. 2012. Sustainable management of water resources and hydrotechnical infrastructure in the light of forecasted climate changes (in Polish). Institute of Meteorology and Water Management – National Research Institute. Warsaw.

9. Malinowski Ł. 2016. Master's thesis: The impact of climate change on the Narew River and its tributaries (in Polish). Białystok.

10. Marcinkowski P., Piniewski M., Kardel I., Szcześniak M., Benestad R., Srinivasan R., Ignar S., Okruszko T. 2017. Effect of Climate Change on Hydrology, Sediment and Nutrient Losses in Two Lowland Catchments in Poland. Water, 9(3).

11. Miler A.T., Frydel K. 2014. Changes in groundwater levels against the background of climatic changes in the Kaliska forest inspectorate (in Polish). Infrastructure and Ecology of Rural Areas, 11/3, 743-755.

12. Okruszko T., Kijańska M. 2009. Climate change and water management. The Institute for Sustainable Development. Warsaw.

13. Olesen J.E. et al. 2011. Impacts and adaptation of European crop production systems to climate change. European Journal of Agronomy, 34, 96-112.

14. Orłow-Gozdowska Ł. 2005. Program for protection and development of water resources in the Mazowieckie Voivodeship in the field of river restoration for bi-environmental fish. Department of Agriculture and Modernization of Rural Areas of the Marshal's Office of the Mazowieckie Voivodeship (in Polish). Warsaw.

15. Oschlies, A., Held H., Keller D.,Keller K.,Mengis N.,Quaas M.,Rickels W.,Schmidt H. 2017, Indicators and metrics for the assessment of climate engineering, Earth's Future, 5, 49-58,

16. Piniewski M, Laizé CL, Acreman MC, Okruszko T, Schneider C. 2014. Effect of climate change on environmental flow indicators in the Narew Basin, Poland. Journal of Environmental Quality Environ, 43(1), 155-67.

17. Pociask-Karteczka J., Limanówka D., Nieckarz Z. 2003. Impact of North Atlantic oscillation on the flows of the Carpathian rivers (1951–2000) (in Polish). Folia Geographica, ser. Geographica Physica, 33–34, 89-104.

18. Romanowicz R.J., Nachlik E., Januchta-Szostak A., Starkel L., Kundzewicz Z.W., Byczkowski A., Kowalczak P., Żelaziński J., Radczuk L., Kowalik P., Szamałek K. 2014. Threats related to excess water (in Polish). Nauka 1, 123-148.

19. Sadowski M., Romańczak A., Dynakowska M., Kalinowska A., Siwiec E. 2013. Sixth Government Report and the first biennial report for the Conference of the Parties to the United Nations Framework Convention on Climate Change. Ministry of the Environment (in Polish). Warsaw.

20. Szczygieł L. 2008. Stopping climate change – ne-

cessity or expensive frights? (in Polish). Energetyka, 12, 799-810.

21. Stagl J., Mayr E., Koch H., Hattermann F.F., Huang S. 2014. Effects of Climate Change on the Hydrological Cycle in Central and Eastern Europe. In: Rannow S., Neubert M. (eds) Managing Protected Areas in Central and Eastern Europe Under Climate Change. Advances in Global Change Research, 58.

22. Trzpil M. 2008. Climate change in the modern world as an element of national security (in Polish). Bezpieczeństwo narodowe. Warszawa, nr.7/8.

23. Tubiello, F.N., J.-F. Soussana, S.M. Howden, and W. Easterling. 2007. Crop and pasture response to climate change. Proc. Natl. Acad. Sci., 104, 19686-19690

24. Wibig J., Jakusik E. 2012. Climatic and oceanographic conditions in Poland and the Southern Baltic – expected changes and guidelines for the development of adaptation strategies in the national economy (in Polish). Institute of Meteorology and Water Management – National Research Institute. Warszawa

25. Wrzesiński, D. 2014. Uncertainty of the system of the outflow of rivers in Poland (in Polish). [In:] Monografie Komitetu Gospodarki Wodnej PAN, z. XX, 191-200.

26. Ziemiański M., Ośródka L. 2012. Climate change and monitoring and forecasting the condition of the atmospheric environment (in Polish). Institute of Meteorology and Water Management – National Research Institute. Warsaw.

2

A Preliminary Study into the Possibility of $\delta^{13}C$ Being used as a Sensitive Indicator of the Trophic and Hydrobiological Status of Aquatic Ecosystems

Lilianna Bartoszek[1*], Piotr Koszelnik[1], Justyna Zamorska[2],
Renata Gruca-Rokosz[1], Monika Zdeb[2]

[1] Department of Environmental and Chemistry Engineering, Rzeszów University of Technology, al. Powstańców Warszawy 12, 35-959 Rzeszów, Poland
[2] Department of Water Purification and Protection, Rzeszów University of Technology, al. Powstańców Warszawy 12, 35-959 Rzeszów, Poland
[*] Corresponding author's e-mail: bartom@prz.edu.pl

ABSTRACT

There is a need to search for additional indicators allowing for more accurate identification of both the trophic status of waters as well as its chemical and biological consequences. The work detailed in this paper involved a preliminary analysis pertaining to the possibility of using an isotopic index in association with the values for trophic and saprobic indicators in describing a dam reservoir experiencing a far-reaching eutrophication. The water samples for the physicochemical analysis were collected from three sites along the axis of the dam reservoir in Rzeszów three times during the spring and summer of 2013. The results sustained the classification of the Reservoir's waters as hypertrophic, irrespective of the particular zone sampled. While phytoplankton blooms characterised by reference to the numbers of organisms per unit volume of water were also similar throughout the Reservoir, diversification in terms of taxonomic composition was to be noted, given the occurrence of cyanobacteria among the dominant diatoms in the area close to the dam. This presence was accompanied by enrichment of the Reservoir's suspended organic matter with carbon of the heavier ^{13}C isotope. On this basis, the $\delta^{13}C$ isotopic index can be regarded as a potentially useful indicator allowing for more accurate identification of both the level and the nature of the ongoing trophic degradation in bodies of water.

Keywords: reservoir, trophic status, Carlson indexes, saprobic index, isotopic index

INTRODUCTION

Verification of the trophic status of ecosystems is carried out in line with two groups of criteria, i.e. concentration or index. Where concentration criteria are concerned, the trophic state is determined by reference to the identified concentration ranges of nitrogen and phosphorus compounds, or else chlorophyll *a* [Forsberg & Ryding 1980, Vollenweider & Kerekes 1982, Nürnberg 2001]. The integrated indicators known as TSI (Trophic State Index) [Carlson 1977, Walker 1979] and ITS (Index of Trophical State) [Neverova-Dziopak et al. 2011] were used as well. In these cases, the developers drew on many years of observation to propose simple models by which the advancement of eutrophication could be estimated.

In both cases, the lack of possibility for more-detailed interpretation, especially the distinguishing between various higher trophic states constitutes the limitation. Advanced eutrophication is accompanied by intense blooms of cyanobacteria and other planktonic algae. The classes of algae present depend on many factors, such as temperature, hydrology or solar radiation [Grabowska et al. 2003]. Hence, the proportions of classes present may differ, notwithstanding generally similar values for generalising parameters like chlorophyll *a* [Ostrowska 2013]. Nevertheless, the occurrence or non-occurrence of certain classes of algae relates closely to the subsequent impact on water quality. Reservoirs, especially those suffering degradation, frequently play host to the cyanobacteria capable of producing such varied toxins as microcystins, nodularin, anatoxins, saxitoxins, etc. These are released into the aquatic environment following the death and breakdown of cells [Chorus & Bartram 1999, Boopathi & Ki 2014], often causing a far greater deterioration in water quality than is indicated by the analysis of trophic level.

A helpful supplement to the trophic indicators in the assessed degradation of bodies of water may be the saprobic system, until recently used to assess and classify the purity of surface waters. The saprobic system is based on the assumption that aquatic organisms differ in their tolerances to organic pollutants, to the extent that each species present or absent may assume some indicator value [Lampert & Sommer 2001, Gorzel & Kornijów 2004]. The system has thus been used to assess the degree of contamination of water with the organic matter capable of causing biochemical decomposition.

The practical difficulty in using the saprobic system lies in the absolute necessity that every aquatic organism found in a sample should be identified, which isan extremely difficult, labour-intensive and expensive activity, given the large number of systematic groups to be considered [Klimaszyk & Trawiński 2007, Czerniejewski & Czerniawski 2008].

Thus, there is a need to look for and hopefully identify additional indicators allowing for more accurate identification of both the trophic status, as well as its chemical and biological consequences. In this regard, the analysis of stable isotopes is a tool which is increasingly often used in assessing the quality of changes in the aquatic environment and their dynamics, among others of carbon (C), expressed as the ratio of heavier to lighter content relative to the standard ($\delta^{13}C$) in individual components of the environment [Koszelnik 2009]. This value depends on the original carbon source, as well as the environmental carbon cycle, including the metabolism engaged in by indicator organisms [Gu et al. 1999]. Thus, for example, Lehmann et al. [2004] report $\delta^{13}C$ enrichment in particulate organic carbon (POC) from Lake Lugano during summer production.

The aim of the work detailed here has been the successful running of a preliminary analysis regarding the possibility of isotopic indices being usefully set against values for trophic and saprobic indicators in a dam reservoir undergoing far-reaching eutrophication.

RESEARCH AREA AND METHODOLOGY

The researched Rzeszów Reservoir came into existence in 1974, thanks to the damming of the River Wisłok at a point 63+760 km along its course. The Reservoir is supplied by two main tributaries, i.e. the Wisłok and the Strug, and its main purpose has been to allow for proper operating of the water supply to the city of Rzeszów. However, given the location on the outskirts of such a large city, a vital role as a sports and recreation lagoon has also been served [Bartoszek et al. 2015]. The morphometric parameters of the Reservoir in 2014 are as shown in Figure 1. The overall volume is seen to have decreased by 0.7 million m^3 over the 40-year period, with major silting having taken place, and gradual development of new land surface in the upper zone in particular. The attempts to improve the reservoir's utility were made in 1986–87 and 1995–97, entailing the work to deepen the part next to the dam, while also achieving a narrowing through partial backfill on the right part of the bank also just by the dam. In each case, approximately 250,000–300,000 m^3 of sediment were dredged [Bartoszek et al. 2015].

The Rzeszów Reservoir has a 2025 km^2 watershed in foothill areas that are largely agricultural, though the upper parts are forested, while the middle part also has industrial centres (glassworks, tanneries and refineries). The reservoir can thus be regarded as under strong anthropogenic pressure associated with local agriculture that causes severe erosion of the land. Wastes of various kinds are also discharged in the area, while other kinds of diffuse pollution also occur [Koszelnik 2007, Gruca-Rokosz et al. 2009].

The water samples were taken from three sites along the axis of the reservoir (Fig. 1), three times during the spring-summer period of 2013. Temperature (T_w), pH and dissolved oxygen (OS) were measured *in situ* with a Hach Lange HQ40D meter. Total organic carbon (TOC) and total nitrogen (TN) was determined using a TOC-VCPN analyzer (Shimadzu), phosphate phosphorus (P-PO$_4^{3-}$) and chlorophyll *a* spectrophotometrically (Aquamate, Thermo Spectronic) using filtered samples of water following reaction with

Figure 1. Locations of the sampling points in the Rzeszów reservoir and its morphometric parameters

ammonium molybdate and hot extraction with ethanol, respectively. Total phosphorus (TP) was determined analogously, but in non-filtered samples of water that were mineralized (with H_2SO_4 and peroxodisulfate). The trophic status of the water was approximated by reference to Carlson indices (TSI_{TP} and TSI_{Chla}) [Carlson 1977].

Total suspension (TS) and $\delta^{13}C$ in POC were analysed following filtration using Whatman glass microfibre filters. The filters were dried at 50°C and – prior to the analysis for stable isotopes – exposed to fuming concentrated HCl for 72 h in an exicator, in order to remove carbonates. The filters were analysed using a DELTAPlus isotopic ratio mass spectrometer (Finnigan Mat, Germany) coupled with an elemental analyser (Flask 1112, ThermoQuest, Italy). The $\delta^{13}C$ values were expressed per mil (‰), and set against PDB standards, as follows:

$$\delta^{13}C = (R_{sample}/R_{standard} - 1) \cdot 1000 \qquad (1)$$

where: R denotes $^{13}C{:}^{12}C$.

Calibration was achieved using the NBS22 standard for $\delta^{13}C$. The standard deviations associated with the isotopic analyses were lesser than 0.1‰ (n = 10).

The water for phytoplankton analysis was collected using a suitable sampler in July 2013. The quantitative and qualitative analyses of phytoplankton were carried out using a reversed microscope and cylindrical sedimentation chambers (Utermöll method) [Picińska-Fałtynowicz & Błachuta 2012]. The hydrobiological classification was based on saprobic zones, which correspond to specific numerical values of the saprobic index (IS). IS phytoplankton was determined using the Pantle and Buck method (1955), based on the assessment of the group of observed indicator species [Czerniejewski & Czerniawski 2008].

RESULTS AND DISCUSSION

Water quality and trophic status

The results of the physicochemical analyses of the waters studied are as presented in Table 1. Only a slight variability in the ranges of all indicators was observed, the calculated coefficient of variance (CVP) each time falling below 8%.

The reservoir water was well-oxygenated during the period of study. The values for oxygen saturation (OS) with respect to Tw indicated levels are close to full saturation and even supersaturation. The measured pH of water was in the range 8.18–8.44. Such slight alkalisation of water may be presumed to reflect the excessive development of phytoplankton [Bartoszek et al. 2017]. The average concentration of $P\text{-}PO_4^{3-}$ was about 0.03 mg · dm^{-3}, while the value for TP was about 0.15 mg ·dm^{-3}. Slight spatial differentiation is to be observed, with an downward trend for the contents of forms of phosphorus in Reservoir water along the axis from the inflow (St. 3) in the direction of the dam (St. 1). However, no significant spatial

Table 1. Selected physicochemical parameters of water in the Rzeszów reservoir

Site	Scope	Tw	pH	OS	P-PO$_4^{3-}$	TP	TN	TOC	TS	Chl a
		°C	-	%	mg·dm^{-3}					µg·dm^{-3}
St. 1	Minimum	16.2	8.18	91.7	0.005	0.127	0.77	5.75	12.5	13.6
	Maximum	26.1	8.42	106.6	0.060	0.141	1.79	11.3	25.0	53.1
	Average	20.9	-	99.8	0.023	0.136	1.35	8.24	20.0	32.9
St. 2	Minimum	16.3	8.31	84.7	0.010	0.145	0.82	6.11	25.0	18.5
	Maximum	26.1	8.44	113.2	0.053	0.166	1.68	9.25	40.0	46.9
	Average	20.8	-	99.0	0.026	0.156	1.32	8.19	33.3	30.0
St. 3	Minimum	16.3	8.30	87.4	0.016	0.147	0.88	7.06	20.0	21.0
	Maximum	26.1	8.43	111.5	0.108	0.176	1.64	9.86	32.5	71.6
	Average	20.7	-	99.8	0.049	0.163	1.33	8.72	25.0	54.3

differentiation was observed in the case of either TN, the concentrations of which varied from 0.77 to 1.79 mg · dm^{-3} or TOC (5.75- 11.3 mg · dm^{-3}). The highest content of suspended matter (TS) was each time present in the transitional zone of the reservoir (St. 2), the average value being 33.3 mg · dm^{-3}. Marked differentiation was observed in the case of Chl a, with the back zone (St. 3) reporting the highest average value of 54.3 µg · dm^{-3}, while the transition zone (St. 2) had the lowest (30.0 µg · dm^{-3}).

The afore-mentioned recorded values, and especially TP concentrations above 0.1 mg · dm^{-3}, result in the classification of the waters under study as hypertrophic [Vollenweider & Kerekes 1982]. This status is also confirmed by the values of the Carlson index, TSI$_{TP}$, which exceed 70 (Table 2). Equally, in line with the TSI$_{Chla}$ index, the obtained values in the range 50–70 are indicative of eutrophic waters.

Previous publications have confirmed the above-mentioned interpretation arrived at on the basis of a short period of study. Evaluations of the trophic status on the basis of concentration criteria or trophic indexes have all been indicative of advanced eutrophication in Rzeszów Reservoir [Gruca-Rokosz 2013, Bartoszek et al. 2018]. Indeed, according to Gruca-Rokosz [2013], the average annual concentrations of nitrogen, phosphorus and chlorophyll a in the Reservoir's wa-

ters over the 2009–2011 period are sufficient to qualify it as hypertrophic. Only the concentration of Chl a at St.3 site was such as to point merely to eutrophy. In turn, the values for Carlson trophic indices determined for Rzeszów Reservoir have indicated hypertrophy in the case of the phosphoric version (TSI$_{TP}$), and eutrophy in the case of the version based on chlorophyll (TSI$_{Chla}$). An assessment made by Bartoszek et al. [2018], based on the phosphoric and chlorophyll indices, also showed hypertrophic and eutrophic status of the water (respectively).

The above-mentioned results indicate that a high content of total phosphorus in the reservoir waters was not accompanied by a commensurately high content of chlorophyll a. Furthermore, the average values of the N:P ratio did not exceed 10 at any of the sites, suggesting that a relative lack of nitrogen might inhibit the progress of the internal production process in this Reservoir. Some of the nutrients were used by luxuriant emergent vegetation in the littoral zone. In fact, competition for phytoplankton here may be provided by water chestnut (*Trapa natans*) – a plant enjoying legal protection in Poland, which covers a large part of the Reservoir surface, most especially near the dam [Kukuła & Bylak 2017].

Isotopic composition of the suspension

The values for δ^{13}C determined in the POC taken from the sites studied were progressively higher along the reservoir, at: -28.4 ‰ (St. 3), -27.9 ‰ (St. 2) and -27.3 ‰ (St. 1) (Table 2). POC in the sediment collected next to the dam (St. 1) was most enriched in the heavier ^{13}C isotope. Such enrichment is to be observed in the context of algal-bloom conditions in summer [Lehmann et al. 2004]. Differences obtained for δ^{13}C may also suggest a slightly different kind of

Table 2. Average values of Carlson's trophic indexes (TSI), N:P ratio for water and δ^{13}C for suspension of the Rzeszów reservoir

Site	N:P	TSI$_{TP}$	TSI$_{Chla}$	δ^{13}C
	-	-	-	‰
St. 1	9.8	75	63	-27.3
St. 2	8.4	77	63	-27.9
St. 3	8.2	78	68	-28.4

primary production processes along the reservoir, i.e. in its riverine and lacustrine zones.

Biological parameters

Table 3 presents the results of hydrobiological tests of water. The results obtained for the saprobic index values are indicative of the β-mezosaprobic zone, i.e. water hardly contaminated by organic matter. Generally, no differences in the IS values were found between the selected water-intake points, which may suggest that – given the short retention time (of 0.8 d) – the flow of water through the reservoir is not associated with any decrease in the level of pollution. Nevertheless, the largest numbers of producers and smallest numbers of reducers were reported from a near-dam zone (Site 1), where the producer/reducer ratio equal to 50.3 contrasted with the values for the remaining sites of 8.3 and 7.5. A greater share of producers in biocoenosis occurs as a result of self-purification processes that should entail a reduction in saprobility [Lampert & Sommer 2001].

Diatoms were the dominant algae at all research sites (Fig. 2). These were mainly species from Cyclotella, Navicula and Nitzschia genera. In the majority of dam reservoirs, cyanobacteria and green algae dominate in the summer period [Ostrowska 2013]. Diatoms are more abundant in waters subject to constant mixing, there from mainly in flowing waters. In the analysed case, the dominance of diatoms may in part reflect the high flow characterising the water, as well as its turbidity, as confirmed by the TS values up to 40 mg · dm^{-3}. Diatoms require less light than other photosynthesising organisms, and this is most likely a decisive factor in the dominance of this type of algae. The studies on bioindicators have shown that

Table 3. Results of hydrobiological research of water

Site	Saprobic index	Numer of organism per 1 cm³	Primary producers
			Decomposers
St. 1	2.08	13244	12986
			258
St. 2	2.13	12829	11453
			1376
St. 3	2.03	13158	11610
			1548

diatoms are among the best indicators of the state of the environment [Panek 2011b]. They react *inter alia* to the changes in trophic conditions and the conditions relating to the contamination with organic matter, which allows them to be used in monitoring surface waters in areas experiencing varying degrees of anthropogenic transformation [Panek 2011a]. Cyanobacteria were observed mainly in the lower part of the Reservoir near the dam (St.1). Mass development of these organisms is considered one of the signs of progressing eutrophication [Szymański et al. 2013]. At St.2 site, the numbers of cyanobacteria in the water were lower, while at the St. 3 site in the vicinity of the inflow, they were not present at all. In order to develop, these organisms require a large amount of biogenic substances (especially phosphorus) to be present in water [Pac 2012].

Further factors favouring cyanobacterial blooms include the sun exposure and a higher temperature of water that lasts for a longer period of time. The most favourable conditions for development could therefore be present in the near-dam zone, given the slower flow of water there. Cyanobacteria may develop in the circumstances of nitrogen deficiency, i.e. where the values for

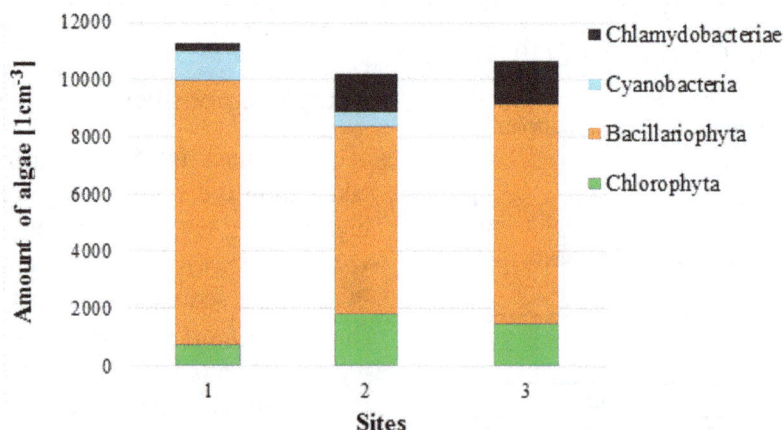

Figure 2. Plankton occurring in water in July 2013 at individual research sites

the ratio of nitrogen to phosphorus in water are low, for example owing to the ability of certain species to bind free atmospheric nitrogen [Kobos 2009]. The cyanobacteria found in the Reservoir belonged to the Microcystis, Oscillatoria, Aphanothece and Romeria types, among which strains capable of producing toxins are present [Mazur-Marzec et al. 2009]. At the same time, these cyanobacteria lack heterocysts, ensuring that their development depends on aquatic concentrations of nitrogen [Bergman et al. 1997].

An isotopic index of trophic status

Graphical analysis of spatial distributions associated with the indicators examined (Fig 3) indicates that the isotopic composition of POC did not reflect variability characterising TN and Chl *a*. TSI_{Chl} also varied across a range not affecting the $δ^{13}C$ parameter (Fig. 3A and B). This was not the case for TP, the N:P ratio, TSI_{TP} and the hydrobiological indicators. It was at site St. 1, identified as having the largest population of blue-green al-

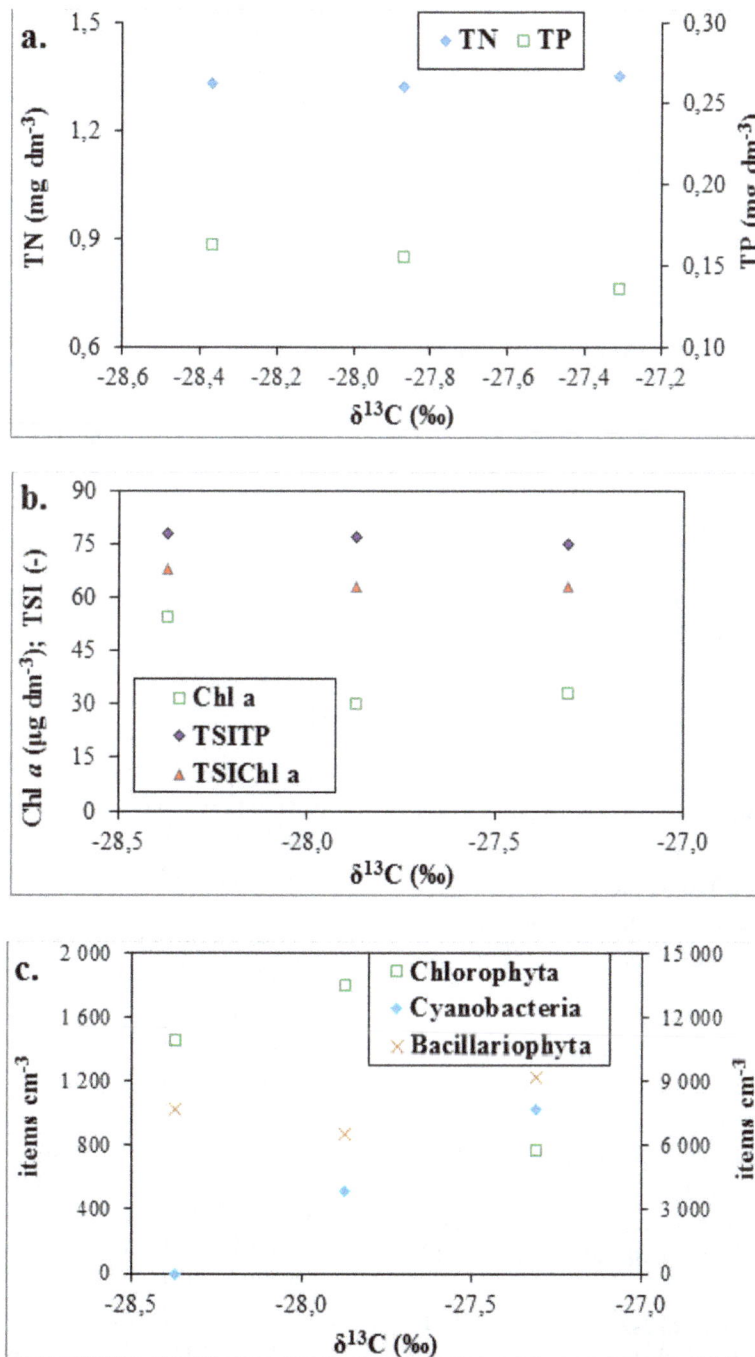

Figure 3. Relations between spatial distribution of selected water parameters and $δ^{13}C$ values; a. with total nitrogen, total phosphorus, b. with chlorophyll *a*, TSI_{TP}, TSI_{chla}, c. with phytoplankton

gae where the highest $\delta^{13}C$ value was reported. At the same time that site had higher noted concentrations of diatoms and lower concentrations of green algae, compared with other sites (Fig. 3C). The decrease in TP and the resulting increase in N:P seems to exert an indirect impact, by stimulating the growth of specific phytoplankton groups. The situation described may suggest that summer blooms of cyanobacteria achieve greater enrichment of matter produced with the heavier ^{13}C isotope, as they do so potentially poisoning waters with toxins.

CONCLUSIONS

The water of Rzeszów Reservoir was classified as hypertrophic irrespective of zones. The phytoplankton blooms expressed in the form of numbers of organisms per unit volume of water were also shaped at similar levels. However, diversification of taxonomic composition was evident, given the presence of cyanobacteria among the dominant diatoms in the zone next to the dam. This effect was accompanied by the enrichment of organic matter suspended in the reservoir by the heavier ^{13}C isotope. On this basis, the $\delta^{13}C$ isotopic index can be regarded as a useful indicator allowing for more accurate identification of the trophic status and follow-up degradation of bodies of water. Further analyses are necessary in this respect, on the basis of a larger sample of data.

Acknowledgements

Financial support was provided by the grant no 2011/03/B/ST10/ 04998 from the Polish National Science Centre. We would like to thank our colleagues from the department laboratory for their support and help in sampling and laboratory analysis.

REFERENCES

1. Bartoszek L., Koszelnik P., Gruca-Rokosz R., Kida M. 2015. Assessment of agricultural use of the bottom sediments from eutrophic Rzeszów reservoir. Annual Set The Environment Protection, 17, 396–409.

2. Bartoszek L., Gruca-Rokosz R., Koszelnik P. 2017. Analysis of the desludging effectiveness of the Cierpisz and Kamionka Reservoirs as an effective method of the eutrophic ecosystems recultivation. Annual Set The Environment Protection, 19, 600–617 (in Polish).

3. Bartoszek L., Miąsik M., Koszelnik P. 2018. Degradation and the ability to sustainable restoration of strongly degraded shallow reservoir. Global NEST Journal, submitted.

4. Bergman B., Gallon J.R., Rai A.N., Stal L.J. 1997. N2 Fixation by non-heterocystous cyanobacteria. FEMS Microbiology Reviews, 19(3), 139–185.

5. Boopathi T., Ki J.S. 2014. Impact of environmental factors on the regulation of cyanotoxin production. Toxins, 6, 1951–1978.

6. Carlson R.E. 1977. A trophic state index for lakes. Limnology and Oceanography, 22(2), 361–369.

7. Chorus I., Bartram I. 1999. Toxic cyanobacteria in water: a guide to their public health consequences, monitoring and management. WHO Publ., E. & F.N. Spon, London-New York.

8. Czerniejewski P., Czerniawski R. 2008. Final report on hydrochemical and hydrobiological works carried out on the Klasztorne Górne lake together with an assessment of reclamation opportunities. Urząd Miasta i Gminy w Strzelcach Krajeńskich (in Polish).

9. Forsberg C., Ryding S.O. 1980. Eurtophication parameters and trophic state indices in 30 Swedish waste-receiving lakes. Arch. Hydrobiol., 89, 189–207.

10. Gorzel M., Kornijów R. 2004. Biological methods of assessing the quality of river waters. Kosmos Problemy Nauk Biologicznych, 53(2), 183–191 (in Polish).

11. Grabowska M., Górniak A., Jekatierynczuk-Rudczyk E., Zielinski P. 2003. The influence of hydrology and water quality on phytoplankton community composition and biomass in a humoeutrophic reservoir, Siemianowka reservoir (Poland). Ecohydrol. Hydrobiol., 3, 185–196.

12. Gruca-Rokosz R., Tomaszek JA., Koszelnik P. 2009. Competitiveness of dissimilatory nitrate reduction processes in bottom sediment of Rzeszów reservoir. Environment Protection Engineering, 35, 5–13.

13. Gruca-Rokosz R. 2013. Trophic state of the Rzeszów reservoir. Journal of Civil Engineering, Environment and Architecture, 60(3), 279–291 (in Polish).

14. Gu B., Schelske C.L., Brenner M. 1999. Relationship between sediment and plankton isotope ratios ($\delta13C$ and $\delta15N$) and primary productivity in Florida lakes. Canadian Journal of Fisheries and Aquatic Sciences, 53, 875–883.

15. Klimaszyk P., Trawiński A. 2007. River status assessment based on benthic macroinvertebrates. INDEX BMWP-PL. Poznań (in Polish).

16. Kobos J. 2009. Water blooms – the diversity of

phytoplankton. IV National Sinic Workshops: Toxic blooms of cyanobacteria in fresh and brackish water. Gdynia, Gdańsk University, Institute of Oceanography, 24–42 (in Polish).

17. Koszelnik P. 2007. Atmospheric deposition as a source of nitrogen and phosphorus loads into the Rzeszow reservoir, SE Poland. Environment Protection Engineering, 33(2), 157–164.

18. Koszelnik P. 2009. Isotopic effects of suspended organic matter fluxes in the Solina reservoir (SE Poland). Environment Protection Engineering, 35(4), 13–19.

19. Kukuła K., Bylak A. 2017. Expansion of water chestnut in a small dam reservoir: from pioneering colony to dense floating mat. Periodicum Biologorum, 119(2), 137–140.

20. Lampert W., Sommer U. 2001. Ecology of inland water. PWN, Warszawa (in Polish).

21. Lehmann M.F., Bernasconi S.M., McKenzie J.A., Barbieri A., Simona M., Veronesi M. 2004. Seasonal variation of the δ13C and δ15N of particulate and dissolved carbon and nitrogen in Lake Lugano: Constraints on biogeochemical cycling in eutrophic lake. Limnology and Oceanography, 49(2), 415–429.

22. Mazur-Marzec H., Błaszczyk A., Toruńska A. 2009. Risk assessment related to the occurrence of cyanobacterial blooms in usable waters. IV National Sinic Workshops: Toxic blooms of cyanobacteria in fresh and brackish water. Gdynia, Gdańsk University, Institute of Oceanography, 9–21 (in Polish).

23. Neverova-Dziopak E., Kowalczyk E., Bartoszek L., Koszelnik P. 2011. Trophic state of the Solina reservoir. Zeszyty Naukowe Politechniki Rzeszowskiej, Budownictwo i Inżynieria Środowiska, 58(2), 197–208 (in Polish).

24. Nürnberg G. 2001. Eutrophication and trophic state. LakeLine, 29(1), 29–33.

25. Ostrowska M. 2013. Indicating algae in a Turawa retention reservoir. JEcolHealth, 17(4), 163- 168 (in Polish).

26. Pac M. 2012. Cyanobacteria in freshwater environment. Water-Environment-Rural Areas, 12, 3(39), 187–195 (in Polish).

27. Panek P. 2011a. Naturalists and engineers, or the assessment of water quality in Poland. Przegląd Przyrodniczy, XXII(1), 3–9 (in Polish).

28. Panek P. 2011b. Biotic indicators used in water monitoring since the implementation of the Water Framework Directive in Poland. Przegląd Przyrodniczy, XXII(3), 111–123 (in Polish).

29. Picińska-Fałtynowicz J., Błachuta J. 2012. Methodological guidelines for conducting phytoplankton research and assessing the ecological status of rivers on its basis. GIOŚ, Warszawa (in Polish).

30. Szymański D., Dunalska J.A., Jaworska B., Bigaj I., Zieliński R., Nowosad E. 2013. Seasonal Variability of Primary Production and Respiration of Phytoplankton in the Littoral Zone of an Eutrophic Lake. Annual Set The Environment Protection, 15, 2573–2590 (in Polish).

31. Vollenweider R.A., Kerekes J.J. 1982. Eutrophication of waters. Monitoring assessment and control. Technical report. Environment Directorate, OECD, Paris.

32. Walker W. 1979. Use of hypolimnetic oxygen depletion as a trophic index for lakes. Water Resour. Res., 15(6), 1463–1470.

3

Machines and Horticultural Implements for the Cultivation of Small-Scale Herbs and Spices

Witold Niemiec[1], Tomasz Trzepieciński[2]*

[1] Department of Water Purification and Protection, Rzeszow University of Technology, Al. Powstańców Warszawy 6, 35-959 Rzeszów, Poland

[2] Department of Materials Forming and Processing, Rzeszow University of Technology, Al. Powstańców Warszawy 8, 35-959 Rzeszów, Poland

* Corresponding author's e-mail: tomtrz@prz.edu.pl

ABSTRACT

Herbal plants and spices, due to their aromatic and medicinal properties, are widely used in pharmacy, cooking, cosmetology and agriculture. The increase in the number of willow plantations for energy crops has enhanced the interest of the salicylic glycosides contained in the willow bark which shows multidirectional pharmacological activity. The small-scale production of herbs and spices requires the application of agricultural machines and horticultural implements adjusted to the area cultivated. In this paper, the comprehensive characteristics of the medicinal properties of willow are presented. Newly developed machines used for the mechanisation of the establishment and cultivation of herbs and spices are also described. The machines presented are a part of the technology developed at the Rzeszow University of Technology which can also be used for the mechanisation of biomass production. In order to improve the physicochemical properties of the soil and increase the amount and quality of yield, natural organic fertilisers can be used. Therefore, the paper presents the correct method of storing manure and a device for the application of solid and liquid fertilizers to the soil as well as a device for monitoring the state of the soil-water environment.

Keywords: cultivation, harvesting, herbal plants, herbs, spices, willow

INTRODUCTION

In recent years, the cultivation of herbs covers an area of about 14 thousand hectares in Poland. Athough the production of herbal food and cosmetics has an increasing economic impact, in the coming years the production of herbal plants will have the greatest impact on the Polish market for herb cultivation [Olewnicki et al. 2015]. Poland is one of the leading European countries as far as the cultivation of herbs and spices is concerned. Field cultivation of herbs on farms can be an important source of income, although those cultivating them must remember to adjust the farm production to the changing demand, consumer needs and prices. The issues of how to set up a plantation of herbal plants, how to grow them and how to obtain a high-quality raw material and a

satisfactory income level are all the relevant challenges [Newerli-Guz 2016].

The cultivation of herbal plants should begin with assessing the possibility of selling the raw material. Then, trial crops should be established to check how the species behaves under specific field conditions. In addition, it is important to check the labour required for cultivation and appropriate handling of the crop after harvest [Leonti and Verpoorte 2017, Newerli-Guz 2016]. The discussion of the possible advantages of using herbal medicines instead of purified compounds as well as the truth and myths about herbal medicines have been provided by Carmonia and Pereira [2013]. The impact of the environmental factors on the size and quality of crops should also be taken into account. Thus, it becomes important to consider the methods of intensifying cultivation in relation to the rational fertilisation adapted to the varieties of plants and how to integrate the protection of the plantation against weeds and pests [Lubbe and Verpoorte 2011, Newerli-Guz 2016]. Another important issue is the process of preserving the raw material obtained using dryers. The raw material obtained should be packed and delivered to the herbal plants purchasing centres or directly to the processing plant. Therefore, it is important to analyse the opportunities for selling the raw material produced in order to avoid the problems related to its sale and distribution.

In the case of herbal plants, the quality of the raw material is more important than the crop yield. The content of active substances determined for a specific plant determines its health properties, use in the culinary arts and cosmetology [Newerli-Guz 2016]. The products from the Polish agritourism farms can compete on the European market due to the low soil pollution in our country and low consumption of artificial fertilisers and pesticides.

Currently, more people tend to improve the quality of their lives by promoting a healthy lifestyle and rational nutrition. The valuable vitamins and minerals contained in vegetables, fruits, spices and herbs positively affect the functioning of the human organism. Herbs are also used as feed additives acting as growth promoters that favourably affect the ecosystem of the alimentary canal by inhibiting the growth of pathogenic microorganisms [Mizak et al. 2012]. The products from ecological farms are devoid of artificial fertilisers and chemicals; thus, they are considered to be healthier and provide many nutrients. The development of ecological farming in Poland is important in the structural and functional transformation of rural areas. It affects the policy of sustainable development, the improvement of the financial situation of farmers and the promotion of agritourism.

An area of land, a crop of herbs, and the use of specialised equipment in the mechanisation of the establishment, cultivation and harvesting of the crop have a significant impact on the economic efficiency of agricultural production. This paper, presents the machines and horticultural implements for the establishment, cultivation and harvesting of herbal plants for small and medium-sized farms. When considering the establishment of plantations, special attention was given to the fertilisation of the land using organic fertilisers.

HARVESTING OF WILLOW BARK FOR HERBAL REMEDIES

The natural ingredients found in the wood and bark of willow bark, are very valuable, especially the active substances in the form of salicylic glycosides, flavonoids and phenolic compounds used for pharmaceutical and cosmetic purposes. There are about 350 basic species and varieties of *Salix* willow in the world. Among the many species of willow, the two species are of the greatest importance for modern phytotherapy: the purple willow (*Salix purpurea L.*) and the white willow (*Salix alba L.*). The willow bark, as herbal material, is considered to be antipyretic, anti-inflammatory and analgesic. Such properties are conditioned by salicylic glycosides, which are one of the most important groups of active compounds contained in the willow bark. These glycosides easily break down in the alimentary tract, releasing salicylic alcohol, which is oxidized in the liver to salicylic acid and thus does not damage the stomach mucosa. Despite using them in viral diseases, the active compounds in the bark inhibit platelet aggregation [Waliszewska and Dukiewicz 2014].

The salicin contained in the willow bark has become the precursor of one of the best-known medicines – aspirin (acetylsalicylic acid) [Mahdi 2014, Mahdi et al. 2006]. The extract from the white willow bark (*Salix alba L.)* acts similarly to aspirin, but as a natural product it is slightly weaker and therefore is administered in higher doses. The most important active compounds of the *Salix purpurea* bark are: salicortin, tremulacin, grandydentatin, flavon glycosides, ellagic acid, populin, isosalipurposide, phenolic glycosides, tannins, salireposide, phenolics and fragilin [Schmid et al. 2001]. The chalcones, being the fraction of flavonoid compounds, exhibit multidirectional pharmacological activities, including: anti-inflammatory, antihistamine and antiulcer effect. The content of chalcones compounds in plants depends on the season and age of the plant [Forster et al. 2008], and their highest amount was found in autumn [Szczukowski et al. 2002]. In the next part of the article, the selected machines and the devices used for establishing and cultivation of the energy willow will be presented. However, the technical solutions described may be useful in the cultivation of other herbal plants and spices.

MECHANIZATION OF CULTIVATION OF HERBS AND SPICES

Fertilization

The soil is the basic source of nutrients in plant growing. In order to improve the physicochemical properties of soil and increase the amount and quality of yield, fertilizers are used. Fertilizers are the products intended for providing nutrients to plants or increasing soil fertility. In organic growing and the cultivation of herbs and spices organic (natural) fertilizers such as manure, liquid manure, straw and compost are a source of nutrients and substances used for soil fertilization. The sewage sludge can be also used as a fertilizer [Niemiec and Wójcik 2015]. One of the factors affecting the quality of herbs is the contamination of heavy metals, which can accumulate during the cultivation, storage and processing of herbs and may have adverse effects on the consumer health [Tripathy et al. 2015]. Therefore, the sewage sludge should not be used in the cultivation of the herbal plants and spices.

Natural fertilizers are a source of nitrogen; hence, their improper storage can lead to the pollution of water reservoirs and groundwater with nitrates. For this reason, special attention is paid to the storage of natural fertilizers. The devices and tanks adapted to technological systems of animal farming should be provided for removal and storage of animal waste. Incorrect storage of animal manure can lead to the contamination of ground and surface water, and contamination of the air as a result of the emission of ammonia, methane, hydrogen sulfide, carbon dioxide or other anaerobic fermentation products.

Farmers are legally obliged to collect manure in maintenance rooms or in appropriate dunghills with side walls and a drainage system. Slurry and liquid manure must be stored only in tightly covered tanks. An example of a properly made manure tank is shown in Fig. 1. Manure loading is one of the most labor-consuming activities associated with the wastewater disposal. The subject matter of the utility model PL W-52082 is a manure loader having a rotating columnar stand with an extension arm (Fig. 1). The loader enables easy one-man operation and is effective when plunging the gripper into a prism. Furthermore, it provides operation in a significant area of the dunghill.

The agricultural use of liquid animal manure for fertilizing purposes is the most common form of its management. Depending on the size of doses and the frequency of application, slurry and liquid manure differently affect (i) the size and quality of crops and (ii) the changes in the soil environment. Liquid manure spread onto fields should be immediately mixed with the soil. The application of fertilizers, including liquid manure, is regulated by the Act of 10 July 2007 on Fertilizers and Fertilization (Journal of Laws of 2007, no. 147, item 1033) and the Regulation by the Polish Minister of Agriculture and Rural Development of 18 June 2008 on the implementation of certain provisions of the Act on Fertilizers and Fertilization (Journal of Laws of 2008, No. 119, item 765).

The richness of easily absorbed nutrients contained in the liquid manure and their high bioavailability make it an excellent natural fertilizer. The

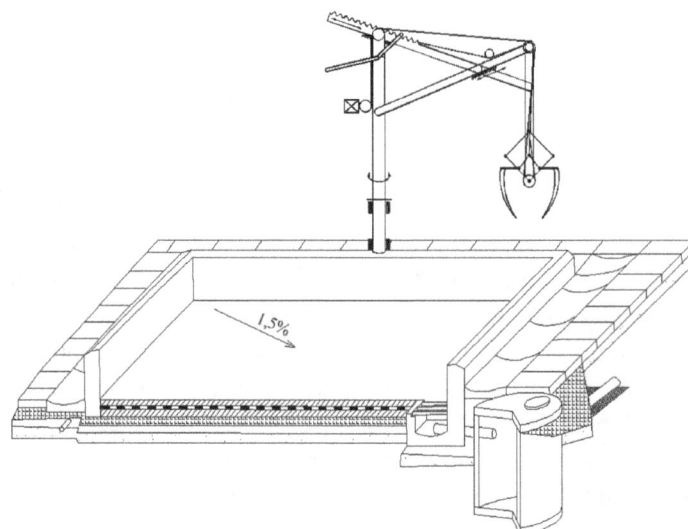

Fig. 1. Dunghill and columnar loader of manure (prepared on the basis of Habel et al. [1990])

bioavailability of ingredients contained in the liquid manure for plants is higher than for ordinary manure. The high content of nitrogen dissolved in water (about 50%) makes slurry a quick-acting fertilizer [Dimitriou and Mola-Yudego 2017]. The application of liquid manure to the soil, however, is associated with the ecological and hygienic-sanitary problems. The injection or at least *soil* management after *liquid manure applications* in arable *soils* is a good agricultural practice because it efficiently reduces the NH_3 losses.

Legally required sanitary conditions related to the application of solid and liquid fertilizers are associated with an immediate covering of the fertilizer with soil. This increases the effectiveness of fertilising. The application of organic liquid fertilisers into the soil is the main advantage of single axle trailer with the tank (Fig. 2). The device for fertilization with solid and liquid fertilizers according to the patent PL P-176563 consists of a tank (1) mounted on the supporting frame of a single axle trailer (2) using quick release couplings (3).

The tank has a rotating mixer (4) driven by the power take-off (PTO) shaft (5) of the tractor. In the rear of the tank (1) there is a liquid manure spreader (6) and a sprinkler (7). The tank can be mounted on the typical manure spreader. The liquid discharged with the use of conventional slurry tanks is poured out over a relatively short distance, which requires a large number of trips to fertilize the whole area of land. In the proposed agricultural solution, there is an impeller pump (8) in the rear of the trailer, which transports the liquid to the sprinkler (7). The rotatable sprinkler allows to control the spraying direction. The advantage of the spreader is the possibility of fertilizing smaller fields from roadways. The description of the patents and utility models used in this paper are shown at the end of the chapter 3, in Table 1.

The device for subsurface application of liquid fertilizer (Fig. 3) meets the requirement of covering the fertilizer with soil. The basic elements of the machine are a frame bearer (1), a wheel (2), a coulter (3), damper springs (4, 6), a pressure roll (5) and pipe (7) supplying liquid. The details of the structure of the device are given in the description of utility model PL W-39050.

In order to control the state of the soil-water environment or the degree of bioavailability of nutrients, a lysimeter construction (Fig. 4) was developed to allow the continuous collection of groundwater samples for specialist tests. The lysimeter is used to measure the amount of actual evapotranspiration which is released by plants. The device consists of a container closed by a perforated lid with a cone-shaped bottom ending with a nozzle. The upper lid and outlet port end with flexible hoses that connect the space inside the device with its surroundings. The water from the container is collected using hoses or a portable vacuum pump.

The advantage of the device is a reinforced structure that allows the use of heavy agricultural equipment in the land without the need to bypass the place of installation of the device. More detailed data are given in the description of utility model PL W-64580.

Establishing and cultivation of plants

The growing demand for biomass, especially various forms of wood, increases the activity in the search for new machines enabling thewillow cultivation on small areas with a low amount of work. In the machinery market, there is a choice in terms of technical facilities intended for the operation in large plantations. These machines are expensive and unsuitable for farming in small crop areas [Trzepieciński et al. 2013]. Hand planting is cumbersome and is characterized by low-performance. However, specialized machines with high efficiency cannot be used on small plantations [Bergante et al. 2016].

The growing interest in wood biomass production in small mountainous areas of southern Poland leads to developing the original machines and devices for planting the energetic willow cuttings at the Rzeszow University of Technology. The planters, which are mounted on a typical agricultural *tractor* with the *three-point suspension system, are* characterized by a simple construction and high reliability.

The frame bearer of the double-row planter (Fig. 5) for the energetic willow is attached to the agricultural tractor. Each of the two working sections of the planter consists of a set of drive wheels (8). The working sections are made in a form of the drums (4) which consist of the five pipes (13), when the pipes are uniformly distributed around the drum. Cuttings (14) are placed manually by the worker in the pipes (10). During the movement of the planter, the cuttings are consecutively placed in the groove (15), and are pre-kneaded in a vertical position. Finally, the cuttings are planted into the soil by pressure wheels (5).

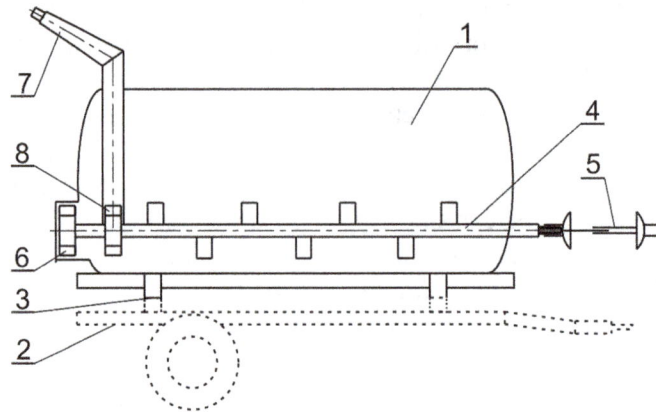

Fig. 2. The tank for fertilization of solid and liquid fertilizers: 1 – tank, 2 – single axle trailer, 3 – quick release coupling, 4 – mixer, 5 – power take-off shaft, 6 – liquid manure spreader, 7 – sprinkler, 8 – impeller pump

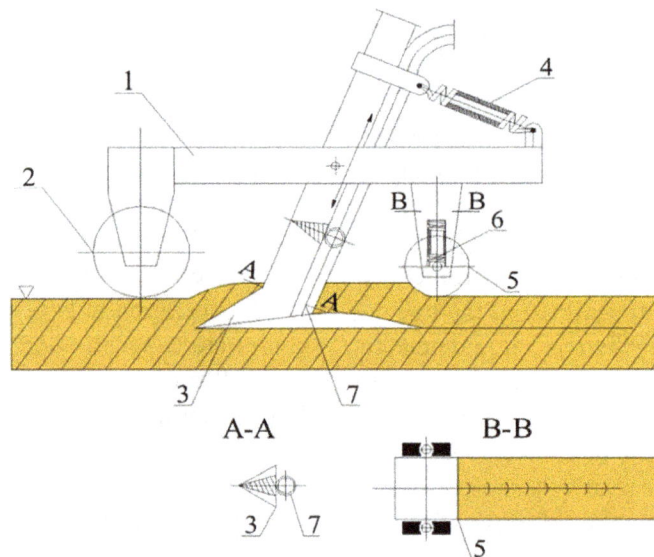

Fig. 3. The device for injecting liquid into the ground: 1 – frame bearer, 2 – roller, 3 – coulter, 4, 6 – damper springs, 5 – pressure roll, 7 – supply pipe

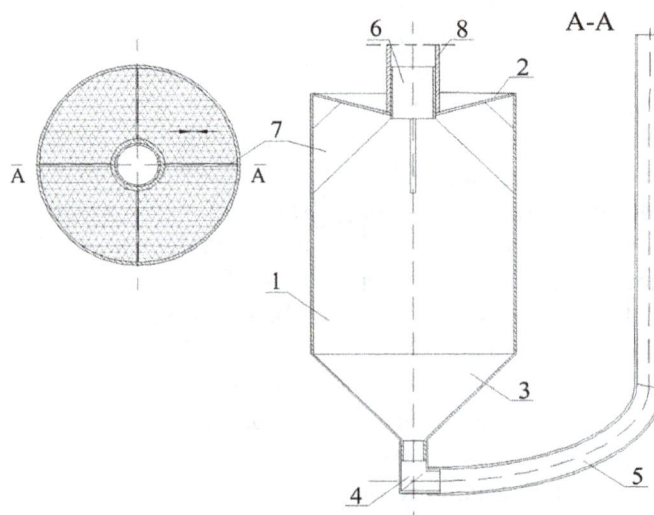

Fig. 4. Scheme of the device for collection and measurement of infiltrating water under field conditions: 1 – tank, 2 – perforated lid, 3 – conical bottom, 4 – outlet port, 5 – flexible pipe, 6 – connector pipe, 7 – brackets, 8 – nozzle cover

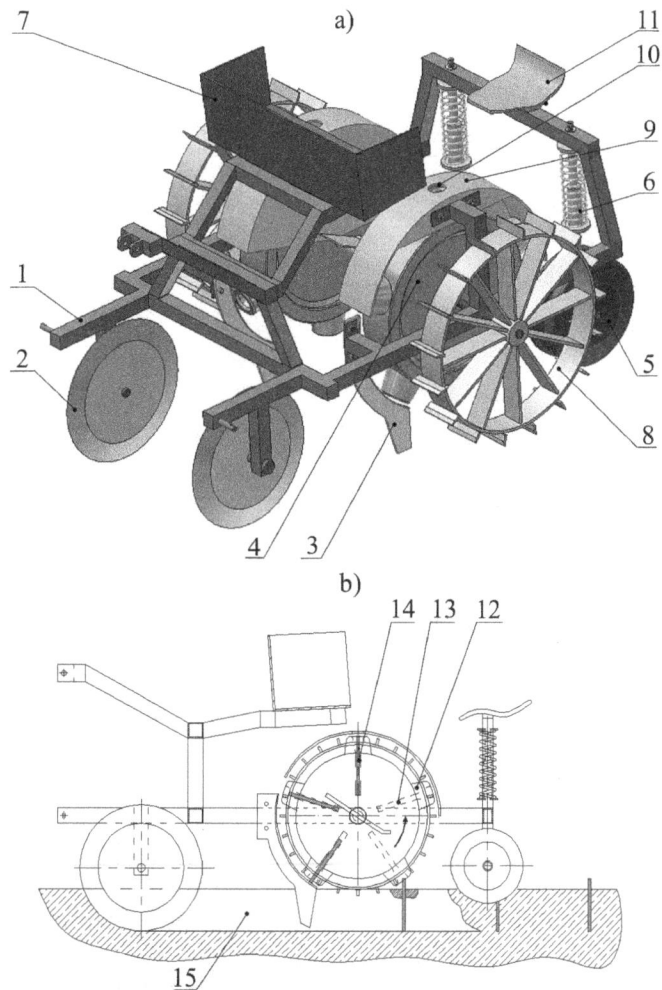

Fig. 5. Planter of plants with lignified shoots; general view (a) and cross-section (b): 1 – supporting frame, 2 – disc coulter, 3 – calibration strip, 4 – drum, 5 – pressure wheels, 6 – spring, 7 – reservoir for cuttings, 8 – drive wheel, 9 – cover, 10 – hole, 11 – seat, 12 – compaction section, 13 – pipe, 14 – cutting, 15 – groove

The presented planter requires a groove for introducing the cuttings, and finally cuttings are planted into the soil by the separate pressure wheels. This disadvantage is overcome by the track planter (Fig. 6) which gradually presses the cuttings into the soil. The planter is mounted on the *tractor* with the *three-point suspension system* (1) and is equipped with two working sections with spacing adapted to the typical spacing of willow planting (approximately 0.75 m). The elimination of the furrow-forming wheels (coulters) and the calibration strips reduced the working resistances during the planting operations. The details of the planter can be found in the description of patent PL P-400110.

Harvesting

In Poland, the plantations of willow have a small area and sometimes are far apart from each other. Therefore, the use of specialized machines, even in the form of a mechanization service, is an economic barrier for the owner of the plantation [Santangelo et al. 2015, Spinelli et al. 2012]. Mechanization plays a key role in the economic implementation of shrubs short rotation. An efficient harvesting operation and proper selection of the machines are the basic assumptions for the implementation of effective crop cultivation [Vanbeveren et al. 2015]. In order to meet the needs of biomass producers, in particular those operating on small and medium-sized farms, a mower (Fig. 7) has been developed under the patent PL P-213402. The machine meets the following criteria:

- shoots should be cut without damaging the stumps, which is important from the point of view of vegetation regeneration ,
- shoots should be cut as close to the soil surface as possible.

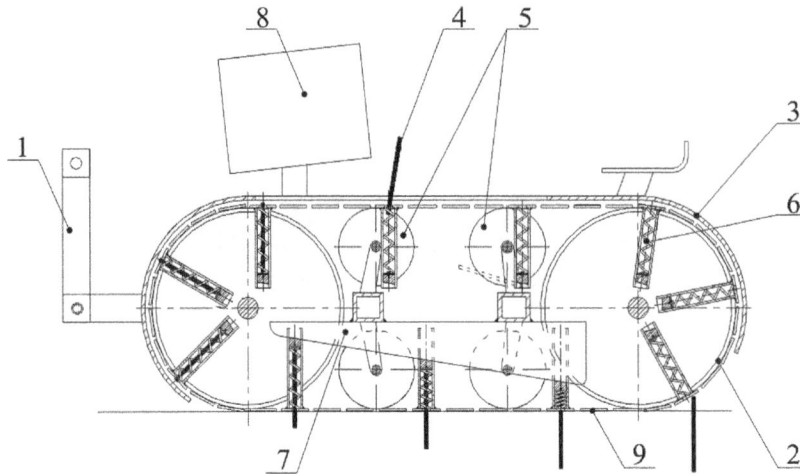

Fig. 6. The planter for cuttings of ligneous plants: 1 – three-point suspension system, 2 – wheel, 3 – housing, 4 – cutting, 5 – rollers, 6 – spring, 7 – wedge, 8 – reservoir for cuttings, 9 – track

The structure of a developed and patented mower for cutting down woody plants (Fig. 7) is composed of the frame (1), the three-point linkage (2) for attaching to the agricultural tractor and a working arm (3) coupled to the frame using an articulated joint (4). At the end of the working arm (3) a circular saw (6) is mounted. In order to provide the required power and efficiency, the saw is PTO driven.

The drive of the circular saw is transmitted from PTO by using a shaft (7), intersecting the axis gear (8) and the belt transmission (9). All movable elements of the machine are protected by safety guards. The prototype of the tractor-mounted mower for woody plants is presented in Fig. 8.

Table 1 presents the details of selected machines and devices used in the technology of establishing and cultivating herbal plants and species, especially willow.

CONCLUSIONS

Poland is a country with large production and processing capacities of herbs and spices. Among the plants cultivated in the country, about 170 species are used in the herbal industry. Cultivation of herbal plants and spices is not widespread, although it remains an alternative to the classical corn growing and allows for diversification of agricultural income sources, especially in small and medium-sized farms and ecological agricultural farms. The decreasing number of the small farms focused on animal production creates opportunities for increasing the share of herb and spices production in the crop share. A significant increase in the social awareness related to the ecological and health policy of European Union countries is also a factor determining the production of healthy food, herbs and spices. The grow-

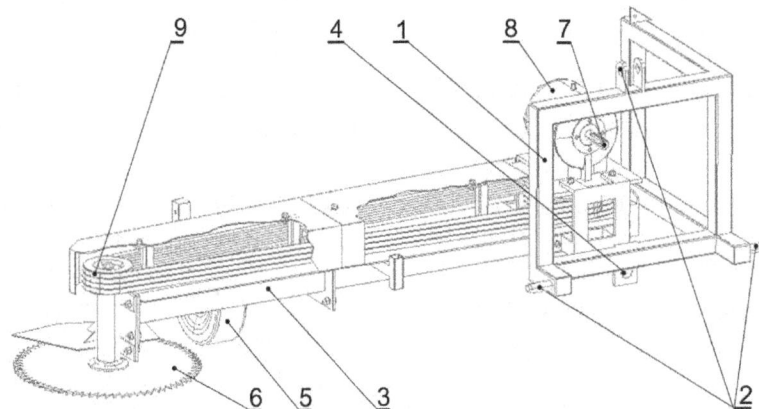

Fig. 7. Model of a tractor mower for woody plants: 1 – frame, 2 – three-point suspension system, 3 – working arm, 4 – articulated joint, 5 – supporting wheel, 6 – circular saw, 7 – shaft, 8 – intersecting axis gear, 9 – belt transmission

Fig. 8. Prototype of the mower for woody plants

Table 1. Machines and horticultural implements in the technology of herbal plants cultivation

Authors	Machine/device	Patent (P)/Utility model (W) no.	Year
Niemiec W., Puchała J.	Device for liquid dosage into soil	W-39050	1983
Niemiec W., Grygiel A., Kaszubski M.	Device for fertilization of solid and liquid fertilizers	P-176563	1995
Niemiec W.	Device for collecting and measurement of infiltrationed water in field research	W-64580	2007
Niemiec W., Skiba S., Ślenzak W.	Mower for tree-like plants	P-213402	2008
Niemiec W., Stachowicz F., Trzepieciński T.	Device for planting ligneous plants	W-66636	2011
Trzepieciński T.	Device for planting ligneous plants	P-400110	2015

ing number of the energy willow plantations may contribute to a greater interest in the use of the willow bark for the medical purposes.

REFERENCES

1. Bergante S., Manzione M, Facciotto G. 2016. Alternative planting method for short rotation coppice with poplar and willow. Biomass Bioenergy, 87, 39–45.

2. Carmonia F., Pereira A.M.S. 2013. Herbal medicines: old and new concepts, truths and misunderstandings. Revista Brasileira de Farmacognosia, 23(2), 379–385.

3. Dimitriou I., Mola-Yudego B. 2017. Nitrogen fertilization of poplar plantations on agricultural land: effects on diameter increments and leaching. Scandinavian Journal of Forest Research, 32(8), 700–707.

4. Forster N., Ulrichs C., Zander M., Katzel R., Mewis I. 2008. Influence of the season on the salicylate and phenolic glycoside contents in the bark of Salix daphnoides, Salix pentandra, and Salix purpurea. Journal of Applied Botany and Food Quality, 82(1), 99–102.

5. Habel A., Jucherski A., Niemiec W., Frankowski A., Kozakiewicz Z., Kaszubski M. 1990. Mechanical loader of manure. Utility Model No. W-52082.

6. Leonti M., Verpoorte R. 2017. Traditional Mediterranean and European herbal medicines. Journal of Ethnopharmacology, 199, 161–167.

7. Lubbe A., Verpoorte R. 2011. Cultivation of medicinal and aromatic plants for specialty industrial materials. Industrial Crops and Products, 34(1), 785–801.

8. Mahdi J.G. 2014. Medicinal potential of willow: A chemical perspective of aspirin discovery. Journal of Saudi Chemical Society, 14(3), 317–322.

9. Mahdi J.G., Mahdi A.J., Mahdi A.J., Bowen I.D. 2006. The historical analysis of aspirin discovery, its relation to the willow tree and anticancer potential. Cell Proliferation, 39(2), 147–155.

10. Mizak L., Gryko R., Kwiatek M., Parasion S. 2012. Probiotics in animal nutrition. Życie weterynaryjne, 89(9), 736–742 (in Polish).

11. Newerli-Guz J. 2016. The cultivation of herbal plants in Poland. Roczniki Naukowe 18(3): 268– 274 (in Polish).

12. Niemiec W., Wójcik M. 2015. The possibilities of utilization of municipal sewage sludge in selected sewage-treatment plants. Zeszyty Naukowe Politechniki Rzeszowskiej – Mechanika, 87(4), 339–347 (in Polish).

13. Olewnicki D., Jabłońska L., Orliński P., Gontar Ł. 2015. Changes in Polish domestic production of herbal plants and in selected types of enterprises that process herbal plants in the context of the global increase in demand for these products. Zeszyty Naukowe SGGW, 15(1), 68–76 (in Polish).

14. Santangelo E., Scarfone A. Del Giudice A., Acampora A., Alfano V., Suardi A., Pari L. 2015. Harvesting systems for poplar short rotation coppice. Industrial Crops and Products, 75, 85–92.

15. Schmid B., Kötter I., Heide L. 2001. Pharmacokinetics of salicin after oral administration of standard willow bark extract. European Journal of Clinical Pharmacology, 57(5), 387–391.

16. Spinelli R., Schweier J., De Francesco F. 2012. Harvesting techniques for non-industrial biomass plantations. Biosystems Engineering, 113(4), 319–332.

17. Szczukowski S., Tworkowski J., Sulima P. 2002. Willow bark with a source of salicylic glycosides. Wiadomości Zielarskie, 44(1), 6–7 (in Polish).

18. Tripathy V., Basak B.B., Varghese T.S., Saha A. 2015. Residues and contaminants in medicinal herbs – A review. Phytochemistry Letters, 14, 67–78.

19. Trzepieciński T., Niemiec W., Stachowicz F. 2013. Selected design problems of mowers for felling woody plants and protection of green areas. Technika Rolnicza Ogrodnicza Leśna, 1, 13–15.

20. Vanbeveren, S.P.P., Schweier, J., Berhongaray, G., Ceulemans, R. 2015. Operational short rotation woody crop plantations: manual or mechanised harvesting. Biomass and Bioenergy, 72, 8–18.

21. Waliszewska B., Dukiewicz H. 2014. The use of willows in pharmacy. Zeszyty Naukowe Wydziału Nauk Ekonomicznych Politechniki Koszalińskiej, 18, 57–66 (in Polish).

4

Influence of Ultrasonic Disintegration on Efficiency of Methane Fermentation of *Sida hermaphrodita* Silage

Magda Dudek[1*], Paulina Rusanowska[1], Marcin Zieliński[1], Marcin Dębowski[1]

[1] University of Warmia and Mazury in Olsztyn, Department of Environmental Engineering, ul. Warszawska 117a, 10-720 Olsztyn, Poland
* Corresponding author's e-mail: magda.dudek@uwm.edu.pl

ABSTRACT

The technologies related to the anaerobic decomposition of organic substrates are constantly evolving in terms of increasing the efficiency of biogas production. The use of disintegration methods of organic substrates, which would improve the efficiency of production of gaseous metabolites of anaerobic bacteria without the production of by-products that could interfere with the fermentation process, turns out to be an important strategy. The methane potential of commercially available biodegradable raw materials is huge and their effective use gives the prospect of obtaining an important renewable energy carrier in the form of biogas rich in methane. Ultrasonic disintegration may play a special role in the pre-treatment of substrates subjected to methane fermentation. The pre-treatment based on ultrasonic sonication has a positive effect on the availability of anaerobic compounds released from cellular structures for microorganisms. The research was aimed at determining the influence of ultrasonic sonication on the anaerobic distribution of the organic substrate used, which constituted the mallow silage along with cattle manure with hydration of 90%. The research was carried out using the UP400S Ultrasonic Processor. The disintegration process was applied in two technological variants. The efficiency of biogas and methane production was determined depending on the technological variant used and the time of disintegration. The influence of sonication time on the effectiveness of anaerobic transformation was demonstrated. The highest biogas yield and methane production potential was recorded at 120s. The prolongation of the action time of the ultrasonic field did not significantly increase the biogas production. The use of disintegration of liquid manure as the only medium for the propagation of ultrasonic waves was sufficient to increase the production of gaseous metabolites of anaerobic bacteria. Subjecting the substrate additionally containing mallow silage to the process to sonication did not significantly affect the efficiency of the fermentation process. The percentage of methane in the biogas produced was independent of the pre-treatment conditions of the substrate and was in the range of 66–69%.

Keywords: ultrasonic disintegration, methane fermentation, *Sida hermaphrodita* silage, biogas production.

INTRODUCTION

A wide range of waste substrates can be processed into a valuable energy carrier that is methane-rich biogas, with the use of anaerobic fermentation systems. The type of waste submitted to anaerobic degradation is mostly derived from agriculture, e.g. animal waste (liquid manure), industrial waste (waste products from slaughterhouses, blood, fish waste products, etc.), waste and residue matter from plant production (maize silage, wheat straw, barley seed cake, grass, clover leaves, sugarcane bagasse, etc.) as well as household waste [Mata-Alvarez et al. 2000, Alagöz et al. 2018].

The anaerobic fermentation processes are continually improved in terms of the implemented technologies, so as to enhance the acquisition of gaseous metabolites from anaerobic bacteria. One of the factors limiting the rate of anaerobic fermentation is the process of hydrolysis, which prolongs the whole technological process [Alagöz et al. 2018]. In order to improve the efficiency of digestion and shorten the duration of fermentation, various disintegration methods are applied, most often to comminute the sludge itself or the raw material used as a feedstock in fermentation chambers. Physical methods are the most frequent ones in the preliminary disintegration process. The available research reports point to the high efficiency of ultrasound disintegration compared to other approaches to biomass conditioning. The action of ultrasounds in aqueous solutions consists of quick cycles of the compression and decompression of sound waves, which promote the formation of microbubbles within the solution. The rapid release of energy during decompression leads to the destruction of the structure of organic substrate matter subjected to disintegration. At first, this accelerates hydrolysis, which facilitates a more rapid course of the subsequent fermentation stages [Zhang et al. 2008, Grönroos et al. 2005].

The studies conducted thus far have focused on the disintegration of activated sludge, the stage which improves the solubility of organic substances in sludge, contributes to the destruction of cells and flocs, and therefore helps to release the intercellular matter. It also improves the rate and efficiency of methane fermentation processes, which translates into greater production of biogas and methane [Bougrier et al. 2006, Wood et al. 2009]. The inclusion of the ultrasound pretreatment processing of activated sludge prior to the anaerobic fermentation can raise the biogas production by 24% up to 140% [Saha et al. 2011, Zhen et al. 2017].

It has been demonstrated that the essential factors which influence the effects obtained in an ultrasound field are the ones which characterize the emission of ultrasounds, i.e. frequency, power and intensity, but also the exposure time. Many researchers draw the attention to the fact that the ultrasounds generated within a relatively narrow frequency range from 16 to 50 kHz should be used for the disintegration of organic substrates [Eder et al. 2002]. The changes in a medium induced by the active influence of an ultrasound field may vary, being dependent on both the aforementioned characteristics of ultrasounds and the physicochemical properties of the medium, such as viscosity of the liquid, presence of electrolytes and polyelectrolytes, the macrostructure and character of the suspension, ambient temperature and many other factors. Hence, the ultimate effect results from the interplay of numerous factors which occur simultaneously. Hydration of the substrate is an extremely important parameter. The higher the hydration, the better the transmission of an ultrasound field deep into the structures

of organic matter, and therefore the better effects of the action produced by the factor. The exposure of organic substrates to an ultrasound field has a positive impact on the destruction of structures and release of the cellular matter, which becomes more readily available to the microorganisms in the subsequent anaerobic fermentation process. This results in a more effective production of gaseous metabolites and a shorter duration of the whole process. When the cell structures are densely compacted, as in lignocellulose substrates, the high effectiveness of the lysis of cells and release of intracellular substrates to the liquid phase is much more difficult to achieve. It is then necessary to apply either a large dose of energy or long exposure to ultrasounds. Sonification of sludge, like most preliminary substrate conditioning processes, is an energy-consuming process, which raises concerns whether this approach could be implemented in industrial practice. Chu et al. (2001) report that an insufficient dose of ultrasounds does not release insoluble substances directly to the suspension, but allows hydrolytic enzymes to attack organic substances more easily [Chu et al. 2001]. Many researchers have discovered that the ultrasound stimulation promotes the activity of enzymes, growth of cells and permeability of the cell membrane [Xie et al. 2009, Liu et al. 2003, Pitt and Ross 2003].

The purpose of this study has been to determine the effect of a dose of ultrasounds on the disintegration of selected substrates prior to the methane fermentation process.

METODOLOGY

The research on the effect of sonification on particular components in a substrate prior to methane fermentation was conducted under laboratory conditions. The analyzed substrate was a mixture of Virginia fanpetals (*Sida hermaphrodita rusby)* silage and bovine slurry, mixed in a 1:1 ratio. In all the variants, the hydration of the substrate was around 90%. The characteristics of the substrate used for the experiments are specified in table 1.

The study was divided into two parts. Stage I consisted in the sonification of the substrate, which was separated between two technological variants, depending on the material subjected to ultrasounds:
- Variant 1 – mixture of Virginia fanpetals silage + bovine liquid manure V1,
- Variant 2 – bovine liquid manure V2.

Table 1. Characteristics of *Sida hermaphrodita* silage used in the studies

Parameters	Value
TS [%]	27.6±0.78
VS [%T.S.]	91.9±0.97
SS [% T.S.]	6.81±0.076
T_N [g/g]	0.005±0.0001
COD [g O_2/g_{TS}]	1.539±0.043
TC [g/g]	0.417±0.054
TOC [g/g]	0.379±0.007

In variant 2, ultrasound-disintegrated bovine slurry was added in the same ratio to Virginia fanpetals silage and then dosed to fermentation chambers at stage II of the experiment.

During the first stage of the research, a UP400S ultrasound homogenizer was employed for the process of disintegrating the substrate. The specification of the device is given in table 2, and its typical range of applications consists of: homogenisation, deagglomeration, lysis and disintegration of cells, protein extraction and emulsification of liquids. Three substrate disintegration time lengths were tested during the study: 60, 120 and 180 s, with the homogenizer set at its maximum power.

The UP400S (400 W, 24kHz) ultrasound homogenizer is the most powerful laboratory device of this type. The samples submitted to ultrasound processing contained 100 g of substrate each.

The second stage of the research aimed to identify the methane potential of the tested organic substrate depending on the technological variants established in stage 1. An automatic methane potential test system AMPTS II Bioprocess Control was applied. This device is used to measure the flows of biomethane produced during the anaerobic fermentation of biodegradable substrates.

The AMPTS II Bioprocess Control system was composed of three sub-units. The main part consisted of bioreactors immersed in a water bath in which constant temperature was maintained. The capacity of each bioreactor was 500 ml. The reaction chambers were connected to stepper motors, which powered slow rotational blades mixing the contents of the reactors. Mixing was repeated cyclically, every 10 minutes for 30 s at 100 rpm. The second component was a system for measuring the amounts of biogas. It was composed of a water container with measuring cells and motion detectors placed inside. Each reactor chamber had a corresponding measuring cell,

which ensured completely automatic measurements of the produced biogas via motion sensors. The last component, a data collection system, enabled the results be seen and controlled throughout the whole experiment.

The research method applied in this study allows the user to determine the activity of anaerobic sludge, biodegradability of substrates and volumes of gaseous products of microorganisms' metabolism. The equipment recorded and analyzed the changes in partial pressure in the measuring chamber caused by the biogas production during anaerobic processes carried out by bacterial microflora.

In each of the experimental variants, the reaction chambers, each of the total capacity of 500 cm³, were inoculated with anaerobic sludge originating from a digester operating on a semi-technical scale, which was fed with Virginia fanpetals and bovine liquid manure. The characteristics of the inoculant are given in table 3.

Depending on the characteristics of the tested substrates and the technological variants, the sludge was dosed in appropriate amounts of the substrate. The tested load of the fermentation chambers with organic compounds was approximately 5.0 g d.m./dm³.

At the start of the experiment, in order to ensure anaerobic conditions, the reactors and their contents were deoxygenated by forceful blowing of nitrogen. The volumes of biogas produced in the reactors were measured for 40 days under mesophilic conditions and at a temperature of 38°C.

Table 2. Characteristics of UP400S Ultrasonic Processor

Parameters	
Model	UP400S
Power	400 W
Ultrasound frequency	24 kHz
Automatic frequency adjustment	Yes
Regulated amplitude	100%
Adjustable pulse	100%

Table 3. Characteristics of the anaerobic sediment used in the experiment

Parameter	Value
pH	7.72±0.14
Hydration [%]	94.98±0.21
TS [%]	5.02±0.18
VS [%$_{T.S.}$]	72.11±2.41
SS [%$_{T.S.}$]	48.96±1.27

ANALYTICAL PROCEDURES

After both the disintegration and fermentation processes, the analyses of the tested substrate were performed, as required. The contents of dry matter, dry and mineral organic matter were determined with the gravimetric method, which consisted of evaporating a sample, drying the residue at 105°C to constant weight and then weighing it. The COD values were obtained using Hach Lange cuvette tests and a UV/VIS DR 5000 spectrophotometer. The ratio of alkalinity to volatile acids was determined using a TitrLab AT 1000 potentiometric titrator. The content of glucose was monitored with a YSI 2700 Select analyzer. Determination of glucose was achieved through the measurements of D-glucose made with an enzymatic probe. Glucose oxydase is the enzyme immobilized in the membrane. Concentrations of total carbon (TC), inorganic carbon (IC) and total organic carbon (TOC) as well as total nitrogen (TN) were assayed with a Flash 2000 Thermo Scientific elemental analyzer.

The samples for an analysis of the ratio of volatile acids to alkalinity F/T were obtained by centrifugation of the material in a Rotina 380 laboratory centrifuge for 3 min at 9000 rotations per minute.

For assays of COD of sugars, TOC, TC, IC, TN, the samples were additionally vortexed in a Mini Spin Eppendorf laboratory vortex mixer for 90 s at 13000 rpm.

The gas for the qualitative analysis was obtained from respirometers using a needle and a gas-proof syringe. The respirometers were fitted with rubber bungs and tubing, which enabled us to take samples of gas. Each time, 5 cm^3 of biogas was sampled, and the total and percentage compositions of the sampled biogas were determined on a GC Agillent 7890 A gas chromatographer. The chromatograph was coupled with a thermoconductive detector. The percentage content of the following biogas components was measured: methane CH$_4$, carbon dioxide CO$_2$ and oxygen O$_2$.

STATISTICAL ANALYSIS

The statistical analysis of the results and the calculations pertaining to coefficients of determination R2, were supported by STATISTICA 10.0 PL software. The cultures of microalgae for each series and variant of the experiment were grown in triplicate. All physicochemical analyses were performed in triplicate as well. The verification of the research hypothesis regarding the distribution of each variable was based on the W Shapiro-Wilk test. One-factor analysis of variance (ANOVA) was performed to identify the significance of differences between the variables. The homogeneity of variance in groups was checked with the Levene's test. The RIR Tukey test served to determine the significance of differences between the analyzed variables. The level of significance in the above-mentioned tests was set at p = 0.05.

RESULTS AND DISCUSSION

The study on the impact of sonification of substrate prior to the methane fermentation was conducted in order to improve the efficiency of biogas production during the methane fermentation. The first stage of the research consisted in disintegration of substrate (variant 1) and liquid manure (variant 2) before fermentation.

During sonification, the changes in temperature and energy supplied to carry out the process of substrate conditioning were observed (Tab. 4). As the time of exposure of substrate to ultrasounds was prolonged, the amount of supplied energy increased proportionally.

The greatest changes in concentrations of the analyzed parameters were noticed between the research variants. When liquid manure was submitted to disintegration, the concentrations of glucose as well as the ratio of volatile acids to alkalinity F/T were nearly twice as high as in the control sample or variant 1 of substrate conditioning. Statistically significant differences were also observed between the variants with respect to the content of organic content expressed through COD (Table 5). Within the individual technological variants, statistically significant differences were observed with respect to the nitrogen compounds. The exposure of substrate to ultrasounds for 60 s did not cause changes in this parameter relative to the control. Continuing sonification for another 60 s, resulted in an increase of the nitrogen content in both technological variants by about 10%. The subsequent dose of ultrasounds did not have such an effect. In variant 2, where bovine liquid manure was conditioned, the content of organic carbon increased significantly compared to the control and variant 1 samples (Table 5).

In the second stage of the study, disintegrated substrate was fed into fermentation chambers to analyze the biogas production potential. The

Table 4. Parameters of the ultrasonic disintegration process

Type	60 s			120 s			180 s		
	E [Ws]	E[Ws/g]	T [°C]	E [Ws]	E[Ws/g]	T [°C]	E [Ws]	E[Ws/g]	T [°C]
V1	7551.3	75.51	28.5	13585.6	135.86	68.2	18247.6	182.48	87.4
V2	10889.58	108.90	37.6	21794.18	217.94	45.2	30124.4	301.24	57.6

Table 5. Characteristics of the substrate after the disintegration process depending on the technological variant

Type		Carbohydrates [mg $C_6H_{12}O_6$/dm³]	FOS/TAC	TOC [mg/l]	TC [mg/l]	IC [mg/l]	N [mg/l]	DOC [g/l]	C [% s.m.]	N [% s.m.]	$C_{org.}$ [% s.m.]	$N_{org.}$ [% s.m.]
	K	45.0	5.13	19167	19350	182	3502	45.0	41.09	3.03	38.64	3.13
V1	60s	47.9	5.07	19535	19710	177	3597	45.6	38.36	2.78	38.42	2.89
	120s	48.1	4.93	19603	19800	197	3867	45.2	39.55	2.98	38.70	2.26
	180s	50.33	5.09	19724	19916	192	3794	45.1	39.42	2.89	38.48	2.54
V2	60s	103.23	9.23	15745	15912	167	3007	33.2	41.79	3.26	41.22	3.16
	120s	95.39	9.65	16797	16985	188	3932	35.0	43.01	3.15	42.34	3.12
	180s	98.41	9.32	16785	16959	174	3872	34.6	42.6	3.11	41.86	3.02

amount of obtained biogas was assessed in relation to the applied technological variant. The highest productivity of gaseous metabolites produced by bacteria was determined in variant 1, at the sonification time equal to 180 s. The biogas gain was then 1456.4±0.7 Nml, which was 23.5% more than in the control process (Fig. 1). Quarmby et al. achieved 15% higher productivity of biogas from sludge submitted to sonification than in the control test, where substrate was not conditioned [Quarmby et al. 1999]. The lowest quantities of biogas were noted in both tested variants when the application of ultrasounds lasted for 60 s. However, no statistically significant differences were observed between the amounts of biogas produced at the sonification time of 120 s and 180 s.

Moreover, there were no statistically significant differences in the amount of produced biogas depending on the material submitted to

sonification. Sonification of either whole substrate or liquid manure alone resulted in comparable amounts of produced biogas. This finding suggests that there is no need to disintegrate the whole biomass prior to feeding it into a fermentation chamber. For the reasons related to the maintenance of a biogas plant and technological aspects of biogas production, sonification of the liquid phase alone is sufficient.

The content of methane in the biogas obtained, irrespective of the tested process variant, was around 66–69% (Fig. 2). Very similar results were reported by Part et al. who tested the fermentation of *Chlorella vulgaris* microalgal biomass. Disintegration of cells in their experiment was achieved with the help of an STH-750S ultrasonic homogeniser (Sonictopia, Korea) the maximum power of which was 750 W. The ultrasound field applied in the cited study ranged from 5 to

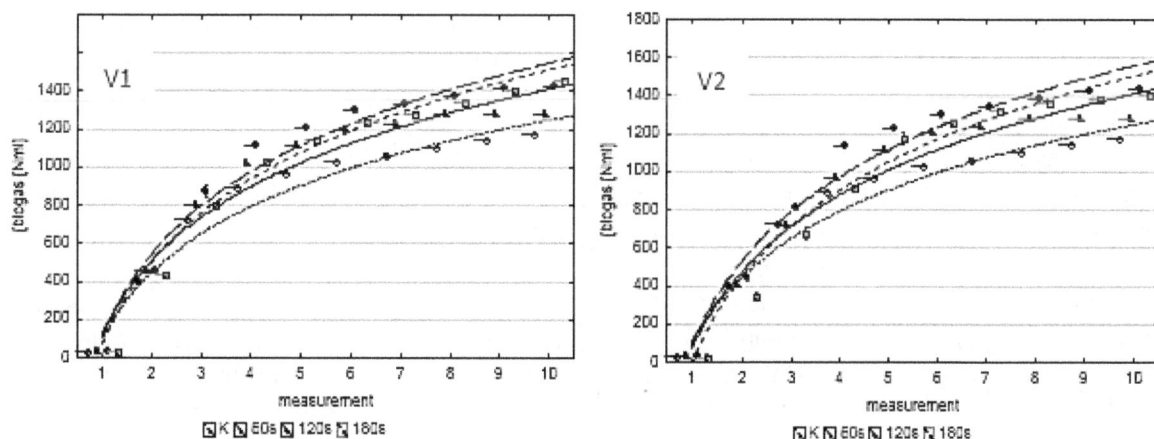

Figure 1. Effectiveness of biogas production in the experiment

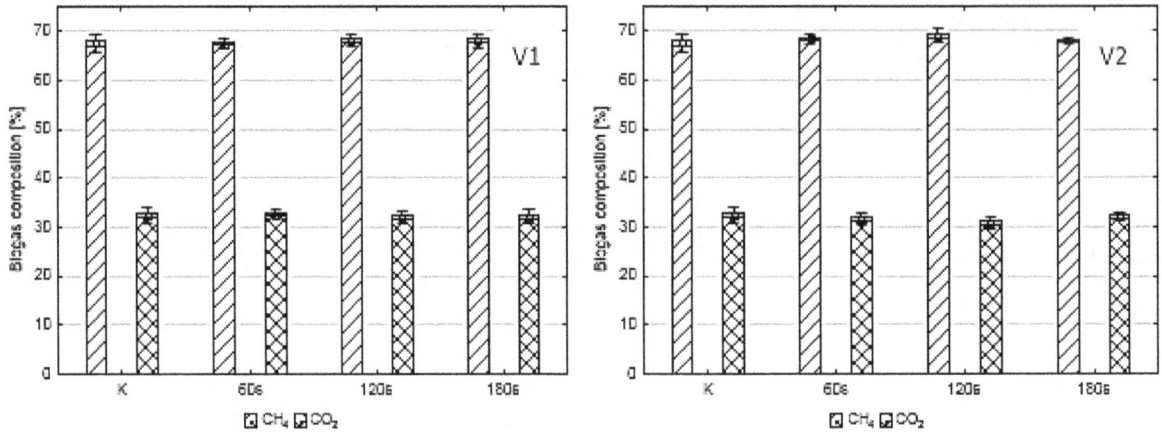

Figure 2. Biogas composition depending on the technological variant.

200 J/ml. The disintegrated biomass was fed to the fermentation chambers which had been inoculated with sludge. The content of methane in the biogas obtained under mesophilic conditions was around [Park et al. 2013].

Alagöz et al. investigated the applicability of preliminary substrate conditioning to improve the productivity of biogas production and to achieve higher biogas production in co-fermentation of pomace from olives and grapes. The use of ultrasonic disintegration enabled the researchers to obtain about 6 000 ml biogas with the methane concentration of 60%. The methane gain per organic matter unit was approximately the same as the one achieved from the fermentation of Virginia fanpetals silage, and equaled *ca* 0.1 Nl/g_{VS} [Alagöz et al. 2018].

The quantities of produced biogas converted per organic matter unit confirmed that the most beneficial disintegration variant was the one where sonification lasted for 120 s. The shortest sonification time only slightly improved the effectiveness of the process, while doubling that time resulted in the production of 1010.67±7.77 Nl/kg_{VS} of biogas (Fig. 3). However, the further extension of the ultrasound pretreatment tested did not lead to significantly higher methane productivity (Fig. 3). The amounts of biogas produced in this experiment are relatively high compared to the previous research on disintegration and fermentation of organic matter. Park et al., mentioned above, who investigated ultrasound disintegration of microalgae *Ch. vulgaris* and their fermentation, reported half the amount of biogas produced in our study, i.e. about 400 ml/g_{VS} [Park 2013]. The productivity of methane for the most productive variant amounted to 687.36±3.18 ml_{CH4}/g_{VS}, which was 30% more methane gain

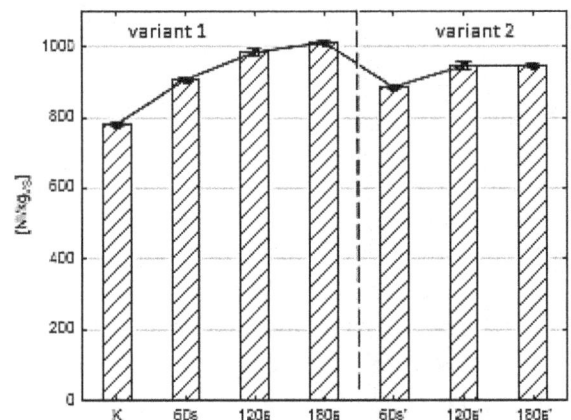

Figure 3. Biogas production depending on the technological variant

than from the control sample, which had not been conditioned prior to fermentation. Similar results were reported by Passos et al., who submitted to sonification mixed biomass of microalage used to pre-treat municipal wastewater and sewage at a secondary settlement tank. The average load of the tanks was about 24 g COD/m^2·d. The biomass of microalgae was exposed to an ultrasound field of the power of 50–70 W. The exposure time was 10, 20 and 30 min. These researchers reported an improvement in the methane productivity ranging from 6 to 33% [Passos et al. 2014].

CONCLUSIONS

The study pertained to the disintegration process applied in two technological variants. The efficiency of biogas and methane production was determined depending on the technological variant used and the time of disintegration. The influence of sonication time on the effectiveness of anaerobic transformation was demonstrated. The highest biogas yield and methane production potential was recorded at 120s. The prolongation of the action time of the ultrasonic field did not significantly increase the biogas production. The use of disintegration of liquid manure as the only medium for the propagation of ultrasonic waves was sufficient to increase the production of gaseous metabolites of anaerobic bacteria. Subjecting the substrate additionally containing mallow silage the process to sonication of did not significantly affect the efficiency of the fermentation process. The percentage of methane in the biogas produced was independent of the pre-treatment conditions of the substrate and was in the range of 66–69%.

Acknowledgment

The study was carried out in the framework of the project under program BIOSTRATEG funded by the National Centre for Research and Development No. 1/270745/2/NCBR/2015 "Dietary, power, and economic potential of *Sida hermaphrodita* cultivation on fallow land".

This work was supported the statutory project 18.610.008–300, University of Warmia and Mazury in Olsztyn, Poland.

REFERENCES

1. Alagöz B.A., Yenigün O., Erdinçler A. 2018. Ultrasound assisted biogas production from co-digestion of wastewater sludges and agricultural wastes: Comparison with microwave pre-treatment. Ultrasonics Sonochemistry, 40(B), 193–200

2. Bougrier C., Albasi C., Delgenés J.P., Carrére H. 2006. Effect of ultrasonic, thermal, and ozone pre-treatments on waste activated sludge solubilisation and anaerobic biodegradability. Chem. Eng. Process., 45, 711–718.

3. Chu C.P, Chang B.V., Liao G.S., Jean D.S., Lee D.J. 2001. Observations on changes in ultrasonically treated waste-activated sludge. Water Research, 35(4), 1038–1046.

4. Eder B., Günthert F.W. 2002. Practical experience of sewage sludge disintegration by ultrasound

5. Grönroos A., Kyllönena , Korpijärvi K., Pirkonen P., Paavola T., Jokela J., Rintala J. 2005. Ultra-sound assisted method to increase soluble chemical oxygen demand (SCOD) of sewage sludge for digestion. Ultrasonics Sonochemistry, 12, 115–120.

6. Liu, Y.Y., Yoshikoshi, A., Wang, B.C., Sakanishi, A., 2003. Influence of ultrasonic stimulation on the growth and proliferation of Oryza sativa Nipponbare callus cells. Colloids and Surfaces B: Biointerfaces 27, 287–293.

7. Mata-Alvarez J., Macé S., Llabrés P. 2000. Anaerobic digestion of organic solid wastes. An overview of research achievements and perspectives. Bioresour. Technol., 274, 3–16.

8. Park K.Y., Kweon J. , Chantrasakdakul P. , Lee K., Cha H.Y. 2013. Anaerobic digestion of microalgal biomass with ultrasonic disintegration. International Biodeterioration & Biodegradation, 85, 598–602.

9. Passos F., Astals S., Ferrer I. 2014. Anaerobic digestion of microalgal biomass after ultrasound pretreatment. Waste Management, 34, 2098–2103.

10. Pitt, W.G., Ross, S.A., 2003. Ultrasound increases the rate of bacterial cell growth. Biotechnology Progress, 19(3), 1038–1044.

11. Quarmby J., Scott J.R., Mason A.K., Davies G., Parsons S.A. 1999. The application of ultrasound as a pre-treatment for anaerobic digestion, Environ. Technol., 20, 1155–1161.

12. Saha M., Eskicioglu C., Marin J. 2011. Microwave, ultrasonic and chemo-mechanical pretreatments for enhancing methane potential of pulp mill wastewater treatment sludge Bioresour. Technol., 102, 7815–7826.

13. TU Hamburg-Harburg Reports on Sanitary Engineering, 35, 173–187.

14. Wood N., Tran H., Master E. 2009. Pretreatment of pulp mill secondary sludge for high-rate anaerobic conversion to biogas. Biores. Technol., 100, 5729–5735.

15. Xie B., Liu H., Yan Y. 2009. Improvement of the activity of anaerobic sludge by low-intensity ultrasound. Journal of Environmental Management., 90, 260–264.

16. Zhang G., Zhang P., Yang J., Liu H. 2008. Energy-efficient sludge sonication: Power and sludge characteristics. Bioresource Technology, 99, 9029–9031.

17. Zhen G., Lu X., Kato H., Zhao Y., Li Y. 2017. Overview of pretreatment strategies for enhancing sewage sludge disintegration and subsequent anaerobic digestion: Current advances, full-scale application and future perspectives. Renew. Sustainable Energy Rev., 69, 559–577.

5

Role of Substrates used for Green Roofs in Limiting Rainwater Runoff

Anna Baryła[1*], Agnieszka Karczmarczyk[1], Agnieszka Bus[1]

[1] Warsaw University of Life Sciences – SGGW, Faculty of Civil and Environmental Engineering, Department of Environmental Improvement, Nowoursynowska 166, 02-787 Warszawa, Poland
[*] Correspondent author's e-mail: anna_baryla@sggw.pl

ABSTRACT

The retention of rainwater is one of the main functions of green roofs in urban areas. One of the elements influencing the variability of rainwater retention on green roofs is the configuration of the roof, i.e. the combination of drainage and vegetation layers and plants. In the article, laboratory studies regarding the influence of the vegetation layer of the green roof on the retention of rainwater were carried out, and the influence of changes in the initial moisture content in extensive and intensive substrates on retention were compared. The analysis of seven randomly selected substrates showed that the runoff coefficients range from 0.59 to 0.71. In the case of the retention, statistically significant differences were observed in terms of the rainfall volume as well as the initial moisture content.

Keywords: green roof, substrates, runoff, retention, moisture

INTRODUCTION

Restoring natural hydrological processes in urbanized areas requires the retention of rainfall in balance with the level prior to urbanization (Burns et al. 2012, Tokarczyk-Dorociak et al. 2017). In areas of forests and meadows, approximately 60–80% of rainfall is used for evapotranspiration and infiltration (Zhang et al. 2001), whereas in urbanized areas, the surface runoff constitutes such a percentage (Geiger and Dreiseitl 1999). Green roofs are considered one of the effective methods of managing rainwater in urbanized areas (Pęczkowski et al. 2016, Baryła et al. 2017). Such roofs have a high potential of stopping runoff, as they can comprise as much as half of the non-permeable area of cities (Mentens et al. 2006). It is estimated that yearly rainfall retention on green roofs can amount to anywhere from 5 to 85% (Li and Babcock, 2014, Cipolla et al. 2016, Elliott et al. 2016, Sims et al. 2016). The variability in rainfall retention can be attributed to, among others, the configuration of the green roof (Berndtsson 2010, VanWoert et al. 2005a, Speak et al. 2013). Construction-wise, green roofs

are made up of three main elements: the vegetative/surface layer, the substrate, and the retention/drainage layer (De Nardo et al. 2005, Teemusk and Mander 2007, Getter et al. 2009, Fioretti et al. 2010, Gwóźdz et al. 2016). In green roofs, substrates are usually designed to retain rainwater and support plant growth; hence, the materials characterized by specific physical, chemical and biological properties are used (these are usually mineral-organic or mineral mixtures) (Karczmarczyk et al. 2017). In accordance with DAFA (2015 guidelines), the vegetation layer has to be characterized by a stable structure and store percolating water, making it available to plants and releasing only excess water to the drainage layer. In order to ensure these properties, substrates are made of absorbent materials such as perlite, volcanic lava, pumice, vermiculite and zeolite, from loose salvaged materials, such as red brick and slag, as well as from the materials obtained artificially, such as LECA or pollytag. Moreover, roof substrates (mainly used on intensive roofs) contain organic substances like peat (low moor peat) or compost (Molineux et al. 2009, Kohler and Poll 2010, Aslup et al. 2011, Toland et al.

2012, Chen 2013). Over the course of their use, the physical properties of substrates undergo changes (Nagase and Dunnett 2011). By mineralization of organic substance, they decrease in volume, and lose their porosity as well as water-holding capacity, which influences the degradation of conditions for plant growth (Bogacz et al. 2013). Research has shown that the structure of a green roof does not appear to be the only factor influencing the rainfall runoff or retention. These factors include the accumulation and intensity of rainfalls (Carter and Ramussen, 2006, Simmons et al. 2008), climatic conditions, seasonality (Mentens et al. 2006), preceding conditions (Bengtsson et al. 2005, Denardo et al. 2005), as well as – though to a lesser degree – the slope of the roof (Villarreal and Bengtsson, 2005, Getter et al. 2007). Despite the studies carried out in many research centres, many questions regarding the influence of different factors on the retention capacity of green roofs still remain.

The aim of the study was: 1. To confirm that the vegetation layer of a green roof is a significant factor influencing the retention of rainwater; 2 to compare the changes in the moisture content (drying up) in extensive and intensive substrates, as well as its influence on the retention capacity

MATERIALS AND METHODS

The comparison of retention as well as the influence of the drying out of substrates on the amount of leachate was determined in a laboratory experiment. Seven columns, 144 mm in diameter, were filled with substrates labelled S 1–7, which had been collected randomly from a local market (Fig. 1). S1 is a substrate of an intensive type, used for the construction of multi-layer roofs. The substrate was taken in the year 2015 from a green roof in the area of Warsaw, built in 2012. The composition of the substrate is: washed

sand, mineral grits (chalcedonite, brick, LECA), low moor peat and compost. S2-S4 (intensive type) and S6 (extensive type) are fresh substrates taken from newly-built green roofs or from "large bags", prior to the application. S2 is an intensive substrate that is to be used under lawns or small shrubs; it contains an artificial aggregate (LECA), a mineral aggregate, sand, compost and low-moor peat, as well as a fertilizer. S5 is a growing medium sampled from a fresh prefabricated *Sedum* mat (Xelo Flor moss-sedum-herbs XF317). Substrates S1-S4 are mixtures of mineral and organic compounds, whereas S6 is a 100% mineral mixture of crushed red brick, gravel, lime and sand. In the case of substrates S3-S4, no specification could be obtained; the only available information is that they conform to FLL (2008) or DAFA (2015). S7 is a substrate created from crushed red brick. The characteristics of the analysed substrates were compiled in Table 1. All columns were filled with 4 cm layers, regardless of the substrate type (Fig. 1). The application of different thicknesses for intensive and extensive substrates would have changed the retention abilities, and thus decrease the runoff volume, making it more difficult to compare the results of the experiment.

Prior to the start of the experiment, the columns were irrigated, with the substrates later being watered with the water from the water supply network for 200 days, in doses and according to a schedule developed based on the atmospheric precipitation observed in a nearby weather station (coordinates 52°16'07.16''N, 21°04'89.84''E) in 2013. The volume of runoff was measured by hand after each simulated rainfall. The measurements of the change in moisture and temperature of the substrates were carried out with a WET probe prior to each watering. In the case of mineral substrate S6, it was not possible to carry out the measurements of the moisture contents and temperature, and thus, only the amounts of runoff were measured. The organic matter – OM (%)

Table 1. Characteristics of substrates used in the study

Substrate	SI	S2	S3	S4	S5	S6	S7
Type	intensive	intensive	intensive	intensive	extensive	extensive	extensive
Age	3 years	fresh	fresh	fresh	fresh	fresh	fresh
Composition	mineral-organic	mineral-organic	mineral-organic	mineral-organic	no data	mineral	mineral
pH	7.50	7.31*	7.19*	7.60*	8.03*	7 74*	–
OM content [%]	1.9	10.4*	7.0*	7.4*	7.2*	0*	0
Bulk density [kg/m³]	1083.6	1054.8'*	105.1*	983.4*	1145.6*	1498.7*	1103.1
* Karczmarczyk et al. 2018.							

RAINFALL

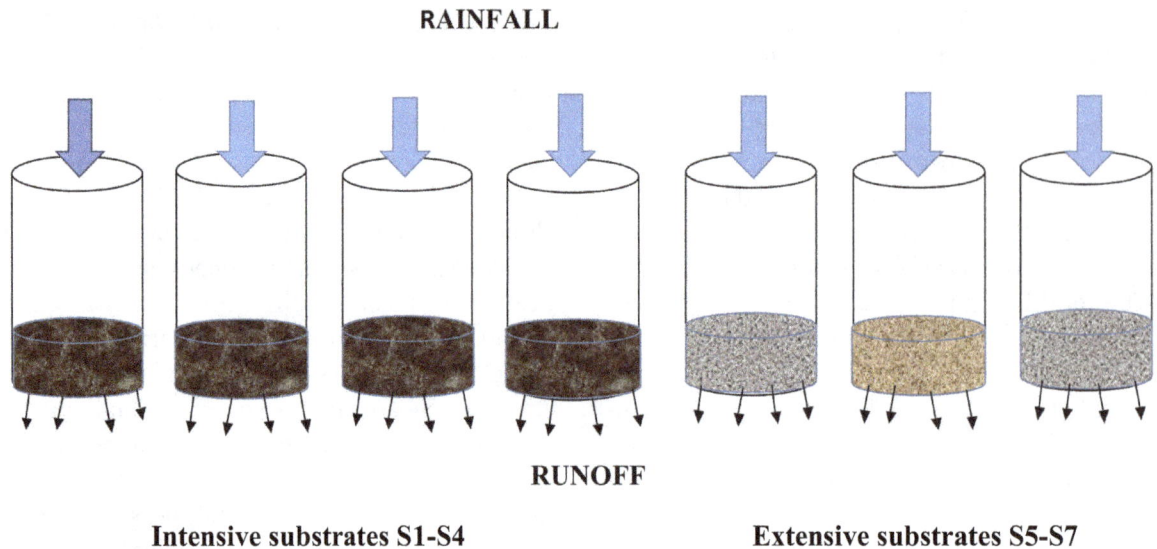

RUNOFF

Intensive substrates S1-S4   **Extensive substrates S5-S7**

Fig. 1 Set up of column experiment

content was defined as the loss of mass following the incineration of samples at a temperature of 550°C. For pH, the PN-ISO 10390:1997 standard was used, which specifies an instrumental method for the routine determination of pH using a glass electrode in a 1:5 (v/v) suspension of soil in water. The pH was measured by means of a Volcraft pH-212 pII mctcr.

Rainfall retention in individual trays was calculated in accordance with the formula:

$$R = \frac{P-Q}{P} \cdot 100\% \qquad (1)$$

where: R is the runoff retention rate (%);
P is the rainfall volume (mm);
Q is the runoff depth of green roof (mm).

The article attempts to identify the environmental factors which may play an important role in maintaining the retention of rainwater on a green roof. Two parameters were tested as explanatory variables for the retention of rainwater on green roofs, R (%), i.e., the rainfall volume

and the initial moisture content of the substrate. In order to assess the differences between individual substrates, the analysis of variance (ANOVA) was carried out, allowing for testing the significance of differences between the average values. Next, Tukey's (HSD) multiple comparison test was carried out (P<0.05). STATGRAPHICS Centurion XIV computer software was used for comparisons and statistical analyses.

STUDY RESULTS

Over the course of 200 days, each of the columns was watered with a dosage of 774.41 mm water from the water supply network (Table 2). A total of 80 rainfall events were simulated, the maximum and minimum dosage of which amounted to 48.47 mm and 3.1 mm, respectively. In the experiment, 48 rainfall events did not exceed 10 mm, 24 fell in the range of 10.1–20.00 mm, 4 in the range of 20.1–30 mm, and 2 events each simulated in the ranges of

Table 2. Experimental data of column leaching experiment

Substrate	SI	S2	S3	S4	S5	S6	S7
Observation oeriod (days)	200						
Precipitation P [mm]	774.41						
Volume of leachate Q [mml	470.58	490.77	455.31	482.45	456.32	553.00	478.40
Retention R [%]: mean (min–max)	55.13 (8.30–100)	52.35 (6.67–100)	53.60 (6.30–100)	49 54 (6.67–100)	55.56 (2.12–100)	47.02 (1.21–100)	51.14 (2.50–100)
Moisture [%]: mean (min–max)	6.83 (2.0–18.4)	13.71 (2.7–24.7)	12.33 (2.1–22.9)	13.27 (2.0–21.6)	8.62 (2.2–15.8)	–	9.71 (2.2–21.6)
Substrate temperature T [°C]: mean (min–max)	21.81 (19.6–25.7)	21.17 (19,6–24,5)	20.74 (19.0–24.5)	20.57 (18.8–24.2)	20.42 (18.7–24,4)	–	20.36 (18.5–24.5)

30.1–40 mm and 40.1–50.0 mm. The value of average retention in the individual trays was similar: S5(55.6%) > S1(55.1%) > S3(53.5%) > S2(52.4%) > S7(51.1%) > S4(49.3%) > S6(47.0). ANOVA showed that there are no statistically significant differences between the groups of retention values obtained for individual columns – $F_{(6.553)} = 0.73$; $p > 0.05$. The Tukey's test which was carried out additionally did not reveal any statistically significant differences between the groups (Fig. 2a).

The initial moisture content varied within the range from 6.83% to 13.13% (Table 2). The lowest average moisture value was noted for S1 substrate. This may be connected with the mineralization of organic substances, which often increases along with the age of the substrate (substrate after 3 years of use). ANOVA showed that there are statistically significant differences between the average initial moisture values in the individual columns – $F_{(5.467)} = 28.61$ at 95.0% level of confidence. In order to determine which groups differed statistically from each other, multiple comparisons using Tukey's test were carried out. On the basis of the test, similarities in S2-S4 columns (fresh intensive substrates) and between S5 and S7 columns (extensive substrates) were confirmed (Fig. 2b). The S1 substrate (intensive substrate after 3 years of use), on the other hand, did not reveal similarities to the remaining columns and was characterized by lower average values (Fig. 2b). Despite the obtained differences

in the initial moisture values in the individual trays, such a relationship was not observed in the case of retention.

The temperature of the substrates in the trays ranged, on average, from 20.36°C to 21.81°C, not showing any large changes over the course of the experiment (Tab. 2). The retention ability of S1-S7 substrates decreased along with an increase in rainfalls (Fig. 3), which confirmed the studies of other authors (Carter and Rasmussen 2006; Teemusk and Mander 2007). Teemusk and Mander (2007) showed 85.7% retention at a rainfall of 2.11 mm. On the other hand, at the rainfalls of over 12 mm, the runoff from green roofs was comparable to that from the reference roof. In Southfield, Michigan (roof thickness of 100 mm), Carpenter and Kaluvakolanu (2011) obtained 98.6% retention for small events (i.e. <12.6 mm), 90.2% retention for medium events (i.e. 12.7–25.4 mm) and 52.7% in the case of big events (i.e.>25.4 mm).

The obtained negative values of correlations, i.e. S1 (-0.53), S2 (-0.60), S3 (-0.59), S4 (-0.48), S5 (-0.47), S6 (-0.54), S7 (-0.61), for $p < 0.05$, indicated an average relationship between the rainfall volume (mm) and retention ability. The linear model between rainfalls and retention ability explained from 16 to 30% variance in retention ability in rainfall function (Fig. 3).

The obtained regression coefficients indicate that along with an increase in initial moisture content, the retention ability of substrates decreases.

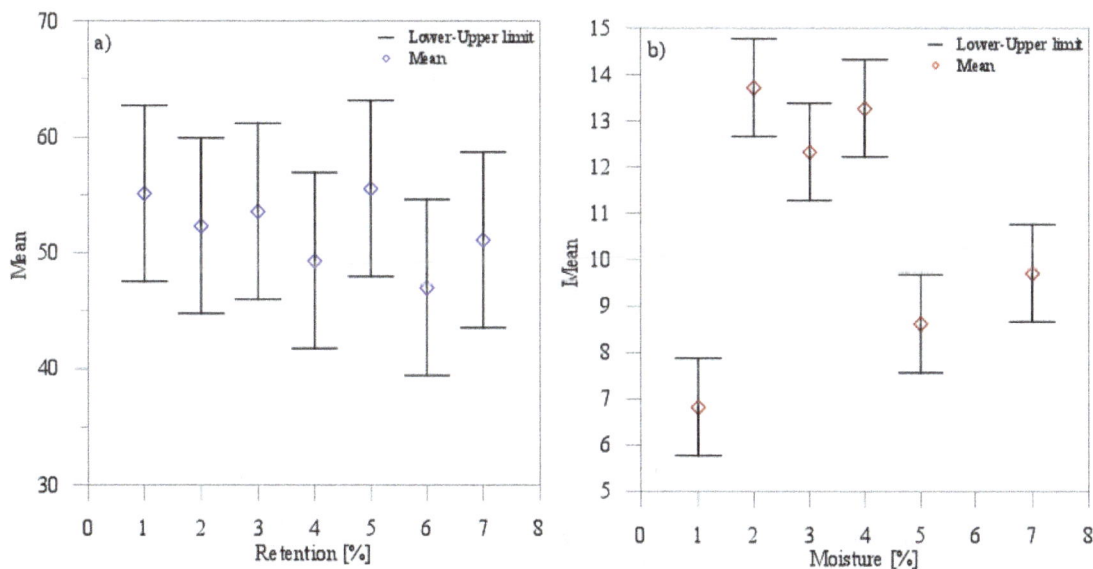

Fig. 2 Result of comparing averages a – left) retention, b – right) initial moisture content in columns with Tukey's interval at 95% level of confidence

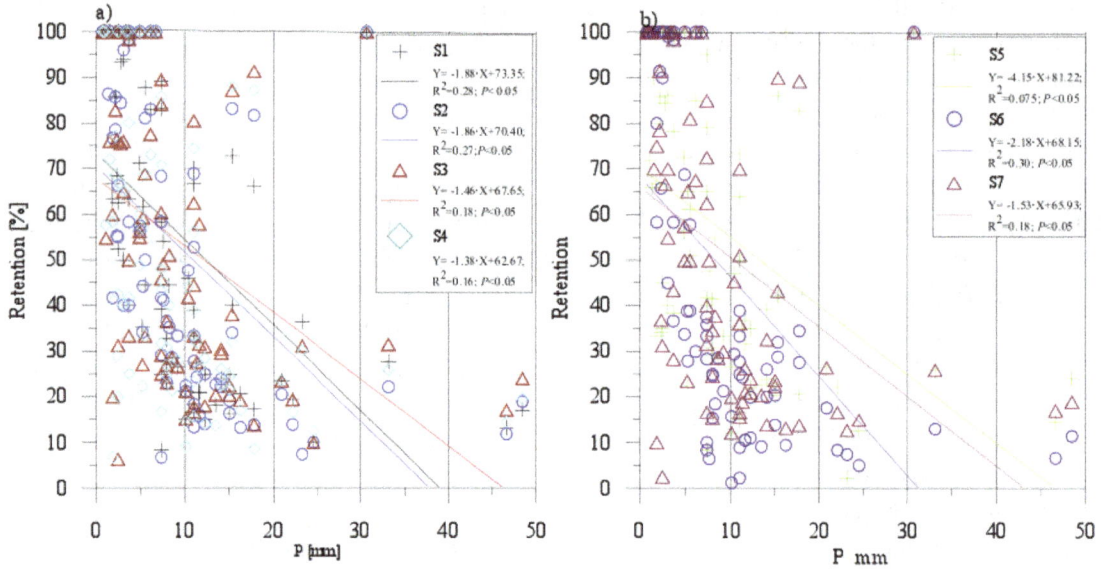

Fig. 3 Relationship between retention R [%] and rainfall P [mm]

Negative correlation coefficients in the linear relationship between retention and moisture content, were lower than in the relationship between the retention and rainfall volume. The calculated correlation coefficients were: S1(-0.42), S2 (-0.32), S3 (-0.53), S4 (-0.53), S4 (-0.56), S7 (-0.46), which signifies an average relationship. For each pair of variables, statistically significant linear correlations at a significance level of 5% were determined.

The linear model of relationships between initial moisture content and retention ability explained from 16 to 31% variance (Fig. 4).

CONCLUSION

The hydrological efficiency of green roofs is influenced by many factors, including the characteristics of rainfalls (rainfall depth, duration and intensity), the roof structure (type and depth of substrate, slope and age of green roof, drainage, and type of vegetation), the length of the dry period, fertilization and maintenance work, local sources of contamination, climate, and seasonal variability (Berndtsson et al. 2006, Morgan et al. 2013, Zhang and Guo 2013, Buffam et al. 2016). Therefore, the results of water retention by green

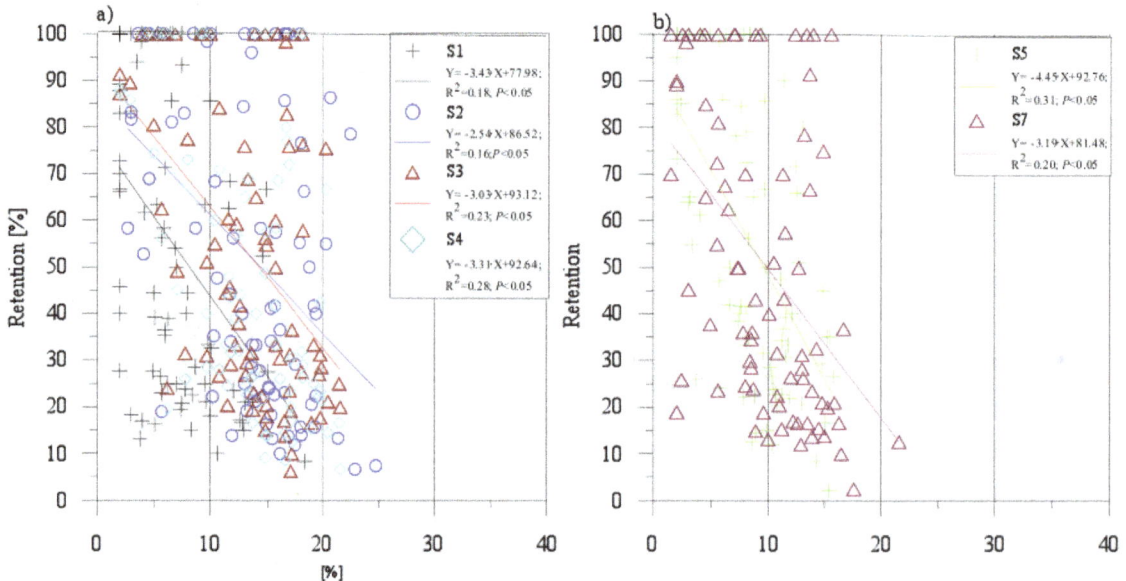

Fig. 4 Relationship between retention R [%] and initial moisture content θ [%].

roofs are characterized by large differences. The authors carried out laboratory studies (controlled environment) on the differences and values of the retention abilities of intensive and extensive substrates characterized by different physical properties. The obtained results confirmed that the vegetation layer plays a significant role in the retention of rainwater. The analysis of seven randomly collected substrates showed that the runoff coefficients ranged from 0.59 to 0.71. Fresh intensive S2-S5 substrates were found to have higher water retention abilities than S6 substrate characterized by a mineral composition. This may stem from the fact that over a longer course of use mineral substrates will be exhibit lower moisture retention abilities, as confirmed by studies carried out by Stovin et al. 2015 and Baryła et al. 2017. In the case of retention, statistically significant differences were observed when it comes to the rainfall volume and initial moisture content. Attention should be drawn to the fact that shallow, extensive systems of green roofs may decrease the risk of flooding, even in the case of torrential rain. Despite all of the analysed substrates showing good retention ability for normal rainfall events, in the case of rainfalls exceeding 20 mm, the maximum observed retention was 39%.

REFERENCES

1. Aslup S.E., Ebbs S.D., Battaglia L.L., Retzlaff W.A. 2011. Heavy Metals in Leachate from Simulated Green Roof Systems. Ecological Engineering 37, 1709–1717.

2. Baryła, A., Karczmarczyk, A., Bus, A., Kożuchowski, P. 2017. Ocena przydatności wskaźnika opadów uprzednich do opisu uwilgotnienia podłoży na zielonych dachach typu ekstensywnego (Assessing the Usefulness of the Previous Rainfall Indicator in Describing the Moisture Content of Substrates on Extensive Type Green Roofs) [in Polish]. Acta. Sci. Pol., Formatio Circumiectus, 16(4), 23–34.

3. Bengtsson L., Grahn L., Olsson J. 2005. Hydrological Function of a Thin Extensive Green Roof in Southern Sweden. Nordic Hydrol. 36 (3), 259–268.

4. Berndtsson, J. Czemiel 2010. Green Roof Performance Towards Management of Runoff Water Quantity and Quality: a Review. Ecological Engineering 36, 351–360.

5. Bogacz, A., Woźniczka, P., Burszta-Adamiak, E., Kolasińska, K. 2013. Methods of Enhancing Water Retention in Urban Areas. Scientific Review Engineering and Environmental Sciences, 22 (1), 27–35 (in Polish).

6. Buffam I., Mitchell M.E., Durtsche R.D. 2016. Environmental Drivers of Seasonal Variation in Green Roof Water Quality. Ecological Engineering 91, 506–514.

7. Burns M.J., Fletcher T.D., Walsh C.J., Ladson A.R., Hatt B.E. 2012. Hydrologic Shortcomings of Conventional Urban Stormwater Mmanagement and Opportunities for Reform. Landsc. Urban Plan. 105, 230–240.

8. Carpenter D., Kaluvakolanu P. 2011. Effect of Roof Surface Type on Stormwater Run-off from Full-scale Roofs in a Temperate Climate. J. Irrigat. Drain. Eng. 137, 161–169.

9. Carter T.L. Rasmussen T.C. 2006. Hydrologic Behavior of Vegetated Roofs. Journal of the American Water Resources Association 42 (5), 1261–1274.

10. Chen C.F. 2013. Performance Evaluation and Development Strategies for Green Roofs in Taiwan: A review. Ecological Engineering, 52, 51–58.

11. Cipolla S.S., Maglionico M., Stojkov I. 2016. A Long-term Hydrological Modelling of an Extensive Green Roof by Means of SWMM. Ecological Engineering 95, 876–887.

12. DeNardo J.C., Jarett A.R., Manbeck H.B., Beattie D.J., Berghage R.D. 2005. Stormwater Mitigation and Surface Temperature Reduction by Green Roofs. Trans. ASAE 48 (4), 1491–1496.

13. DAFA. Dachy zielone. 2015. Wytyczne do Projektowania, Wykonywania i Pielęgnacji Dachów Zielonych–Wytyczne dla Dachów Zielonych (Guidelines for Designing, Constructing and Caring for Green Roofs – Guidelines for Green Roofs) [in Polish]; Stowarzyszenie Wykonawców Dachów Płaskich i Fasad (DAFA): Opole, Poland.

14. Elliott R.M., Gibson R.A., Carson T.B., Marasco D.E., Culligan P.J., McGillis W.R. 2016. Green Roof Seasonal Variation: Comparison of the Hydrologic Behavior of a Thick and a Thin Extensive System in New York City. Environ. Res. Lett. 11, 074020.

15. Fioretti R., Palla A., Lanza L. G., Principi P. 2010. Green Roof Energy and Waterrelated Performance in the Mediterranean Climate. Building and Environment, 45, 1890–1904.

16. FLL 2008. Forschungsgesellschaft Landschaftsentwicklung Landschaftsbau (FLL). Guidelines for the Planning, Construction and Maintenance of Green Roofing – Green Roofing Guideline; FLL: Bonn, Germany.

17. Geiger W. Dreiseitl H. 1999. Nowe sposoby odprowadzania wód deszczowych (New Means of Rainwater Drainage) [in Polish] Poradnik retencjonowania i infiltracji wód deszczowych do gruntu

na terenach zabudowanych. Oficyna Wydawnicza Projprzem-EKO, Bydgoszcz.

18. Getter K.L., Rowe D.B., Andresen J.A. 2007. Quantifying the Effect of Slope on Extensive Green Roof Stormwater Retention. Ecological Engineering, 31, 225–231.

19. Gwóźdz K., Hewelke E., Żakowicz S., Sas W., Baryła A. 2016. Influence of Cyclic Freezing and Thawing on the Hydraulic Conductivity of Selected Aggregates Used in the Construction of Green Roofs. Journal of Ecological Engineering 17(4), 50–56.

20. International Organization for Standardization (ISO). PN-ISO 10390:1997 Equivalent to ISO 10390:1994 Soil Quality – Determination of pH; ISO: Geneva, Switzerland, 1998.

21. Karczmarczyk A., Baryła A., Kożuchowski P. 2017. Design and Development of Low P-Emission Substrate for the Protection of Urban Water Bodies Collecting Green Roof Runoff. Sustainability 9, 1795.

22. Karczmarczyk A., Bus A., Baryła A. 2018. Phosphate Leaching from Green Roof Substrate – Can Green Roofs Pollute Urban Water Bodies?. Water 10, 199.

23. Köhler M., Poll P.H. 2010. Long-Term Performance of Selected Old Berlin Green Roofs in Comparison to Younger Extensive Green Roofs in Berlin. Ecological Engineering 36, 722–729.

24. Li, Y., Babcock Jr., R.W. 2014. Green Roof Hydrologic Performance and Modeling: a Review. Water Sci. Technol. 69, 727–738.

25. Mentens J., Raes D., Hermy M., 2006. Green Roofs as a Tool for Solving the Rainwater Runoff Problem in the Urbanized 21st Century? Landscape Urban Plann. 77 (3), 217–226.

26. Molineux Ch. J., Fentiman Ch. H., Gange A.C., 2009. Characterising Alternative Recycled Waste Materials for Use as Green Roof Growing Media in the U.K. Ecological Engineering 35, 1507–1513.

27. Morgan S., Celik S., Retzlaff W. 2013. Green Roof Storm-Water Runoff Quantity and Quality. J. Environ. Eng. – ASCE 139 (4), 471–478.

28. Nagase A., Dunnett N. 2011. The Relationship Between Percentage of Organic Matter in Substrate and Plant Growth in Extensive Green Roofs. Landsc. Urban Planning, 103, 230–236.

29. Pęczkowski G., Orzepowski W., Pokładek R., Kowalczyk T. Żmuda R. 2016. Retention Proper-

ties of the Type of Extensive Green Roofs as an Example of Model Tests. Acta Sci. Pol., Formatio Circumiectus, 15(3), 113–120 (in Polish).

30. Simmons M.T., Gardiner B., Windhager S., Tinsley J. 2008. Green Roofs are Not Created Equal: the Hydrologic and Thermal Performance of Six Different Extensive Green Roofs and Reflective and Non-reflective Roofs in a Sub-tropical Climate. Urban Ecosyst. 11(4), 339–348.

31. Speak A.F., Rothwell J.J., Lindley S.J., Smith C.L. 2013. Rainwater Runoff Retention on an Aged Intensive Green Roof. Science of the Total Environment, 461–462, 28–38.

32. Sims A.W., Robinson C.E., Smart C.C., Voogt J.A., Hay G.J., Lundholm J.T., et al. 2016. Retention Performance of Green Roofs in Three Different Climate Regions. J. Hydrol. 542, 115–124.

33. Stovin V., Poë S., DeVille S., Berretta C. 2015. The Influence of Substrate and Vegetation Configuration on Green Roof Hydrological Performance. Ecological Engineering 85, 159–172.

34. Teemusk A., Mander Ű. 2007. Rainwater runoff quantity and quality performance from a green roof: The effects of short-term events. Ecological Engineering 30, 271–277.

35. Toland D.C., Haggard B.E., Boyer M.E., 2012. Evaluation of Nutrient Concentrations in Runoff Water from Green Roofs, Conventional Roofs, and Urban Streams. Transactions of the ASABE 55(1), 99–106.

36. Tokarczyk-Dorociak K., Walter E., Kobierska K., Kołodyński R. 2017. Rainwater Management in the Urban Landscape of Wroclaw in Terms of Adaptation to Climate Changes. J. Ecol. Eng. 18(6):171–184.

37. Villarreal, E.L., Bengtsson, L. 2005. Response of a Sedum Green-Roof to Individual Rain Events. Ecological Engineering 25 (1), 1–7.

38. VanWoert N.D., Rowe D.B., Andresen J.A., Rugh C.L., Fernandez R.T., Xiao L. 2005a. Green Roof Stormwater retention. J. Environ. Qual. 34, 1036–1044.

39. Zhang L., Dawes W.R., Walker G.R. 2001. Response of Mean Annual Evapotranspiration to Vegetation Changes at Catchment Scale. Water Resour. Res. 37, 701–708.

40. Zhang S., Guo Y. 2013. Analytical Probabilistic Model for Evaluating the Hydrologic Performance of Green Roofs. J. Hydrol. Eng. 18, 19–28.

6

Invasive Species and Maintaining Biodiversity in the Natural Areas – Rural and Urban – Subject to Strong Anthropogenic Pressure

Beata Fortuna-Antoszkiewicz[1*], Jan Łukaszkiewicz[1*], Edyta Rosłon-Szeryńska[1], Czesław Wysocki[2], Piotr Wiśniewski[3]

[1] Department of Landscape Architecture, Warsaw University of Life Sciences – SGGW, Nowoursynowska 166 St., 02-787 Warsaw, Poland

[2] Department of Environment Protection, Warsaw University of Life Sciences – SGGW, Nowoursynowska 166 St., 02-787 Warsaw, Poland

[2] Department of Environmental Protection of Mokotów District, City of Warsaw, Poland

* Corresponding author's e-mail: beata_fortuna@op.pl

ABSTRACT

Expansion of invasive species can be clearly seen all over Poland. Foreign tree and herbaceous plant species are effectively taking over more and more habitats competing with native vegetation. This phenomenon is strongly pronounced in the areas subject to strong anthropogenic pressure. The presence of invasive plants replacing the native vegetation is a threat for biodiversity and ecological balance. The research carried out by the authors between 2011 and 2017 on selected sites (comparatively: urban and open spaces, including a 600 ha park and a 10 km long forest strip along a river) confirms the pressure exerted by invasive species irrespective of the natural conditions of a particular site or its type – in each case it is most prominent in areas where vegetation is not properly maintained or where it is not maintained at all. The research was based on the dendrological inventories and phytosociological assessments. The inventories were used for a detailed assessment of both the condition and structure of treestands, including accounting for invasive species. Phytosociological assessment can, among others, form a basis for forecasting ecological stability of individual plant communities. Uncontrolled expansion of invasive species, especially in the areas of strong anthropogenic pressure, may cause unfavourable natural succession and in consequence – destabilisation of ecological system in a given area.

Keywords: anthropogenic pressure, alien plant species, biodiversity, vegetation maintenance

INTRODUCTION

Invasive plant species have been introduced into environment by man – either deliberately (by introducing utility plants) or inadvertently (accidentally introduced species). Currently, approximately 12 thousand foreign species have been identified, 10–15% of which are believed to have a negative influence of varying degree (Regulation (EU) No 1143/2014 of the European Parliament and of the Council of 22, October 2014). Such plants are mainly species cultivated in a controlled manner (agricultural cultivation, botanical gardens) which have found their way into the environment, such as *Impatiens parviflora* DC. – imported into botanical gardens in the 19th century from central Asia, now commonly seen in European forests and parks (Gwiazdowski, 2014). Some species, such as *Prunus serotina* (Ehrh.) Borkh., initially planted for practical reasons (acquiring precious wood, recreating tree-stands e.g. in industrial areas)[1], then to improve biotic communities (enriching species pool of lower forest strata), have successfully acclimatized in new areas (Sudnik-Wójcikowska, 1987). Similarly, *Quercus rubra* L. – resistant to pollution, was commonly used, among other, as a fore-crop in poor habitats, including industrial idle lands (such as afforestation in sand pits) (Strzelecki & Sobczak, 1972). Both species turned out to be extremely

---

[1] Its typical use include recreating forest-stands in mining break-downs, reinforcing stockpiles, ravines and their slopes, due to its resistance to dust and fumes and low soil requirements (dry, acidy soils) (Strzelecki & Sobczak, 1972; Seneta & Dolatowski, 2012).

expansive – black cherry, which in European forests has a bush form, grows into thick scrubs effectively blocking the development of other native tree species, thus hindering the forest renewal (Seneta & Dolatowski, 2012). A large number of black cherry patches have been found with a high number of specimens in each patch. This species keeps on taking over new plots and areas, easily invading the natural, semi-natural and anthropogenic vegetation communities. The areas where this species is a threat include forests, protected areas or habitats disturbed by anthropogenic pressure (Tokarska-Guzik et al., 2012). In a few decades after introduction of red oak into cultivated forests (in the beginning of 20th century [Sudnik-Wójcikowska, 1987]), its spontaneous spreading has been reported in numerous plots in various regions, as well as a large number of specimens in newly created patches (as a result of its fecundity – the species grows faster than other oaks and bears fruit already at a young age). Red oak continues expanding onto new plots and areas[2]. It may be harmful in forests and protected areas, since it easily penetrates into natural communities (Sudnik-Wójcikowska, 1987; Seneta & Dolatowski, 2012; Tokarska-Guzik et al., 2012). Another tree species – *Acer negundo* L. – was brought to Europe in the 17th century and in the 18–19th century to Poland (in Warsaw – approx. 1880). Initially, it was considered valuable due to its quick growth, and as such used in gardens (in the beginning of the 20th century, often used in parks and planted in the country). Since 1940s, its strong expansion can be seen; in the 1980s it had already become one of the most common plants in Warsaw, often seen in anthropogenic habitats, acclimated in forest communities (such as riparian forests along Vistula River, border area of oak-hornbeam forests, pine and oak forests) (Sudnik-Wójcikowska, 1987). The presence of *Acer negundo* L. in riparian forests along the Vistula river should be considered a stage of secondary replacement succession (Matuszkiewicz & Roo-Zielińska, 2000).

Foreign herbaceous plants include *Reynoutria japonica* Houtt. – found in Europe since the middle of the 19th century; first reports of this taxon in Warsaw appeared in 1964.; in 1980s

it became common in anthropogenic ruderal and semi-natural habitats (along roads, at waste dumps, idle lands), but also at borders of willow riparian forests, alder forests, pine-birch-oak stands, in shrubs near water reservoirs (Sudnik-Wójcikowska, 1987). Another example includes *Echinocystis lobata*, cultivated after 1945 in Cracow; in 1980s it was commonly seen (e.g. in Warsaw) in anthropogenic habitats (along roads, waste dumps, near allotment gardens etc.) but also in semi-natural and natural communities (mainly in willow and willow-poplar riparian forests) (Sudnik-Wójcikowska, 1987).

Since the middle of the 20th century, a clear increased expansiveness of many foreign herbaceous plant species (e.g. *Reynoutria japonica, Echinocystis lobata, Solidago gigantea, Impatiens parviflora*) as well as tree species (e.g. *Acer negundo, Prunus serotina, Robinia pseudoacacia*) can be observed. Other species, such as *Quercus rubra* successfully reappear in anthropogenic habitats and forests (Sudnik-Wójcikowska, 1987).

Expansion of cities and industrial areas, exerting anthropogenic pressure over greater areas of land contributes to the transformation of habitats and their plant communities. This results in an increased number of sites with disturbed soil and water conditions, often polluted, which are suitable for highly tolerant species that continue to take over a particular ecosystem. Highly expansive foreign species may replace the local populations, leading to a reduction in the species count and finally to a change of ecosystem structure and destabilisation of the entire natural system in a given area (Gwiazdowski, 2014). As a result of introduction of foreign species into forests[3], entire forest sub-compartments became dominated by one "exot" or tree-stands composed of native species with an addition of foreign taxons with a varying degree of mix. Some foreign species may become invasive if they spontaneously spread and infiltrate natural biocenoses (Gazda & Augustynowicz, 2014). The presence of invasive vegetation replacing the native vegetation is a

---

[2]   Red oak currently is present in approx. 3% of forest sub-compartments (nearly 5% of national forests' area). It is a dominant species in approx. 0.5% of forest sub-compartments, so 3 900 ha compared to approx. 80 ha in mid 20th century – a 50-fold increase in area (Gazda & Augustynowicz, 2014).

[3]   Over 30 foreign tree species have been introduced into Polish forests, including 22 coniferous species and 9 deciduous species, some of which are more numerous than others, such as Douglas fir (*Pseudotsuga menziesii* (Mirb.) Franco), red oak (*Quercus rubra* L.), white pine (*Pinus strobus* L.) (Seneta & Dolatowski, 2012; Gazda & Augustynowicz, 2014). Other species previously cultivated as ornamental species include: box alder (*Acer negundo* L.), black cheery (*Prunus serotina* (Ehrh.) Borkh), black locust (*Robinia pseudoacacia* L.) (Gazda & Augustynowicz, 2014).

threat for biodiversity and ecological balance at species level (change of ecosystem species compositions) and super-species level (threat to habitat and ecosystem diversity). Additionally, some species, especially in the locations where humans are present, pose a health hazard (such as low ragweed pollen, which causes strong allergic reactions, or Sosnowsky's hogweed causing painful burns) (Gwiazdowski, 2014)[4].

MATHERIALS AND METHODS

In order to identify the invasive plants appearing in the environment, the authors have conducted research in selected sites between 2011 and 2017. These sites included managed urban sites and rural areas, recreational (5 parks, including one with an area of approx. 600 ha) and technical (2 stripes of forests near water, inducing one 10 km long), as well as unmanaged sites (2 areas). The selected areas are subject to varying degrees of anthropogenic pressure. For the last 40 to 80 years, the vegetation succession (limited human interference) took place in whole or part of these areas. The research carried out in the sites has been preceded with the analysis of natural conditions (climate, soil, water conditions, habitat etc.) as well as functional and spatial analyses. The detailed research is based on the dendrological inventories and phytosociological assessments. The inventories were used for a detailed assessment of both condition and structure of tree-stands (spatial structure, species composition, health, age) including accounting for invasive and expansive species (e.g. Fortuna-Antoszkiewicz & Łukaszkiewicz 2017; Łukaszkiewicz & Fortuna-Antoszkiewicz 2017). These comprised: identification of taxons, their frequency and distribution, dendrometric measurements of trees/shrubs. Phytosociological assessment (phytosociological photographs using a 5-point Braun-Blanquet scale) was carried out to identify the vegetation communities and formed a basis for forecasting

ecological stability of individual phytocenoses (Wysocki & Sikorski, 2009).

RESULTS

The research carried out in sites in various regions of Poland (Figure 1) resulted in identification of the following invasive species (Table 1): trees (4 taxons) – *Prunus serotina* Ehrh., *Quercus rubra* L., *Acer negundo* L., *Robinia pseudoacacia* L.; herbaceous plants (7 taxons) – *Reynoutria japonica* Houtt., *Reynotria xbohemica* Chrtek et Chyrtkova, *Impatiens glandulifera* Royle, *Impatiens parviflora* DC., *Solidago canadensis* L., *Solidago gigantea* Aiton, *Echinocystis lobata*[5].

The most common invasive taxons in the researched area include *Robinia pseudoacacia* L., *Acer negundo* L. and *Reynoutria japonica* Houtt. – each found in 4 sites; the least common – *Echinocystis lobata* – found in one site. In decreasing frequency: in 3 sites – *Quercus rubra* L., *Prunus serotina* Ehrh., *Impatiens parviflora* DC.; in 2 sites – *Reynotria xbohemica* Chrtek et Chyrtkova, *Impatiens glandulifera* Royle, *Solidago canadensis* L., *Solidago gigantea* Aiton.

In the existing tree-stands, invasive tree species can be found mainly around mother specimens, but they also invade open areas (unused park interiors, mini-interiors created after felling of dead large tree specimens, at fringes of the tree-stand). In the researched sites, invasive herbaceous plants can be found in open and well sunlit patches of land, mainly in synanthropic habitats (e.g. near communication routes).

Frequency and numbers of individual invasive taxons are varied:
- they are more numerous in the intensively utilised areas with a stronger anthropogenic pressure, unmaintained or maintained only to a minimum degree (such as Chorzów – expansive part, on a hill; Warsaw: Ursynów park – reservation part at the foot of the escarpment, Żerań Canal – in the areas with high penetra-

---

[4] In Poland, the management of foreign species is regulated by the Act of 16 April 2004 on protection of the environment as amended and Resolution of the Minister of Environment of 9 September 2011 on a list of alien plants and animals which if released into environment may threaten native species or habitats [Journal of Laws no. 210, item 1260], and at European level – Regulation (EU) No. 1143/2014 of the European Parliament and the Council of 22, October 2014 on the prevention and management of the introduction and spread of invasive alien species (European Union Journal of Law L 317/35).

[5] Clearly dominant herbaceous expansive species forming patches of dense monocultures have been identified in sites: they are an indication of high nitrogen (nitrite) concentration and a proper distribution of humus in the soil. These include: common nettle (*Urtica dioica*) – typical to mesophilic herbaceous communities; ground elder (*Aegopodium podagraria*) – growing among other in elm and ash forests; hemp-agrimony (*Eupatorium cannabinum*) – a rhizome plant growing in bogs formed by water containing calcium compounds and in spring water communities.

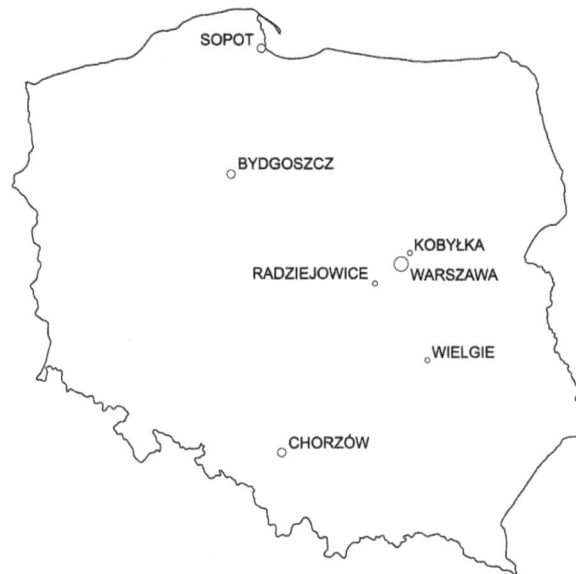

Figure 1. Location of research polygons in Poland (prepared by: P. Wiśniewski)

tion, Exhibition Canal; Radziejowice – eastern periphery near expressway; Sopot park);

- in the areas with weaker anthropogenic pressure, the invasive plants are less numerous (such as Southern escarpment in Bydgoszcz, Kobyłka pond and scrub complex, Żerań Canal – in less used locations) and are less varied in terms of species (e.g. Wielgie park – only one taxon: *Echinocystis lobata*; grows seasonally in a depression, within the water way area on a patch with full sun exposition, covering 100% of available space);

- in the heavily used sites with strong anthropogenic pressure in plots subject to maintenance activities (such as mowing, clearing/correction of small trees in lower forest strata), the invasive plants are basically absent (e.g. Ursynów park in escarpment crown) or are present only in controlled spots (e.g. Silesia Park – western intensive zone).

SILESIA PARK IN CHORZÓW – CASE STUDY

Silesia Park is a special site, it has an area of 600 ha, and was established (in 1950–1968) on post-industrial and degraded lands within the Silesia agglomeration. It is an area subject to strong anthropogenic pressure (approx. 3 million users annually). The current species structure of the park is a consequence of 60 years of often spontaneous growth, i.e. secondary succession.

In the initial stage of Park establishment, the tree and shrub species were introduced for the purpose of reclamation and phytoremediation. This decision was made because most of the Park area had poor soil (e.g. podzol), additionally degraded by mining and heavy industry. Forests were planted using pioneering species, hoping that as they grew, favourable conditions (habitat transformation) for more demanding and long living species would be created. The used native species included: birch, some poplars (aspen, black, white) as well as willows, hazel and elder. Two foreign species, already present in the area were also planted: *Quercus rubra* L. and *Prunus serotina* Ehrh. (Łukaszkiewicz & Fortuna-Antoszkiewicz 2017). In this period, these species were commonly used in western Poland for afforestation and as a forecrop for poor and degraded soils.

In 2013–2014, the authors conducted an evaluation of Silesia Park vegetation. The tree-stand structure analysis carried out within its current boundaries covered: spatial and species structure, age and health. Additionally, phytosociological evaluation was carried out in selected plots (Figure 2); the species composition and spatial structure of communities was analysed with special attention given to undergrowth strata (herbaceous plants). In general, the following phytocenoses are present in Silesia Park: a/ xerothermic turfs; b/ pasture-like and near pasture meadows – in the areas of expansive park lawns; c/ wet meadows, e.g. in terrain depressions or near water reservoirs; d/ water and near water communities

Table 1. Invasive plants in the investigated objects (elaborated: B. Fortuna-Antoszkiewicz, J. Łukaszkiewicz, E. Rosłon-Szeryńska, P. Wiśniewski, 2011–2017)

Object / location / surroundings / type of object / characteristics / years of research / area covered by research		
Phytosociological communities (dominant)	Invasive plants occurring in the area	Maintenance of vegetation
PARKS AND GARDENS		
1/ Silesia Park (The Gen. George Ziętek Voivodship Park of Culture and Recreation) / Chorzów, Upper Silesian region, in the triangle of three large industrial cities: Chorzów, Katowice and Siemianowice Śląskie / in the neighborhood: urban areas with multi-family housing + main traffic routes / area recultivated – city park (formerly: the so-called folk park) of a supralocal (regional) character (1950s - 1960s) / **the south-west part -** composed classical forms of vegetation; **east part** - with a compact, dense stand / **2013-2016** / ca. **600 ha**		
Actual vegetation: xerothermic grasslands; meadows with pasture and semi-pasture features - in the area of extensive park lawns; wet meadows - eg in depressions of the area, or in the vicinity of water reservoirs; by-water and water communities (peripheral zone of park ponds); communities having the features of alder-ash carrs, riparian forests, oak-hornbeam-linden forest, oak woods and forest fringe communities - in the area of extensive park stands; synanthropic or semisynanthropic communities - in areas developed extensively or completely abandoned to secondary plant succession.	**Woody species:** *Quercus rubra* L., *Prunus serotina* (Ehrh.) Borkh. **Herbaceous species:** *Reynoutria japonica* Houtt., *Reynotria xbohemica* Chrtek et Chyrtkova, *Impatiens glandulifera* Royle, *Impatiens parviflora* DC., *Solidago canadensis* L., *Solidago gigantea* Aiton	**Not occurring** **on a larger part of the area** (in the eastern extensive zone, on the hill - an area with a dense forest stand)
2/ Park in Radziejowice / central Mazovia, open areas (agricultural) / the neighborhood : rural areas, on fragments - single-family housing / a historic palace and park complex (landscape park - beginning of the 19th century) / **west part -** composed classical forms of vegetation; eastern part – naturalistic / **2015** / ca. **25 ha**		
Potential vegetation: subcontinental oak-hornbeam forests of the Middle Polish variety. **Actual vegetation:** [forest communities] *Ribeso nigri-Alnetum; Fraxino-Alnetum; Tilio cordatae-Carpinetum betuli caricetosum remotae; Tilio cordatae - Carpinetum betuli typicum*; [meadow communities] *Cirsietum rivularis; Caricetum gracilis; Arrhenatheretum elatioris.*	**Woody species:** *Prunus serotina* (Ehrh.) Borkh. - in open areas of the terrain (mini interiors); *Quercus rubra* L. - occurs singly on peripheral fragments (East); **Herbaceous species:** *Reynoutria japonica* Houtt., *Impatiens glandulifera* Royle, *Impatiens parviflora* DC	**Not occurring** **on a peripheral zone of park** (naturalistic zone, eastern)
3/ Park in Wielgie / southern Mazovia / surrounded by: open and agricultural areas / **2012** / ca. **8 ha** / historic landscape park (middle of 19th century) / garden's composition forms in decline; throughout the area - natural succession		
Potential vegetation: *Fraxino-Alnetum* - in the lowest part; *Tilio-Carpinetum* - in the upper part. **Actual vegetation:** domination of plants from *Circeo-Alnetum* i *Tilio-Carpinetum;* on a part of the area - monoculture of herbaceous species, e.g. *Urtica dioica* (from mesophilic herbs), *Aegopodium podagraria* (occur. among others in riparian forests), *Eupatorium cannabinum* (occur. on swamps with water incl. Ca compounds and in communities accompanying w spring areas).	**Herbaceous species:** *Echinocystis lobata* - 1. position	**Not occurring in the whole area** (compact tree stand) **/** neglected since the 1940s.
4/ Park Ursynów - SGGW (WULS) / Warsaw / on top of Warsaw Escarpment, the southern part / from the south - intense multi-family housing; from north - open areas with investment pressure / historical palace- park ensemble (18th century) / headquarters of the university / on the top of escarpments - composed garden forms; at the foot of the escarpment - a naturalistic part **(nature reserve) / 2011-2012** / ca. **8 ha**		
Potential vegetation: top of the escarpment: *Tilio-Carpinetum typicum;* an escarpment area (slopes and lower terrace): stand similar to the stand of potential vegetation. **Actual vegetation:** communities of eutrophic deciduous forests (class *Querco-Fagetea*); meadow and pasture communities (class *Molinio-Arrhenatheretea*); ruderal communities (class *Artemisietea vulgaris*) + associated species.	**Herbaceous species:** *Reynoutria japonica* Houtt., *Impatiens parviflora* DC. - occur at the foot of the escarpment	**Continuous maintenance** care on the crown of the escarpment / **the part of the area without maintenance -** on the slope and at the foot of the escarpment
5/ Historical palace- park ensemble in Sopot / a part of the North Park - a clearing and woodland around the lime pleaching (so called "*bindage*") / the Skarpa Sopocka Upland, Franciszka Goyka str. 1-3 / surrounded by: Skarpa Sopocka forests, residential development / historical park by estate of Wilhelm Jüncke (1903 r.) / garden's composition in landscape style / **2011** / ca. **0,5 ha**		

Table 1 cont.

Object / location / surroundings / type of object / characteristics / years of research / area covered by research		
Phytosociological communities (dominant)	**Invasive plants occurring in the area**	**Maintenance of vegetation**
Potential vegetation: *Galio odorati-Fagetum* in the lower part and on slopes of the escarpment; *Stellario-Carpinetum* - in the part of the upland. **Actual vegetation:** Overgrowing of glades with a degenerative form of substitute forest and shrub communities, mainly of features of oak-hornbeam forests.	**Woody species:** *Robinia pseudoacacia* L.	**No maintenance** for about 40 years
BELT PLANTINGS AND UNDEVELOPED AREAS		
6/ Area along the Wystawowy Canal [Exhibition Canal] / Warsaw, Saska Kępa / in neighborhood: allotments (East) and multi-family housing (West) / water-side woodlots / spontaneous with the remaining composed plantings from the 1970s / **2017** / strip of land ca. **5,0 ha** (length 1170 m, wide 27-60 m)		
Potential vegetation: communities of deciduous forests with maples and robinia similar to reach oak-hornbeam forests with elements of *Carpinion* (All.) i *Robinietea* (Cl.). **Actual vegetation:** a complex of poorly developed segetal and ruderal communities (with the domination of *Galinsogo-Setarietum*) in the allotment gardens.	**Woody species:** *Acer negundo* L.	**Sporadic / periodic** along both banks of the Canal
7/ South Escarpment in Bydgoszcz / Bydgoszcz, South Escarpment, Toruńska str./ surroundings: multi-family buildings / undeveloped land / spontaneous plants communities - habitat of slope oak-hornbeam / **2017** / ca. **0.6 ha** (slope with a length of 200 m and a width of 30 m)		
Potential vegetation: subcontinental oak-hornbeam forests from the belt of the great valleys of the Wielkopolska-Kujawska Region, from the Kujawski District. **Actual vegetation:** *Tilio-Carpinetum typicum* sub-continental forest of the slope variety with a small amount of invasive plants in the undergrowth and undergrowth.	**Woody species:** *Robinia pseudoacacia* L., *Acer negundo* L.)	**Lack** of **maintenance/** limited penetration (the path along escarpment's slope)
8/ The area along the Żerański Canal / fragment - west bank / Warsaw, Żerań / municipality Nieporęt / open areas; development of single and multi-family housing / / water-side, protective woodlots / spontaneous vegetation with the remaining composed plantings from the 1960s / **2015 -2016** / ca. **25 ha** (strip of land length ca. **10 km** / average wide 20,0 [30,0] m)		
Potential vegetation: communities of deciduous forests with robinia similar to poor oak-hornbeam forests and mixed coniferous forests with elements of classes: *Querco-Fagetea, Vaccinio-Piceetea* and *Robinietea* (Cl.); communities of thermophilic deciduous forests with robinia, similar to bright oakwood forests (with elements of *Quercetalia pubescetis* order (O.) and *Robinietea* class (Cl.). **Actual vegetation:** degenerative forms of substitute forest and shrub communities, including subcontinental and mixed coniferous forest (*Querco roboris-Pinetum* sensu lato, *Pino-Quercetum / Vaccinio-Picetea* class) with a significant proportion of neophytes; locally: anthropogenic robinia forest (association *Chelidonio–Robinietum,* class *Robinietea*); locally: communities of compact sandy grasslands with sheep's fescue (mainly associations: *Diantho-Armerietum, Sileno-Festucetum* and others).	**Woody species:** *Quercus rubra* L., *Prunus serotina* (Ehrh.) Borkh., *Robinia pseudoacacia* L., *Acer negundo* L. **Herbaceous species:** *Reynoutria japonica* Houtt., *Reynotria xbohemica* Chrtek et Chyrtkova, *Solidago canadensis* L., *Solidago gigantea* Aito	**Not occurring in the whole area /** negligence since the 1980s.
9/ Ponds - thicket complex in Kobylka (wasteland) / Kobylka, Napoleona street - green in the industrial zone / surrounded by: industrial plants, railway track / undeveloped land / by-water vegetation in the area of ponds formed in the former channel of the Długa river in Kobyłka / **2015** / ca. **2 ha**		
Potential vegetation: sedge and swamp communities of the class *Phragmitetea;* riparian forest (*Salici-Populetum*) - in the lower part; oak-hornbeam forests of low variety (*Tilio-Carpinetum*) - in the upper part. **Actual vegetation:** water and by-water communities (peripheral zone of ponds); nitrophilous shrub-thicket communities (*Sambuco-Salicilion*), being a further stage of succession in the forest regeneration process; communities with the characteristics of riparian forests and low oak-hornbeam forests.	**Woody species:** *Robinia pseudoacacia* L., *Acer negundo* L	**Lack of maintenance -** no penetration

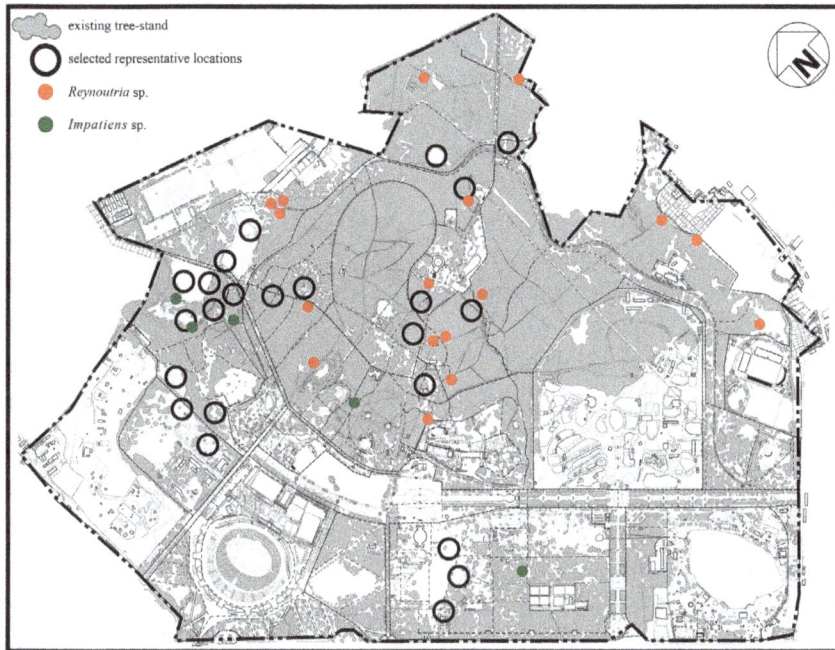

Figure 2. Silesia Park – phytosociological evaluation and location of selected invasive herbaceous plants (prepared by: B. Fortuna-Antoszkiewicz, J. Łukaszkiewicz, P. Wiśniewski, 2013/2014)

(bank area of park ponds); e/ communities similar to alder riparian, oak-hornbeam, oak forests and forest fringe communities – within expansive park tree-stands; f/ synanthropic and semi-synanthropic communities – in the areas with an expansive management plan or where secondary succession was allowed.

Currently, among the pioneering tree species used for recultivation and as a fore-crop in Silesia Park, red oak and black cherry can be described as expansive. Numerous species grow in large parts of the area, especially in the extensive part of the Park (on the hill, where vegetation maintenance is limited) – in tree (A) as well as in other strata, and what is important – in the underbrush strata (B) – especially near older mother species (e.g. representative red oak specimens with breast height circumference of: 120/ 126/150/ 152/ cm and approx. 25.0 m high and representative specimens of black cherry with trunk circumference of: 130 / 132 cm and approx. 20.0 m high) or their clusters; they are also intensively renewing in the undergrowth strata (C) practically in the entire tree-stand area (Figure 3, 4). Highly concerning is that locally, the seedlings of both taxons cover 100% of area with no seedlings of other native tree species, which are also growing nearby.

The phytosociological evaluation (Figure 1) shows that there are 6 species of herbaceous plants, within the Park which are considered as highly

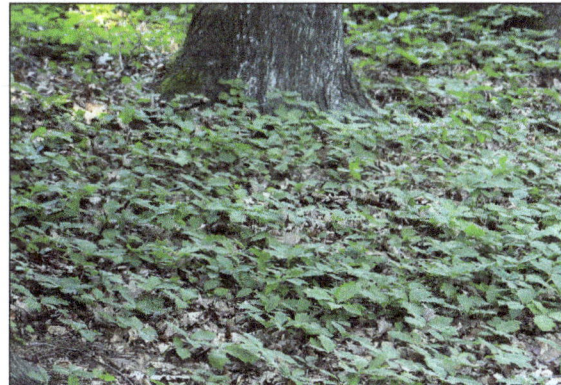

Figure 3. Monoculture of *Quercus rubra* L. seedlings in the herb layer – Silesia Park, Chorzów (photo: J. Łukaszkiewicz, 2014)

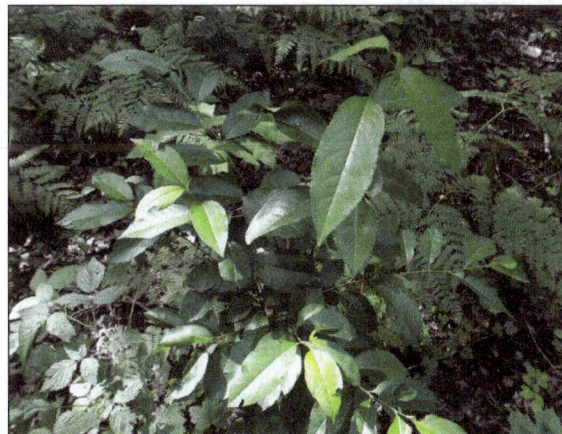

Figure 4. An unrestrained renewal of *Prunus serotina* Ehrh. – Silesia Park, Chorzów (photo: P. Wiśniewski, 2013)

invasive country-wide (Tokarska-Guzik et al., 2012; Regulation of Minister of Environment of 9 September, 2011): *Reynoutria japonica* Houtt., *Reynotria xbohemica* Chrtek et Chyrtkova, *Impatiens glandulifera* Royle, *Impatiens parviflora* DC., *Solidago canadensis* L., *Solidago gigantea* Aiton. The identified invasive species are expansive; they are highly competitive and oust other plant species which had occupied a particular spot in a given phytocenosis (Figure 5, 6). They are present mainly in the western part of the Park (extensive part on a hill).

Generally, the phytocenoses in Silesia Park have reached a level of certain self-regulation and ecological stability, among other due to large area and compactness of the Park (approx. 600 ha). Due to subsoil recultivation (renewal of physical and chemical properties) and formation of a particular phytoclimate within the Park, secondary succession takes place, which involves substitution of pioneering tree species (such as birch, robinia, poplars etc.), planted as a forecrop for more demanding trees. Generally, the tree species typical to oak-hornbeam, sometimes riparian and oak forest communities (mezo- and eutrophic deciduous forests – *Querco-Fagetea* class) arc rcncwcd. Simultaneous succession of expansive and invasive species is an undesired phenomenon; this pertains to both tree and herbaceous plants (such as red oak, black cherry, knotweeds, balsams, goldenrods). This issue is problematic and casts doubt on the optimistic forecast of tree-stand development via succession (towards natural, stable and undisturbed phytocenoses).

DISCUSSION

High number of the researched taxons: *Robinia pseudoacacia* L., *Acer negundo* L., *Quercus rubra* L., *Prunus serotina* Ehrh., *Impatiens parviflora* DC. found within researched sites are consistent with the results of research carried out in forest areas in entire Poland. For example: in the eastern part of Opoczyńskie Hills, the same species have been found to have highest frequency and expansiveness potential (*Quercus rubra* L. – 99 plots, *Prunus serotina* Ehrh. – 98, *Robinia pseudoacacia* L. – 95, *Impatiens parviflora* DC. – 84, *Acer negundo* L. – 78). are mixed and deciduous forests (Trojecka-Brzezińska, 2014) are most vulnerable to the invasion by antropophytes,

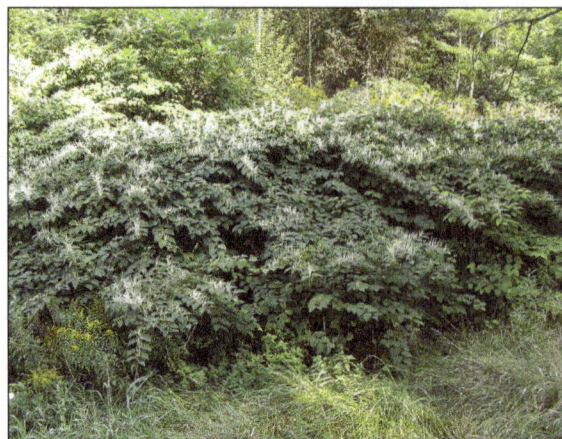

Figure 5. Extensive clusters *Reynoutria japonica* Houtt.- Silesia Park, Chorzów (photo: B. Fortuna-Antoszkiewicz, 2014)

Figure 6. *Impatiens glandulifera* Royle displacing native herbaceous plants in the border zone of the stand – Silesia Park, Chorzów (photo: B. Fortuna-Antoszkiewicz, 2014)

similarly to the tree-stands in researched sites (deciduous and mixed).

The research on the spreading of invasive species e.g. in Ladzka primeval forest (bordering Białowieża primeval forest on south-west) also confirm that the most numerous species include *Impatiens parviflora* DC. and *Prunus serotina* Ehrh. (the species showing preference for non-fresh and poor habitats and pine tree-stands aged from 20 to 60 years growing in the area). In context of anthropogenic factors, the identified species showed preference for the areas with a dense road network (>50 m/ha) and areas less than 0.5 km away from buildings and forest border (Fyałkowska et al., 2014). The research on the expansion of *Prunus serotina* Ehrh. in Kampinos National Park (covering the entire Park with an area of 385 km²) shows that the species is frequently[6] found in the ecosystems which underwent anthropogenic transformation of soil conditions and vegetation. The disturbances caused by human interference may be a decisive factor for the vulnerability of a given ecosystem to invasive species (Otręba & Kondras, 2014).

The analysis of invasive species in 9 research sites (Table 1), in the context of natural conditions and anthropogenic pressure shows the necessity of taking preventive actions (vegetation maintenance, tree-stand management) – following many years of negligence or no maintenance whatsoever – to protect native phytocenoses and maintain local biodiversity.

Other examples of areas which require adapting a rational vegetation maintenance and tree-stand management plan are the areas protected by law (such as thermophilous oak forest *Potentillo albae-Quercetum* in King John Sobieski reservation in Warsaw – under recession [Ciurzycki et al., 2014]), or locations were secondary succession develops towards forest communities – due to the abandoning of land (e.g. secondary forest succession in the meadows in Małe Pieniny [Frączek & Dziepak, 2014]). In such cases, a permanent destruction of cultural landscape and local decrease of biodiversity may result, similarly as in the case of uncontrolled invasive plant introduction.

CONCLUSIONS

- Generally, the presence of invasive plant species is symptomatic of an unfavourable vegetation succession in a given area. To a large degree, it is connected with increasing anthropogenic pressure on the environment.
- Excessive expansion of invasive species may disturb the ecological balance of an ecosystem in a given area, replacing less expansive and less competitive species. This may lead to impoverishing of the species structure and decreasing biodiversity. The current research proves (Tokarska-Guzik et al., 2012) that the environmental functions are best achieved by afforestation composed of native species, especially if they are of a free and non-schematic structure.

- In the selected sites which are subject to anthropogenic pressure, the presence of invasive species was detected to a varying degree and was most prominent in intensively utilised areas. This resulted in alarger area taken over by individual species and greater species variety. In the case of tree species, all tree-stand strata have been taken over, with intensive renewal in the undergrowth strata.
- The authors' own research carried out in the selected sites confirms invasive species' pressure is present irrespective of location, natural conditions of terrain and site type – in each case, it is most prominent in the plots where vegetation is not properly maintained or is not maintained at all.
- In order to maintain optimum vegetation systems (stable native phytocenoses) in the areas subject to strong anthropogenic pressure (urban, recreational and tourist areas) it is necessary to introduce systemic supporting activities: constant monitoring of succession and a rational, planned maintenance of vegetation to reduce negative impact of environmental changes, and in the case of invasive species – elimination in the early stages of expansion.

REFERENCES

1. Ciurzycki W., Stępniewski L., Marciszewska K. 2014. The bright oakwood Potentillo albae-Quercetum recession in the reservation of the name of King Jan Sobieski in Warsaw. In A. Obidziński, K. Marciszewska (Eds.), Lasy wobec zmieniającej się presji człowieka [Forests in the face of changing human pressure]. Wyd. Samodzielny Zakład Botaniki Leśnej SGGW w Warszawie, 38 [in Polish].

2. Fortuna-Antoszkiewicz B., Łukaszkiewicz J. 2017. Valorization and the condition of preservation of bank-side linear woodlots' forms on the example of the selected part of the Żerański Canal. In B. Świderek (ed.), Special Element in its Surroundings – traces, Wyd. Oficyna Wydawnicza WSE iZ w Warszawie, 89–116.

3. Frączek M., Dziepak M. 2014. Secondary forest succession on the glades in Małe Pieniny. In A. Obidziński, K. Marciszewska (Eds.), Lasy wobec zmieniającej się presji człowieka [Forests in the face of changing human pressure]. Warszawa, Wyd. Samodzielny Zakład Botaniki Leśnej SGGW w Warszawie, 39 [in Polish].

4. Fyałkowska K., Wroniewski M., Obidziński A. 2014. Natural and anthropogenic determinants of the spread of invasive alien species of plants

---

[6] Spreading of black cherry lasts relatively short (approx. 60 years) and its dispersion is closely tied to the locations where it had been previously introduced (Otręba & Kondras, 2014).

in Ladzka Forest. In A. Obidziński, K. Marcisze-wska (Eds.), Lasy wobec zmieniającej się presji człowieka [Forests in the face of changing human pressure] (pp. 29). Wyd. Samodzielny Zakład Botaniki Leśnej SGGW w Warszawie, [in Polish].

5. Gwiazdowicz M. 2014. Invasive alien species. Infos. BAS Biuro Analiz Sejmowych, No 11(171), 5th of June 2014. Wydawnictwo Sejmowe dla Biura Analiz Sejmowych, [in Polish].

6. Gazda A., Augustynowicz P. 2012. Alien species of trees in Polish production forests. What do we know about the pool and the distribution of selected taxa? Studia i Materiały CEPL w Rogowie, R. 14. Issue 33 / 4 / 2012.

7. Łukaszkiewicz J., Fortuna-Antoszkiewicz B. 2017. Silesia Park in Chorzów / Poland – the successful re-naturalization of industrial landscape after 60-years. Miškininkystė Ir Kraštotvarka [Forestry and Landscape Management] 2017 1 (12), 25–34.

8. Matuszkiewicz J.M., Roo-Zielińska E. (Eds.). 2000. Embankments betwixt of Vistula as a kind of natural system (section Pilica-Narew). Seria: Dokumentacja Geograficzna No 19, Instytut Geografii i Przestrz. Zagosp. im. Stanisława Leszczyckiego PAN, [in Polish].

9. Otręba A., Kondras M. 2014. Can the mass occurrence of the alien species be an indicator of anthropogenic strains of forest ecosystems? In A. Obidziński, K. Marciszewska (Eds.), Lasy wobec zmieniającej się presji człowieka [Forests in the face of changing human pressure]. Wyd. Samodzielny Zakład Botaniki Leśnej SGGW w Warszawie, 50 [in Polish].

10. Resolution of the Minister of Environment of 9 September 2011 on a list of alien plants and ani-mals which if released into environment may threaten native species or habitats, Dz.U. no. 210, item 1260] (2011).

11. Regulation (EU) No. 1143/2014 of the European Parliament and the Council of 22 October 2014 on the prevention and management of the introduction and spread of invasive alien species, European Union Journal of Law L 317/35 (2014).

12. Seneta W., Dolatowski J. 2012. Dendrology. PWN, [in Polish].

13. Strzelecki W., Sobczak R. 1972. Afforestation of wasteland and land difficult to renew. PWRiL, [in Polish].

14. Sudnik-Wójcikowska B. 1987. Flora of the city of Warsaw and its changes during the nineteenth and twentieth centuries. Wydawnictwa Uniwersytetu Warszawskiego, [in Polish].

15. Tokarska-Guzik B., Dajdok Z., Zając M., Urbisz A., Danielewicz W. 2012. Plants of foreign origin in Poland, with particular reference to invasive species. Generalna Dyrekcja Ochrony Środowiska, [in Polish].

16. Trojecka-Brzezińska A. 2014. Synanthropization of forest phytocoenoses of the eastern part of Opoczno Hills. In A. Obidziński, K. Marciszewska (Eds.), Lasy wobec zmieniającej się presji człowieka [Forests in the face of changing human pressure]. Wyd. Samodzielny Zakład Botaniki Leśnej SGGW w Warszawie, 28 [in Polish].

17. Environmental Protection Act Dz.U. of 2004, no. 92, item 880 as amended (2004).

18. Wysocki Cz., Sikorski P. 2009. The outline of applied phytosociology. Wyd. SGGW, [in Polish].

7

Kinetics of the Photocatalytic Decomposition of Bisphenol A on Modified Photocatalysts

Piotr Zawadzki[1*], Edyta Kudlek[1], Mariusz Dudziak[2]

[1] Silesian University of Technology, Faculty of Energy and Environmental Engineering, Institute of Water and Wastewater Engineering, Konarskiego 22B, 44-100 Gliwice, Poland

[2] Silesian University of Technology, Faculty of Energy and Environmental Engineering, Institute of Water and Wastewater Engineering, Division of Water Supply and Sewage Systems, Konarskiego 18, 44-100 Gliwice, Poland

* Corresponding author's e-mail: zawadzki.piotr@onet.eu

ABSTRACT

This paper presents the evaluation of the photocatalytic kinetics of bisphenol A decomposition in the presence of commercial titanium(IV) oxide and modified photocatalysts (composites). The following modification methods were used: mechanical mixing, calcination and impregnation. The decomposition process was carried out with the addition of photocatalysts and activated carbon at doses of 100 mg/dm^3 and 25 mg/dm^3, respectively. The photocatalytic process was performed in a reactor from the Heraeus Company (Warsaw, Poland) with a volume of 0.7 dm^3. The reactor was equipped with an immersed medium-pressure mercury lamp with a power of 150 W ($\lambda = 200$–580 nm). The degree of bisphenol A decomposition was determined by chromatographic analysis preceded by solid-phase extraction SPE. The qualitative-quantitative analysis was performed using a high-performance liquid chromatograph HPLC (UV detector, $\lambda = 218$ nm) from Varian (Warsaw, Poland). The dependence of the BPA decomposition on the duration of irradiation was found, wherein the modified photocatalysts were the most effective (from 75 to 90% after 15 minutes). The order of photocatalyst efficiency has been proposed as follows: $TiO_2 < TiO_2/AC < C_{dextran}-TiO_2/AC < C_{methanol}-TiO_2/AC < C_{ethanol}-TiO_2/AC < TiO_2-AC$. The highest degree of decomposition was observed in the presence of TiO_2/AC (99%). Numerous studies suggest that the results of the TiO_2 photocatalytic oxidation of organic substances fit well with the Langmuir–Hinshelwood (L–H) kinetic model. The kinetic parameters of the photocatalysis process were carried out according to the L-H model. According to the pseudo-first-order parameters, the results showed that the decomposition of bisphenol A was most intensive in the first 15 minutes of the process.

Keywords: kinetics, photocatalysis, modified photocatalysts, bisphenol A

INTRODUCTION

The 2,2-Bis(4-hydroxyphenyl)propane, widely known as bisphenol A (BPA), is a pollutant classified as a contaminant of emerging concern (CEC). BPA is an organic compound used to produce polymers such as epoxy resins. It commonly appears in various everyday products, particularly in products that are in contact with food (infant feeding bottles, microwave ovenware, coatings on metal lids, etc.). BPA is considered an endocrine-disrupting compound (EDC) and can cause adverse endocrine disruptive effects. In the literature, bisphenol A is described as generating toxic effects in living organisms after short exposure times and chronic effects with long-term exposure. The described xenoestrogen is an anthropogenic impurity with a low acute toxicity, and poisoning from BPA is very rare [Careghini et al. 2015].

Bisphenol A is emitted into the environment mainly via untreated industrial wastewater during manufacturing processes, and it leaches from products stored in poorly maintained landfills. Therefore, many sources result in a continuous release of this compound and an increase in its concentration in aquatic ecosystems. Analysis of samples collected from surface water from all over the world has shown the presence of BPA in aquatic ecosystems in Portugal, Italy, Denmark, USA and China (Table 1). In the ranking of the five largest manufacturers of BPA, China is the largest manufacturer, and Asian countries account for a major share of the global production of BPA (approximately 2.4 million tons). Environmental studies in China have shown that the concentrations of BPA detected in surface waters were comparable to global levels (less than 1 $\mu g/dm^3$) except for several areas with median concentrations higher than 4 $\mu g/dm^3$ [Ma et al. 2006; Dong et al. 2009].

The presence of organic compounds in surface waters and the necessity of their removal is one of the most important aspects of water treatment technology. Considering the low susceptibility of micro-impurities to biological degradation and the difficulty of removal them with conventional water treatment and wastewater treatment systems, new methods should be sought to eliminate pollutants from aquatic ecosystems. An alternative to classical technologies are advanced oxidation processes (AOPs). Their common chemical feature is the formation of a hydroxyl radical (OH·), which drives the oxidation processes of organic compounds. AOPs include methods such as ozonation, photolysis and photocatalysis. These processes do not transfer the substances to another phase (e.g., an activated sludge) that requires further processing and elimination but they enable the complete removal of pollutants from water [Pirila 2015]. Numerous studies have shown that AOPs are highly effective in the elimination of pharmaceutical substances [Bohdziewicz et al. 2013], endocrine-active compounds [Dudziak et al. 2014] and heavy metals [Lenoble et al. 2003].

In recent years, photocatalytic oxidation using titanium(IV) oxide (TiO_2) powder or other semiconductors has received considerable attention. The photoactivation of TiO_2 requires enough energy to activate the semiconductor. A major drawback of commercial photocatalysts is their low activity un-der visible light (Vis). It forces the use of expensive, artificial light sources such as ultraviolet radiation (UV). Currently, second-generation visible-light-active photocatalysts that can absorb the radiation in the visible range ($\lambda > 400$ nm) are under investigation. Therefore, many works have been focused on the modification of titanium(IV) oxide nanoparticles. The modification of TiO_2 has been used for various purposes, including increasing both the photocatalytic activity of the semiconductor and its efficiency in adsorbing pollutants, enhancing the separation performance of the TiO_2 powder from the aqueous phase after process and neutralizing the intermediates oxidation products [Inagaki et al. 2005].

The aim of this work was to evaluate the photocatalytic kinetics of bisphenol A decomposition in the presence of commercial titanium(IV) oxide and modified photocatalysts (composites) based on TiO_2, activated carbon and elemental carbon.

MATERIALS AND METHODS

In this experiment the experimental solutions were prepared using deionized water and an analytical standard of bisphenol A obtained from Sigma-Aldrich (Poznan, Poland) with a purity of 99.0%. A constant concentration of xenobiotics equal to 1.0 mg/dm^3 was used. The pH of the model solution was adjusted to 7 using 0,1 mol/dm^3 HCl and 0.2 mol/dm^3 NaOH. The physicochemical characteristics of BPA are presented in Table 2.

The commercial TiO_2-P25 (approximately 75% of anatase) with a specific surface area S_{BET} of 50 ± 15 m^2/g and mean particle size of 21 nm was doped with carbonaceous materials. The sources of carbon were commercial activated carbon CWZ-30 from Gryfskand, methyl alcohol of a purity of 99.5% and ethyl alcohol of a purity of 96.0% from Avantor Performance Materials Poland S.A. (Gliwice, Poland) and dextran 110 000 with 98.5% purity by Polfa KUTNO (Kutno, Poland). The physicochemical characteristics of the base materials are summarized in Table 3.

SYNTHESIS OF PHOTOCATALYST SAMPLES

Six types of photocatalysts were tested, namely a commercial titanium(IV) oxide, TiO_2-activated carbon mixture (TiO_2-AC), com-

Table 1. Medium level of bisphenol A in the surface waters of the world

Country	Level, $\mu g/dm^3$	References
Portugal	70.00 ÷ 4000.00	[Avzedo et al. 2001]
Netherlands	< 900.00 ÷ 1000.00	[Vethaak et al. 2005]
China	1.50 ÷ 26.20	[Fu et al. 2007]
USA	< 100.00 ÷ 800.00	[Staples et al. 2000]
Poland	< 2.88	[Kotowska et al. 2014]

Table 2. Physicochemical characteristics of bisphenol A

Chemical structure	Physicochemical properties	
	Molecular formula	$C_{15}H_{16}O_2$
	Molecular weight [g/mol]	228.28
	Water solubility (25°C)	Insoluble (200–300 mg/dm³)
	Vapor pressure (20°C) [mmHg]	4.0×10^{-8}
	$logK_{OW}$	2.20–3.82
	$logK_{oc}$	2.53–4.23
	Dissociation constant in 25°C (pK_a)	9.60
	Odor	Mild phenolic
	Form	White flakes or crystals

Table 3. Physicochemical characteristics of TiO_2 and activated carbon

	Activated carbon	Photocatalyst
Symbol, origin	CWZ-30, Gryfskand	P-25, Evonik Degussa GmbH
Surface area [m²/g]	1134.0	35.0–65.0
Particle size [nm]	-	21.0
Granulation [%]	90.0	-
Denisty [g/cm³]	0.28–0.36	4.26

posite consisting of activated carbon and titanium dioxide (TiO_2/AC) and composites consisting of activated carbon, titanium dioxide and three sources of elemental carbon (methanol ($C_{methanol}$-TiO_2/AC), ethanol ($C_{ethanol}$-TiO_2/AC) and dextran ($C_{dextran}$-TiO_2/AC)). The photocatalysts doped with carbonaceous materials were prepared using a hybrid method based on the following preparation techniques: mechanical mixing, calcination and impregnation.

The modification of commercial TiO_2 was started by mixing the semiconductor nanoparticles and activated carbon in deionized water with a ratio of 80:20 (w/w). The suspension was

magnetically stirred at room temperature for 30 minutes in the dark and then calcined in a furnace at 300°C for 8 hours. The resulting photocatalyst is referred to as TiO_2/AC. To prepare the alcohol-modified photocatalysts, the same procedure was repeated; however, methanol or ethanol was added dropwise to the previously prepared composite (1:1, w/v). The C-TiO_2/AC photocatalyst was magnetically stirred in the dark for 30 minutes and then placed in the drying oven at 80°C for 6 hours. The following symbols were used: $C_{methanol}$-TiO_2/AC and $C_{ethanol}$-TiO_2/AC. The composite marked as $C_{dextran}$-TiO_2/AC was prepared by thoroughly dissolving sugar in a TiO_2-AC suspension with a ratio of 80:20 (w/w). Details of each calcination process are shown in Table 4.

INSTRUMENTS AND ANALYTICAL METHODS

Deionized water spiked with analytical standards of the studied compounds at a constant concentration of 1000 mg/dm³ was irradiated with an immersed medium-pressure

Table 4. Characteristics of photocatalysts preparation methods

Catalyst	Carbon source	Conditions/preparation			Symbol
		Calcination		Preparation	
		Temperature [°C]	Time [h]		
TiO_2-AC	commercial AC	-	-	mixing TiO_2 with AC	TiO_2-AC
TiO_2/AC	commercial AC	300	8	mixing TiO_2 with AC, calcination	TiO_2/AC
$C_{methanol}TiO_2$/AC	commercial AC, methanol	300	8	mixing TiO_2 with AC, calcination	$C_{methanol}$-TiO_2/AC
		80	6	addition of methanol into TiO_2/AC, drying	
$C_{ethanol}TiO_2$/AC	commercial AC, ethanol	300	8	addition of methanol into TiO_2-AC, calcination	$C_{ethanol}$-TiO_2/AC
$C_{dextran}TiO_2$/AC	commercial AC, dextran	300	8	dissolution of dextran in TiO_2-AC, calcination	$C_{dextran}$-TiO_2/AC

150 W mercury lamp. The lamp irradiation ranged from 300 to 580 nm. The process of heterogenic photocatalysis was conducted in a laboratory batch reactor (volume of 700 cm³) by Heraeus (Figure 1). The temperature of the reaction mixture was approximately 20–21°C owing to a cooling jacket. An aeration pump with a capacity of 4 dm³ air for 1 minute was introduced into the reactor.

The photocatalytic process was investigated in the presence of a commercial photocatalyst, activated carbon and modified photocatalysts. Photocatalytic experiment was carried out at a catalysts dose of 100 mg/dm³. In the assisted photocatalysis, the dose of activated carbon applied to the reaction mixture reached 25 mg/dm³. The solutions of bisphenol A were irradiated continuously for 45 minutes. In the preliminary study, a 15-minute contact time was set to ensure the adsorption of micropollutants on the surface of the catalyst (adsorption). The adsorption process was carried out in the dark. In this way, the sorption efficiency of the photocatalysts was determined. The samples for analysis were collected after 5, 10, 15, 20, 30 and 45 min of the reaction. The heterogenic photocatalysis in the presence of pure titanium dioxide is denoted as TiO_2, whereas the

mixture of titanium(IV) oxide and activated carbon is labelled as TiO_2-AC. For the modified photocatalysts, the following determinations were used: TiO_2/AC, $C_{methanol}$-TiO_2/AC, $C_{ethanol}$-TiO_2/AC and $C_{dextran}$-TiO_2/AC.

The removal rates of bisphenol A before and after the photocatalytic process were determined by the chromatographic analysis preceded by solid-phase extraction (SPE). The activated carbon and catalyst particles were separated from the treated solution using a 0.45 μm glass filter filtration kit (Merck Millipore Company, Poznan, Poland) connected to vacuum pump by AGA Labor (Warsaw, Poland). The samples were extracted with solid-phase extraction in single-use C-18 Supelclean™ ENVI-18 columns with a volume of 6 cm³ and 1.0 g of the solid phase. The bed was conditioned with 5 cm³ of acetonitrile and 5 cm³ of methanol and washed with 5 cm³ of water to remove any residue. The analyte was eluted from the column with a 3 cm³ mixture of acetonitrile and methanol (60:40, v/v). The qualitative-quantitative analysis was performed using a high-performance liquid chromatograph HPLC (UV detector, λ = 218 nm) equipped with a Hypersil GOLD column by Thermo Scientific (length – 25 cm, diameter – 4.6 mm and granulation – 5.0 μm) with methanol as the mobile phase.

Numerous studies of TiO_2 photocatalytic oxidation of organic pollutants have shown that the corresponding data fit the Langmuir–Hinshelwood (L–H) kinetic model. Therefore, for very low concentrations of micropollutants in aquatic environment, e.g., bisphenol A, the L-H equilibrium simplifies to a pseudo-first-order kinetic model [Asenjo et al. 2003; Kumar et al. 2008]. The Langmuir-Hinshelwood kinetic model (Equation 1) is assumed as follows:

$$r = -\frac{dC}{dt} = \frac{k \cdot K \cdot C}{1 + K \cdot C} \qquad (1)$$

The pseudo-first order rate constant k was determined for each photocatalyst since it enables one to estimate the photocatalytic activity of the catalyst. The integral form of the pseudo-first-order rate equation is usually expressed by Equation 2:

$$ln\frac{C_0}{C} = kt \qquad (2)$$

where C_0 and C are the initial and final concentrations of BPA at time,

$t = 0$ and $t = t$, where t is the irradiation time.

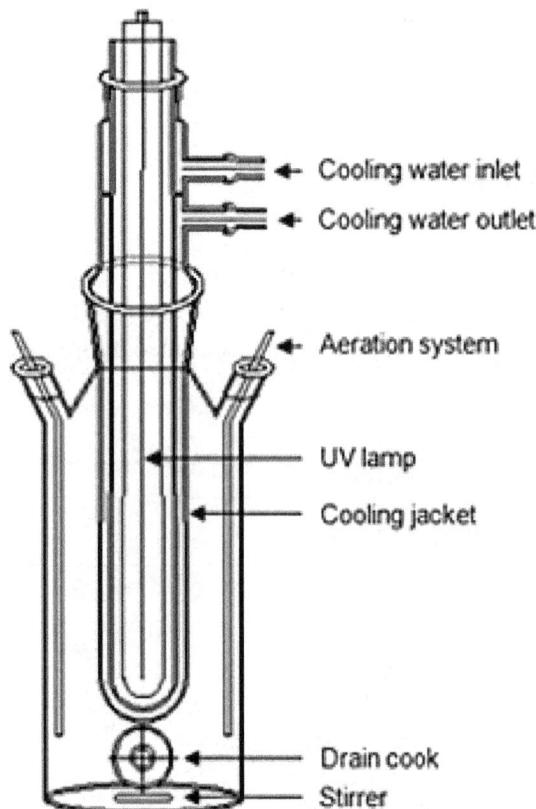

Figure 1. Scheme of the photocatalytic batch reactor [Kudlek et al. 2015]

- ← Cooling water inlet
- ← Cooling water outlet
- ← Aeration system
- — UV lamp
- — Cooling jacket
- — Drain cook
- — Stirrer

RESULTS AND DISCUSSION

The adsorption behaviour and the pollutant decomposition efficiency increase as the surface area of the photocatalyst increases. Ao et al. [Ao et al. 2003] observed that a higher active surface area provides more hydroxyl radicals and, by extension, a higher degree of decomposition of pollutants. The elimination of bisphenol A was the most effective during the sorption process of contaminants on the photocatalyst nanoparticles, which is evidenced by the results from "0" min, i.e., before the radiation source was switched on (Figure 2). It was observed that the compound was adsorbed in the range from 50 to 97%. The highest value of this parameter was observed in the system referred to as TiO_2-AC. A highly developed surface area of activated carbon in the range from 700 to 1800 m^2/g plays a crucial role in the sorption of bisphenol A. However, in this case the decomposition of the studied compound was dominated by the adsorption process, and the particles of the macromolecular substance were only transferred to another phase. While the chromatographic analysis showed a reduction in the concentration of bisphenol A, it could be adsorbed deeply inside the micro and mesopores of the activated carbon (Figure 3). Thus, it was not decomposed. Despite the important role of the surface area of photocatalysts, Matsunaga et al. [Matsunaga and Inagaki 2006] showed a negative effect of carbon on the activity of the photocatalysts. Large amounts of carbon coating the surface of TiO_2 reduced its ability to absorb UV light.

Based on the chromatographic analysis results, the decomposition of bisphenol A was higher for modified photocatalysts than for commercial TiO_2. Additionally, the increased time of exposure to UV radiation showed a decrease in the BPA levels. The results are presented in Figure 2 in consideration of the catalyst used. The greatest reduction in the concentration of bisphenol A was observed for the experimental solutions irradiated in the presence of the considered materials. The initial concentration of the tested micropollutant was reduced by 65 to 98% over 15 minutes of irradiation. After 45 minutes of irradiation, the highest degree of decomposition of the organic compound was found in the TiO_2-AC

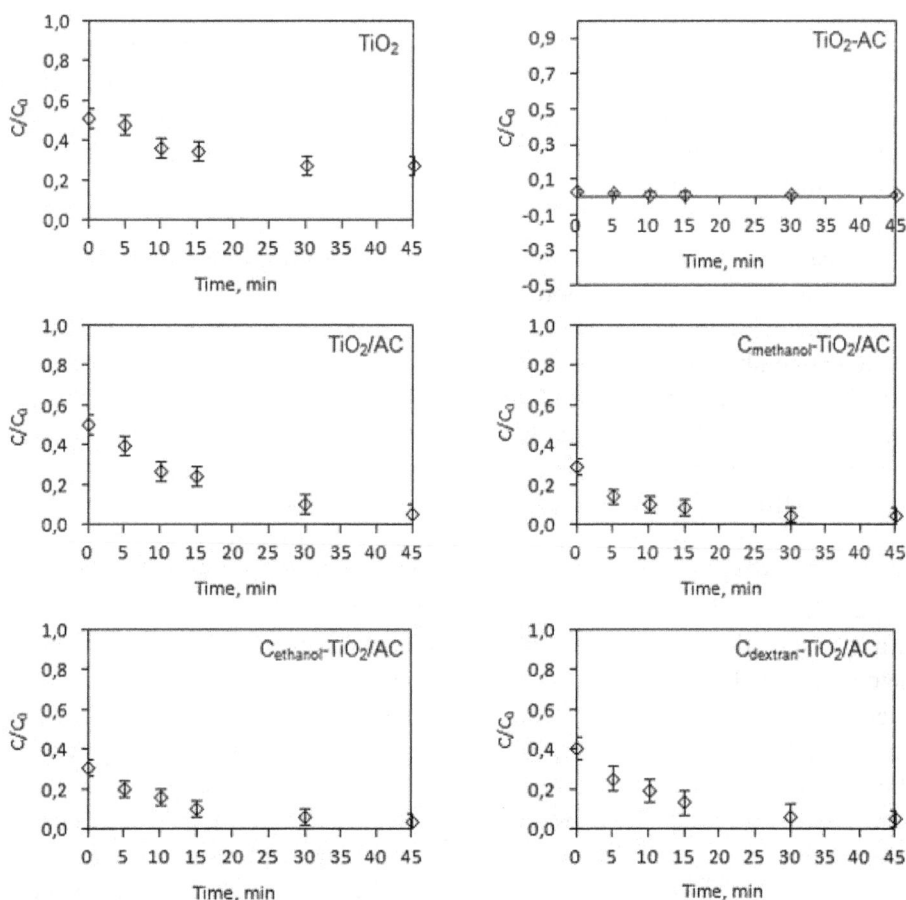

Figure 2. Decomposition of bisphenol A in the photocatalytic process at different photocatalyst configurations

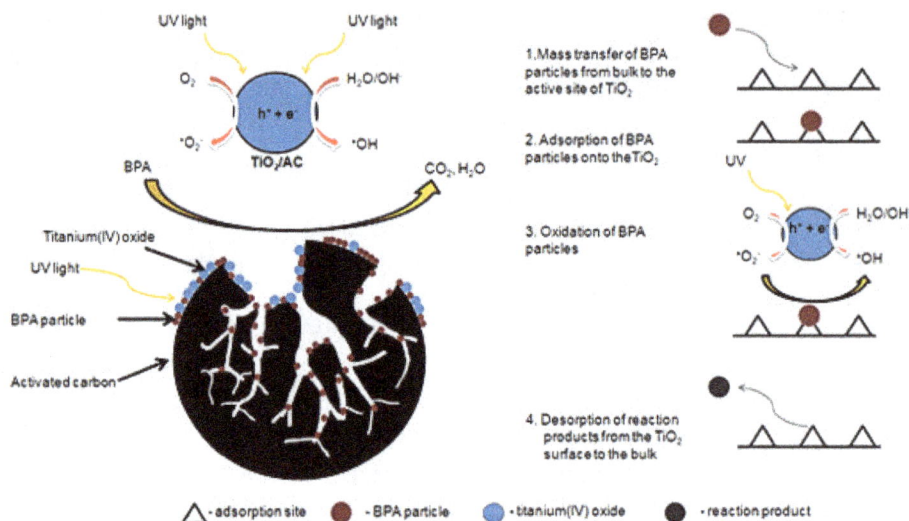

Figure 3. Proposed mechanism of adsorption and decomposition of bisphenol A on modified photocatalysts. Based on [Xing et al. 2016; Singh et al. 2016]

process (99%). The photocatalysts doped with elemental carbon were equally as effective. For the elemental carbon-doped photocatalysts, the initial concentration of bisphenol A was reduced by approximately 95%.

The effect of the number of elemental carbon atoms on the efficiency of bisphenol A decomposition was observed for the following configurations: $C_{methanol}$-TiO_2/AC, $C_{ethanol}$-TiO_2/AC and $C_{dextran}$-TiO_2/AC. The concentration ratio C/C_0 increased as the number of carbon atoms increased. After irradiation of experimental solution in the presence of the methanol-modified catalyst (one carbon atom), the decomposition level of bisphenol A was greater compared to that of the $C_{dextran}$-TiO_2/AC photocatalyst, which was modified using a compound of six carbon atoms.

Based on the Langmuir–Hinshelwood model, the following kinetic parameters of the photocatalysis process were carried out: pseudo-first-order rate constant k, the coefficient of determination R^2 and the half-life $t_{1/2}$. The results are described in Table 5. The dependence of the $-\ln(C/C_0)$ function on the time of irradiation ($t_{1/2}$) is shown in Figure 4.

From earlier observations, the pseudo-first-order parameters indicated that the bisphenol A oxidation reaction was much faster in the presence of modified photocatalysts. One of the parameters indicating the dynamics of the ongoing processes is the pseudo-first-order rate constant k. Analysis of the k parameter in the photocatalytic process indicates rapid decomposition during first minutes of the photocatalytic processes. The re-

sults showed that the decomposition of bisphenol A was most intense in the first 15 minutes of irradiation. Within this time, the data fitted to the linear equation showed very good correlation coefficients R^2. However, lowering the k constant from 15 to 30 minutes of the photooxidation process may be due to the reduction of the photocatalyst surface due to the agglomeration of the intermediate particles. The two-stage oxidation process is mainly attributed to the formation of intermediate products, which, after adsorbing on the photocatalyst surface, block the active sites of the semiconductor. Their quantity depends on the amount of output pollutants adsorbed on the photocatalyst surface and, thus, on the success of the pollutant decomposition [Ao et al. 2003]. In elemental carbon-doped photocatalysts, the correlation between the degree and rate of recombination of the electron-hole pairs (e^-–h^+) and the number of elemental carbon atoms in the structure of pure TiO_2 has been observed. When the concentration of foreign elements is too high, it causes the distance between the e^-–h^+ pair to be too low. The effect is a lack of the separating force [Carp et al. 2004]. Photocatalysts doped with elemental carbon exhibit a pseudo-first-order rate constant, k, that is more than three times higher than that of commercial titanium(IV) oxide. The greatest value of the k parameter was noted in the catalyst referred to as $C_{methanol}$-TiO_2/AC ($k = 0,078$ min^{-1}). No intermediate bisphenol A degradation was found for only the TiO_2/AC composite. This phenomenon could be explained by the inhibition of the electron-hole pair recom-

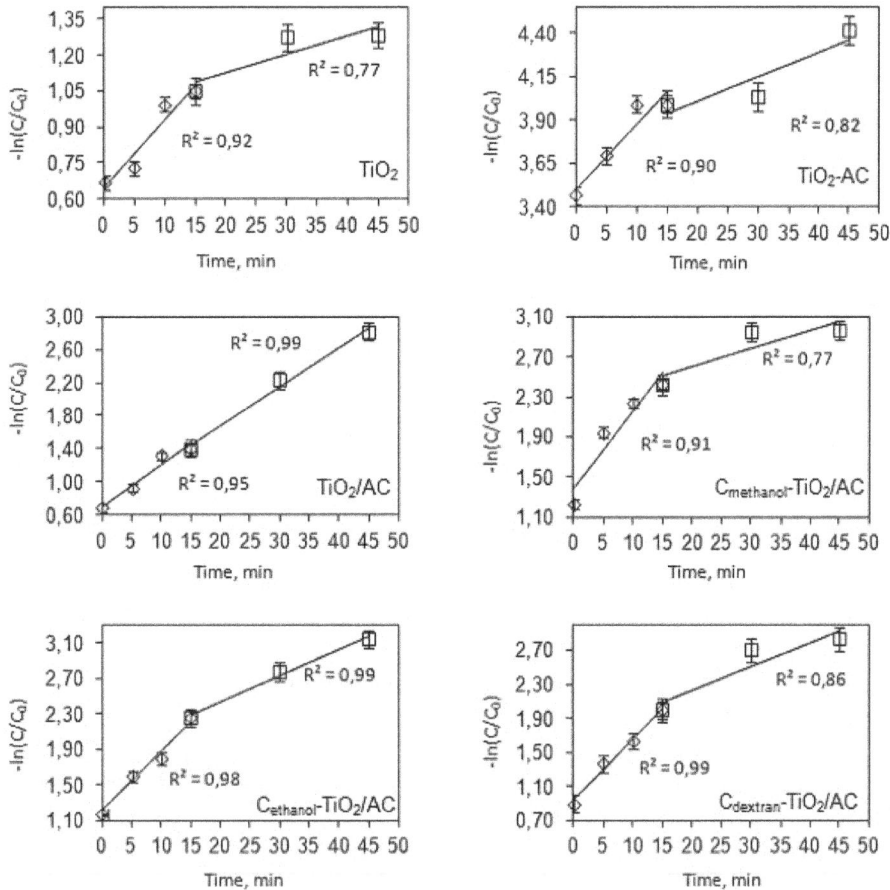

Figure 4. Kinetics of the bisphenol A decomposition in the photocatalytic process with different photocatalyst configurations.

bination process, which in turn leads to the inhibition of the pollutant decomposition.

Above observations are confirmed by the analysis of the half-life $t_{1/2}$ results, which are presented in Table 5. Statistical indices, such as the standard deviations (SD), were used to evaluate the model adequacy. It is defined in Equation 3:

$$SD = \sqrt{\frac{\sum(X_i - X)^2}{N - 1}} \qquad (3)$$

where: X_i is the experimental values of parameters,

X is the mean value of the parameter,
N is the number of experimental data.

Data analysis indicates satisfactory accuracy in the estimation of the kinetic constant for the measured parameters (pseudo-first-order rate constant k, coefficient of determination R^2, and half-life $t_{1/2}$). During the first period of the process (up to 15 minutes), the appointed half-life of the model was higher compared to the decomposition of bisphenol A from 15 to 45 minutes of photocatalysis. This was due to the increasing level of intermediate compounds with the elongation of

Table 5. Pseudo first-order parameters during photocatalysis process

Catalyst	Pseudo first-order parameters					
	k, min^{-1}		R^2, -		t/2, min	
	0–15	15–30	0–15	15–30	0–15	15–30
TiO$_2$	0.028±0.006	0.007±0.002	0.92±0.03	0.77±0.03	47.9±9.27	237.7±38.95
TiO$_2$-AC	0.037±0.006	0.014±0.002	0.90±0.01	0.82±0.05	113.5±17.77	315.4±42.26
TiO$_2$/AC	0.051±0.008	0.047±0.005	0.95±0.03	0.99±0.01	27.3±2.10	30.3±2.49
C$_{methanol}$-TiO$_2$/AC	0.078±0.005	0.017±0.004	0.91±0.03	0.77±0.01	26.5+0.25	172.5±34.48
C$_{ethanol}$-TiO$_2$/AC	0.069±0.005	0.029±0.005	0.96±0.03	0.98±0.01	27.2±2.13	87.2±8.18
C$_{dextran}$-TiO$_2$/AC	0.071±0.001	0.027±0.002	0.99±0.01	0.86±0.01	22.9±0.81	87.9±3.03

Table 6. Classification of reaction rates based on half-lives [Wright 2004]

Half-life, s	Time span for near-completion	Rate classification
$10^{-15} - 10^{-12}$	ps or less	ultra fast rate
$10^{-12} - 10^{-6}$	µs or less	very fast rate
$10^{-6} - 1$	seconds	fast rate
$1 - 10^3$	minutes or hours	moderate rate
$10^6 - 10^3$	weeks	slow rate
$> 10^6$	weeks or years	very slow rate

the irradiation time. The half-life $t_{1/2}$ of bisphenol A was reduced from 48 min and 113 min for TiO_2 and TiO_2-AC, respectively, to approx. 26 minutes for the modified photocatalysts. TiO_2 doped with dextran had the shortest half-life of all the tested photocatalysts. Table 6 presents the classification of the chemical reaction rates based on their half-lives. It was observed that in the first 15 minutes the bisphenol A photodegradation process proceeded at a moderate rate. Recombination of the electron-hole pairs slowed the reaction, whereby the continued irradiation of the solutions allowed the reaction to be classified as slow.

process was found. Half-lives of the compound in the first 15 minutes of the process ranged from 22 to 27 minutes. The $t_{1/2}$ value was half that of commercial titanium(IV) oxide. The analysis of the photochemical oxidation kinetics revealed the two-stage oxidation process of the bisphenol A decomposition. No intermediate bisphenol A degradation was found only for the TiO_2/AC composite, indicating a slowdown in the recombination process of the electron-hole pairs, which contributes to the inhibition of the pollutant decomposition reaction.

CONCLUSIONS

In this paper, the kinetics of bisphenol A decomposition in the presence of commercial titanium(IV) oxide and photocatalysts modified with carbonaceous materials were evaluated. It was found that the level of bisphenol A decomposition increases with irradiation time wherein the modified photocatalysts were the most effective compared to commercial TiO_2. The proposed hybrid method involved the modification of titanium dioxide to produce materials with enhanced properties for removing BPA in comparison to the properties of commercial titanium dioxide. Therefore, the addition of *carbon sources* such as alcohols, sugars and activated carbon generated photocatalysts characterized by higher activity. It was observed that the photocatalysts adsorbed the tested compound in the range from 50 to 97%. A significant effect of activated carbon on this process was found. Modification of the pure photocatalyst intensified the decomposition process of the organic compound, demonstrating a removal rate that exceeded 98%. Based on the Langmuir–Hinshelwood model, a significant difference between the rate of micropollutant degradation in the first 15 minutes of irradiation and between 15 and 45 minutes of the

REFERENCES

1. Ao C.H., Lee S.C., Yu J.C. 2003. Photocatalyst TiO2 supported on glass fiber for indoor air purification: effect of NO on the photodegradation of CO and NO2. J. Photochem. Photobiol. Chem. 156(1–3), 171–177.

2. Asenjo N.G., Santamaria R., Blanco C., Granda M., Alvarez P., Menendez R. 2013. Correct use of the Langmuir–Hinshelwood equation for proving the absence of a synergy effect in the photocatalytic degradation of phenol on a suspended mixture of titania and activated carbon. Carbon. 55, 62–69.

3. Azevedo D.A., Lacorte S., Viana P., Barcelo D. 2001. Occurrence of nonylphenol and bisphenol A in surface waters from Portugal. J. Brazil Chem. Soc. 12(4), 532–537.

4. Bohdziewicz J., Kudlek-Jelonek E., Dudziak M. 2013. Analytical control of diclofenac removal in the photocatalytic oxidation process. Arch. Civil Engin. Environ. 6(3), 71–75.

5. Careghini A., Mastorgio A. F., Saponaro S., Sezenna E. 2015. Bisphenol A, nonylphenols, benzophenones, and benzotriazoles in soils, groundwater, surface water, sediments, and food: a review. Environ. Sci. Pollut. Res. 22(8):5711–5741.

6. Carp O., Huisman C.L., Reller A. 2004. Photoinduced reactivity of titanium dioxide, Prog. Solid State Chem. 32(1–2), 33–177.

7. Dong J., Li X L., Liang R. J. 2009. Bisphenol A pollution of surface water and its environmental factors. J. Ecol. Rural. Environ. 25(2), 94–97.

8. Dudziak M., Burdzik-Niemiec E. 2014. Comparative studies on elimination of estrogens and xenoestrogens by the oxidation processes. Ecol. Chem. Eng. A. 21(2), 189–198.

9. Fu M., Li Z., Gao H. 2007. Distribution characteristics of nonylphenol in Jiaozhou Bay of Qingdao and its adjacent rivers. Chemosphere. 69(7), 1009–1016.

10. Inagaki M., Kojin F., Tryba B., Toyoda M. 2005. Carbon-coated anatase: the role of the carbon layer for photocatalytic performance. Carbon. 43(8), 1652–1659.

11. Kudlek E., Bohdziewicz J., Dudziak M. 2015. Elimination of pharmaceutical compounds from municipal wastewater by photocatalysis, microfiltration and nanofiltration. Acta Innov. 16, 12–19.

12. Kumar K.V., Porkodi K., Rocha F. 2008. Langmuir–Hinshelwood kinetics – A theoretical study. Catal. Comm. 9(1), 82–84.

13. Lenoble V., Deluchat B., Serpaud J., Bollinger J. 2003. Arsenite oxidation and arsenate determination by the molybdene blue method. Talanta. 61(3), 267–276.

14. Ma X.Y., Gao N.Y., Li Q.S., Xu B., Le L.S., Wu J.M. 2006. Investigation of several endocrine disrupting chemicals in Huangpu River and water treatment units of a waterworks. China Water and Wastewater. 22(1–4), 10–16.

15. Matsunaga T., Inagaki M. 2006. Carbon-coated anatase for oxidation of methylene blue and NO. Appl. Cat. Environ. 64(1–2), 9–12.

16. Pirilä M. 2015. Adsorption and photocatalysis in water treatment: active, abundant and inexpensive materials and methods. Acta Universitatis Ouluensis, University Of Oulu, Oulu.

17. Staples C.A., Dorn P.B., Klecka G.M., O'block S.T., Branson D.R., Harris L.R. 2000. Bisphenol A concentrations in receiving waters near US manufacturing and processing facilities. Chemosphere. 40(5), 521–525.

18. Wright M.R. 2004. An introduction to chemical kinetics. John Wiley and Sons, England..

19. Vethaak A.D., Lahr J., Schrap S.M., Belfroid A.C., Rijs G.B.J., Gerritsen A., de Boer J., Bulder A.S., Grinwis G.C.M., Kuiper R.V., Legler J., Murk T.A.J., Peijnenburg W., Verhaar H.J.M., de Voogt P. 2005. An integrated assessment of estrogenic contamination and biological effects in the aquatic environment of The Netherlands. Chemosphere. 59(4), 511–524.

8

Kitchen Organic Waste as Material for Vermiculture and Source of Nutrients for Plants

Joanna Kostecka[1], Mariola Garczyńska[1*], Agnieszka Podolak[1],
Grzegorz Pączka[1], Janina Kaniuczak[2]

[1] Department of Natural Theories of Agriculture and Environmental Education, Faculty of Biology and Agriculture, University of Rzeszów, 35-601 Rzeszów, Ćwiklińskiej 1A Str., Poland

[2] Department of Soil Science, Environmental Chemistry and Hydrology, Faculty of Biology and Agriculture, University of Rzeszów, 35-601 Rzeszów, Ćwiklińskiej 1A Str., Poland

* Corresponding author's e-mail: mgar@ur.edu.pl

ABSTRACT

Departure from waste storage and maximisation of its utilization is currently the basis of modern waste management. This is favoured by the requirements defined in numerous legal instruments, including both EU directives and local regulations of member states. This also applies to organic waste, especially kitchen waste, which, with adequate education of the public, may constitute waste resources of very good quality to produce e.g. vermicomposts. It is very important, since soils of most European countries require continuous supply of organic matter to replenish humus and nutrients for the plants. The paper describes current trends in the production of kitchen organic waste. Since such waste has been vermicomposted for many years, advantages of this biotechnology have been presented and features of the produced vermicomposts have been characterised.

Keywords; kitchen organic waste, earthworms, vermicompost

BACKGROUND

Organisation of sustainable development and circular economy requires departure from waste storage and maximisation of its recycling. In Poland, this is favoured by the requirements specified in numerous legal instruments, starting from the Thematic Strategy on Soil Protection [Thematic Strategy on Soil Protection...], through the Regulation of the Minister of the Environment of 20 January 2015 on the process of R10 recovery (Journal of Laws of the Republic of Poland [Dz. U. RP], item 132), Regulation of the Minister of Economy of 16 July 2015 on landfill waste disposal (Journal of Laws of the Republic of Poland [Dz. U. RP], item 1277), and ending with the National Waste Management Plan [The National Waste Management Plan] [KPGO 2022]. This also applies to organic waste, especially kitchen waste, which, with adequate education of the public and its participation in the creation and

operation of waste management, may constitute waste resources of very good quality to produce composts and vermicomposts. This is even more important due to the fact that agricultural soils in Europe require continuous supply of organic matter to replenish humus and nutrients for the plants. In Poland, this is especially important due to a high proportion of light soils, since water and air relations in these soils cause a naturally lower content of organic matter than in heavy soils.

Increase in agricultural productivity of weak soils and improvement in the physico-chemical properties of marginal soils should be primarily directed towards soil enrichment in organic material. Production of nitrogen and phosphorus fertilisers is highly energy-using and also causes ecological problems. Therefore, with regard to economy and ecology, the source of nutrients for plants should be various types of organic waste. Thus, one of the objectives of the National Waste Management Plan [KPGO 2022] is to depart from waste storage and to maximize utilization of the macro- and microelements included in the waste, e.g. N, P, Ca, Mg, Cu, Zn, Mn, Mo. Apart from including precious nutrients, it should also meet the requirements specified for mineral and organic-mineral fertilisers [The Fertilisers and Fertilisation Act.]. The Research Centre of the European Commission recommends the following procedure: to avoid biowaste disposal, but the choice of an appropriate way of its processing should result from specific local factors (Krutwagen et al. 2008). This necessitates constant research on various methods of its final disposal.

In Europe, the highest amount of municipal waste produced annually per one person (including biodegradable kitchen waste) is produced by highly-developed countries [Eurostat 2016, FUSIONS 2016]. The conclusion is obvious: organic waste of household origin should be considered as an appropriate component for improvement of soil qualities. Household organic waste, if properly segregated by aware citizens, is a very good source of material to produce fertilisers, although this is in contrast with a still preferred form of its disposal, i.e. combustion. It must be emphasised that although we can then produce our own energy (which would be called renewable by some people), if we deal with clean waste of e.g. potential food, then, from the point of view of sustainable development and circular economy, this is highly disadvantageous. It causes an ultimate loss of resource and a gap in the system of potential recirculation.

The aim of the study was to present selected qualities of the kitchen organic waste production. Since the authors of the publication have been vermicomposting such waste for many years, advantages of this biotechnology have been exposed and features of the produced vermicomposts have been characterised.

METHOD

The method included analysis of selected literature and results of studies conducted on vermicomposting of kitchen waste with the use of 2 worm species: *Eisenia fetida* Sav. and *Dendrobaena veneta* Rosa.

Two mixtures of waste have been vermicomposted: mixture (a) apple, carrot and beetroot juice pomace, potato residues + cellulose at a ratio of 4:1 and mixture (b) apple peelings + potatoes + pasta + bread + cellulose at a ratio of 2:1.

In the stratification layer, waste and resulting vermicompost, determination of carbon was conducted by the Turin's method; N – by the Kjeldahl method, pH in H_2O – was determined by a potentiometric method, conductance (mS) and salinity (g $NaCl \cdot dm^{-3}$) – by a conductometric method, phosphorus – by a vanadium-molybdenum method, colorimetrically using UV-VIS spectrophotometer from Shimadzu UV-2600. Potassium was analysed with an atomic absorption spectrophotometer Hitachi Z-2000, with the use of emission method (EAAS), while calcium and magnesium – using an atomic absorption spectrophotometer Hitachi Z-2000, with the use of flame method (FAAS). C/N ratio was calculated.

RESULTS

Kitchen organic waste

With regard to household waste, biodegradable waste constitutes the highest proportion (30-60%). It is produced during preparation of meals and also comes from food, paper and cardboard residues which are thrown away. FAO warns that about one third of the food produced in the world is wasted (Gustavsson et al. 2011). Collection and analysis of data from the whole Europe regarding the route of food delivery (from production to consumption) shows that in 27 EU member states, food wasting reaches about 89 million Mg waste a year, out of which 12.3 million Mg is restaurant waste. In 2009, this corresponded to 179 kg of waste per year/person (Monier et al 2010; European Commission (DG ENV)). During the same period, e.g. Poland produced about 9 million Mg of food waste (about 6.5 million Mg of waste from food industry, about 2 million Mg of household waste and about 350 thousand Mg of waste from food services) (Sapek 2013).

Data from another period – year 2012 from 28 EU member states showed the same trend towards food wasting. During that period, this negative social and environmental phenomenon was at a level of 88 million Mg of food waste, which corresponded to 173 kg of food waste per person. This estimate included both kitchen waste produced at consumption and inedible parts related to food production (FAO 2014; FUSIONS 2016).

Wasteful consumption of food is currently perceived as a problem with serious ethical, ecological and economic consequences. Therefore, the European Commission promotes continuous reduction of food wasting and assumes reduction of this phenomenon by half by 2020. However, implementation of effective preventive measures is very complicated. Such activities require knowledge of the causes and scale of food waste production in the entire chain of food delivery. The available data base for Europe is highly varied and there are justified reservations to its reliability(Bräutigam et al. 2014).

However, certain qualities of biodegradable kitchen waste are useful for processing in vermiculture (Kostecka et al. 2018). Most of all, it is important that such waste is accumulated gradually, so it may be administered to worms as food, e.g. at the site of its production.

Vermicomposting

Vermicomposting, next to fermentation and composting, is one of the examples of pro-environment biotechnology allowing processing of organic waste (Adi and Noor 2009; Domínguez and Edwards 2011a; Pączka and Kostecka 2012). While composting involves cooperation of macroorganisms and microorganisms, for which the processed matter is a source of energy, *vermicomposting* is a process using concentrated worm populations to decompose the supplied organic matter. The product of this process is vermicompost fertiliser and biomass of worm epitheliomuscular tubes (Sherman 2003; Garg et al. 2005; Domínguez and Edwards 2011b).

Due to a high content of microflora and nutrients facilitating plant growth, vermicompost is called "black gold" (Adhikary 2012). It is a homogeneous, granular structure with a pleasant odour and dark brown colour. The resulting fertiliser contains components which are easily available for plants, including nitrogen, phosphorus and potassium. This is due to the fact the alimentary tract of worms contains microorganisms converting insoluble forms of elements into forms available for plants. Vermicompost improves soil characteristics and affects the quality of crops; therefore it is a part of pro-environment practices (Adhami et al. 2014; Padmavathiamma et al 2008; Song et al 2015). The use of vermicompost has been studied with regard to fertilisation of such plants as tomato, cucumber, potato, celery, leek, pepper, spinach, strawberry, bean or pea (Kostecka and Błażej 2000; Arancon et al. 2003; Singh et al. 2008; Pączka and Kostecka 2013; Kadam and Pathade

2014; Kashem et al. 2015; Song et al. 2015). Nevertheless, this issue requires further studies, since the quality of the produced fertiliser, its chemical composition and effect on plant growth and development depend on the type of waste which is converted by the worms.

Another product of vermicomposting is worm biomass. It may be used, for example, as food for fish, including aquarium fish, which may contribute to the reduction of costs related to fish cultivation and cause faster growth of the fish (Kostecka and Pączka 2006).

A so called "worm ecological box" is an example of a practical and pro-environment smallscale waste recycling. This is a container with a capacity adequate for the planned amount of processed waste including worms which are regularly supplied with organic waste. Such culture may function in the cellar, on the balcony or in the family garden. Such processing of waste may be an inexpensive, but effective, tool of everyday utilization of kitchen organic waste (Kostecka 2000; Kostecka et al. 2018).

Vermicompost productivity depends of the conditions created for the worms in the conducted culture (Kostecka 1994; Sherman 2003). Successful vermicomposting also depends on the structure and composition of organic waste administered to the worms. This is confirmed by studies on potential vermiculture utilization of such waste as e.g. banana tree leaves adequately mixed with bovine manure (Padmavathiamma et al. 2008), horse manure (Sangwan et al. 2008), goat manure (Loch et al. 2005), grape pomace (Paradelo et al. 2009), coffee grounds (Adi and Noor 2009), sediments from waste treatment plants (Parvaresh et al. 2004), kitchen organic waste (Kostecka et al. 1999), paper industry waste (Gajalakshmi et al. 2002; Gupta and Garg 2009), plant residues (Bansal and Kapoor 2000), mushroom waste (Kostecka 2000) or ground textile waste (Kaushik and Garg 2003). Not all organic waste in its pure form guarantee process productivity. Waste of animal origin, such as meat residues or bones, as well as certain kitchen waste, e.g. garlic, onion or chilli are not willingly processed (Adhikary 2012).

Worm cultures may include accompanying fauna which competes for space and food. A combined activity of these animals may accelerate or slow down the undergoing transformations. Vermiculture may also be endangered by excessive concentration of worms, as well as by moles, birds, ants or rats (Kostecka 2000; Adhikary 2012).

Chemical composition of the obtained vermicomposts

Processing of mixture a (apple, carrot and beetroot juice pomace, potato residues + cellulose at a ratio of 4:1) of kitchen organic waste by the population of *E. fetida* and *D. veneta* resulted in vermicomposts of dark colour, characteristic fine structure and scent of garden soil.

These vermicomposts differed in pH, conductance and the content of potassium and magnesium. They did not differ in the content of phosphorus or calcium. All the above parameters were characterised with significantly increased values in the produced fertilisers in comparison with the stratification medium layer (Table 1).

The comparison of pH (*p<0.001*) and conductance (*p<0.001*) of the obtained vermicomposts showed differences depending on the species of worms producing them. The vermicompost resulting from decomposition of waste by *D. veneta* had higher pH, and lower conductance. It was also characterised with a higher potassium content (*p<0.001*) as compared with the vermicomposts produced with the use of *E. fetida*, and it was also characterised with a higher magnesium content (*p<0.001*).

The examined vermicomposts did not differ from one another (*p>0.05*) with regard to nitrogen, phosphorus or calcium content, although the values were higher in the vermicomposts produced by *D. veneta*.

Processing of mixture b (apple peelings + potatoes + pasta + bread + cellulose at a ratio of 4:1) by the population of both worm species re-sulted in vermicomposts of similar colour with a fine structure and no residues of unprocessed waste. Vermicomposting did not involve odour production (i.e. was odourless). Again, a comparison of the resulting vermicomposts showed that the worm species affected both pH (*<0.001*), and conductance of the obtained vermicomposts (*p<0.01*) (Table 2). Similarly to mixture a, the vermicompost obtained from transformation of waste mixture b by *D. veneta* had higher pH and was characterised with lower salinity (and lower conductance) than the vermicomposts produced by *E. fetida*.

A comparison of other features of vermicomposts resulting from mixture b showed that the vermicomposts differed in the content of total nitrogen (p<0.01) and calcium (p<0.01). Unlike with mixture a, the vermicomposts coming from the containers cultured with *E. fetida* had a higher nitrogen content, and those produced by *D. veneta* had a higher calcium content. Both vermicomposts obtained from mixture b did not differ in the content of the other nutrients (phosphorus, potassium and magnesium) (p>0.05) (Table 2).

DISCUSSION

Waste is a problematic issue, since it affects matter circulation, causes loss in the agricultural and forest production space and requires increased financial resources to create landfill sites, costs of their operation and then recultivation. That is why new ways to limit those problems are currently needed, also with regard to organic

Table 1. Comparison of mean chemical composition of the stratification layer and vermicomposts obtained by processing of mixture (a) of kitchen organic waste by populations of *E fetida* and *D. veneta* ($\bar{x} \pm$ SD)

Parameters*		Stratification layer	Vermicompost	
			E. fetida	*D. veneta*
pH in H$_2$O		5,02 a ± 0,09	7,17 b ± 0,22	7,63 c ± 0,13
conductance	mS	0,65 a ± 0,06	3,81 b ± 0,43	3,44 c ± 0,32
salinity	g ·dm^{-3}	0,98 a ± 0,09	5,71 b ± 0,64	5.17 c ± 0,48
C		33.28 a ± 0,33	22.65 b ± 0,75	23.24 b ± 1.01
N		0.83 a ± 0,01	1.50 b ± 0,05	1.56 b ± 0,12
P		0.10 a ± 0.01	0.29 b ± 0.02	0.31 b ± 0.05
K	%	0.09 a ± 0.01	1.59 b ± 0.07	1.88 c ± 0.13
Ca		1.37 a ± 0.13	2.19 b ± 0.20	2.30 b ± 0.21
Mg		0.08 a ± 0.01	0.19 b ± 0.01	0.20 c ± 0.01
C/N		40.1	15.1	14.9

* dry matter.

a, b, c – significant differences between the stratification layer and vermicomposts.

Table 2. Comparison of mean chemical composition of vermicomposts obtained by processing of mixture (b) of kitchen organic waste by populations of *E fetida* and *D. veneta* ($\bar{x} \pm SD$)

Parameters		Wermicompost	
		E.fetida	*D.veneta*
pH in H_2O		5,24 a ±0,18	6,19 b ±0,01
conductance	mS	3.36 a ±0.77	2.65 b ±0.05
salinity	g •dm^{-3}	4,68 a ±1,15	3,98 b ±0.08
C		44.98 a ±9.88	27.19 b ±3.13
N		2.79 a ± 0.68	1.65 b ± 0.19
P		0.05 a ±0.01	0.05 a ±0.01
K	%	0.25 a ±0.03	0.28 a ± 0.01
Ca		0.40 a ±0.02	0.50 b ±0.01
Mg		0.05 a ±0.01	0.06 a ±0.01
C/N		16.1	16.5

a, b – statistically significant differences between vermicomposts produced by two worm species.

waste. Specially interesting and simple solutions are related to vermicomposting, including vermicomposting at the site of waste production (in homes, restaurants and at marketplaces) (Kostecka et al. 2018). In the future, this may result in a smaller amount of waste thrown away to disposal chutes and sewage systems, and may also reduce costs of waste transport to distant places, i.e. landfill sites or large composting plants. Although, in fact, such ideas are currently too innovative and time-consuming for many citizens, the Fertilisers and Fertilisation Act (2018) provides for vermicomposting (chapter 1, article 2.1 section 5)), and accepting such method of disposal requires development of special educational programmes. Spreading those ideas will be of great ecological and economic importance. Organic kitchen waste is accumulated gradually, so it may be administered to worms as food in "ecological boxes" (Kostecka et al. 2014, 2018).

Currently conducted studies showed that the quality of the produced vermicompost depends on the worm species, which is consistent with previously conducted studies by Padmavathiamma et al. (2008). Kostecka and Paczka (2011) analysed differences in the vermicompost composition depending on the concentration of *E. fetida*. They showed that a technology based on frequent thinning of the population of vermicomposting worms had a positive effect on 20% reduction of mean vermicompost salinity (p<0.001). With regard to the content of basic nutrients for plants, they showed a significantly higher increase in the content of assimilable potassium (p<0.001), nitrate nitrogen (p<0.001), phosphorus (p<0.001),

magnesium (p<0.001) and calcium (p<0.05). The pH values in $_{H2O}$, in the resulting fertilisers did not change. Regarding the whole picture of plant nutrition requirements, the use of the technology based on frequent separation of medium and worm population was much more advantageous.

Kostecka et al. (1999) noticed differences in the content of ash, organic matter, nitrogen, potassium, calcium, iron, copper, manganese, zinc, sodium and nickel between the vermicomposts produced by groups of species dominated by the Enchytraeidae or worms in the utilization media. This might result from different nutrient demand in both groups, which naturally formed biocenosis in the vermiculture media. The present differences may also be explained in a similar way. Nevertheless, this provokes further research.

Another issue worth further research is a problem of determining the vermicompost maturity. In much earlier studies, Kołodziej and Kostecka (1994) proposed a method to assess vermicompost directly in the culture site, which is practical for breeders. The method involved determination of assimilable plant nutrients in the produced vermicomposts and was to indicate a satisfactory level of the waste medium mineralization by worms. These authors, based on the analysis of vermicomposts from 121 field cultures and a vermicompost from an experimental and didactic culture conducted in the University of Agriculture in Krakow, Branch in Rzeszow (in 1992–1993), performed determinations of the contents of nitrate nitrogen and assimilable forms of phosphorus, potassium, calcium and magnesium in acetic acid by Spruvay method modified by

Nowosielski, and determinations of the total content of these components in sulphuric acid. They proposed minimum values indicating a sufficient level of mineralization. Specific values referred to the vermicomposts from bovine manure and with determination of the content diluted in 0.03 n acetic acid (from manure of 75–80% humidity) they amounted to: 250 mg $NO_3 \cdot dm^{-3}$ for nitrate nitrogen; 800 mg·dm^{-3} for assimilable phosphorus; 1400 mg·dm^{-3} for assimilable potassium; 1000 mg·dm^{-3} for assimilable calcium and 5000 mg $NO_3 \cdot dm^{-3}$ for assimilable magnesium.

Currently most authors applies the method of vermicompost maturity determination to assessment of changes in the C/N ratio in the obtained fertilisers. With regard to kitchen waste, depending on the mixture of vermicomposted waste and the worm species (e.g. kitchen wasted mixed with bagasse at a ratio of 1:1 vermicomposted by *E.fetida* (Babaei et al. 2016), household waste (30%) + bovine manure (70%) – *E.fetida* (Gupta et al. 2014), household waste + rice straw (1:1) – *E.fetida* (Hussain et al. 2018), kitchen waste + bovine manure (1:1) *E.fetida* i *Lampito mauritii* (Tripathi and Bharadway 2004); the proportion of C/N changes described by these authors was in the range between 12 and 53% – which corresponds to the final C/N value of the vermicomposts of 27–21.

It seems that the biggest C/N changes occur with vermicomposting of pure kitchen waste. Albasha et al. (2015) after 60 days of vermicomposting with *Eudrilus eugenigae* , reported a decrease in waste C/N from 30.8±0.12 to 5.45±0.13 (change by 82.3%). These assessments, however, also require collection of more abundant data and finding the principle governing the changes depending on the process duration, worm species and composition of the waste mixture.

CONCLUSION

The whole world faces the serious problem of an increasing amount of waste, including organic waste. This serious threat to the natural environment is a price for rapid civilisation progress. Therefore, more and more attention is directed to improving the waste management system; however, these activities must necessarily involve participation of the public. Vermiculture involves a process in which kitchen organic waste is converted into organic fertiliser containing precious nutrient for plants. This allows for their effective replenishment in soils, and at the same time for sustainable management of European soil resources, the more so that vermicomposts supply the soils with specific organic carbon. Vermicompost production is also a part of solving the problem of organic waste management. The vermicomposts produced from household kitchen waste by means of biotechnological methods using *E. fetida* and *D. veneta*, constitute wholesome organic fertiliser of good chemical composition thanks to the contents of macroelements. As shown by research, the production of vermicomposts and their chemical composition may answer the nutritional needs of plants.

REFERENCES

1. Adhami E., Hosseini S., Owliaie H. 2014. Forms of phosphorus of vermicompost produced from leaf compost and sheep dung enriched with rock phosphate. International Journal of Recycling of Organic Waste in Agriculture, 3, 68-73.

2. Adhikary S. 2012. Vermicompost, the story of organic gold: A review. Agricultural Sciences. 3, 7, 905-917.

3. Adi A.J., Noor Z.M. 2009. Waste recycling: Utilization of coffee grounds and kitchen waste in vermicomposting. Bioresource Technology, 100, 1027-1030.

4. Albasha M.O., Gupta P., Ramteke P.W. 2015. Management of kitchen waste by vermicomposting using earthworm Eudrilus Eugeniage. 2015. Conference on Advances in Agricultural, Biological & Environmental Sciences. [Available at: http://iicbe.org/upload/1869C0715011.pdf (accessed 4 september 2018)].

5. Arancon N.Q., Edwards C.A., Bierman P., Metzger J.D., Lee S., Welch C. 2003. Effects of vermicomposts on growth and marketable fruits of field-grown tomatoes, peppers and strawberries. Pedobiologia, 47, 731-735.

6. Babei A.A., Goudarzi G., Neisi A., Ebrahimi Z., Alavi N. 2016. Vermicomposting of cow dung, kitchen waste and sewage sludge. Journal of Advances in Environmental Health Research, 4(2), 88-94.

7. Bansal S., Kapoor K. K. 2000. Vermicomposting of crop residues and cattle dung with E*isenia foetida*. Bioresource Technology, 73, 95-98.

8. Bräutigam K.R., Jörissen J., Priefer C. 2014. The extend of food waste generation across Eu-27: Different calculation methods and the reliability of their results. Waste Management & Research, 32(8), 683-694.

9. Dominguez J., Edwards C.A. 2011a. Biology and Ecology of Earthworm Species used for Vermicomposting. In: Edwards C.A., Arancon N.Q., Shreman R. (Eds.) Vermiculture technology. Earthworms, organic wastes and environmental management. CRC, Taylor and Francis Group Press. Boca Raton, London, New York, 3, 27-40.

10. Dominguez J., Edwards C.A. 2011b. Relationships between Composting and Vermicomposting. In: Edwards C.A., Arancon N.Q., Shreman R. (Eds.) Vermiculture technology. Earthworms, organic wastes and environmental management. CRC, Taylor and Francis Group Press. Boca Raton, London, New York, 2, 11-25.

11. European Commission (DG ENV), Directorate C – Industry 2010. Preparatory study on food waste across EU 27. [Available at: http://ec.europa.eu/environment/eussd/pdf/bio_foodwaste_report.pdf (accessed 4 september 2018)].

12. Eurostat 2016. Waste generation and treatment. [Available at: http://epp. eurostat.ec.europa.eu/cache/ITY_SDDS/EN/env_wasgt_esms.htm#stat_pres (accessed 5 June 2015)].

13. FAO (2014) Food balance sheets. [Available at: http://faostat3.fao.org/faostatgateway/go/to/download/FB/*/E (accessed 2 September 2018)].

14. FUSIONS 2016. Reducing food waste through social innovations. Fusions EU project 311972 [Available at: http://www.eufusions.org/phocadownload/Publications/Estimates%20of%20European%20food%20waste%20levels (accessed 2 August 2018)].

15. Gajalakshmi S., Ramasamy E.V., Abbasi S.A. 2002. Vermicomposting of paper waste with the anecic earthworm Lampito mauritii Kingburg. Indian Journal of Chemical Technology, 9, 306-311.

16. Garg V. K. Chand S., Chhillar A., Yadav A. 2005. Growth and reproduction of Eisenia foetida in various animal wastes during vermicomposting. Applied Ecology and Environmental Research, 3, 51-59.

17. Gupta R., Garg V. K. 2009. Vermiremediation and nutrient recovery of non-recyclable paper waste employing Eisenia fetida. Journal of Hazardous Materials, 162, 430-439.

18. Gupta R., Yadav A., Garg V. K. 2014. Influence of vermicompost application in potting media on growth and flowering of marigold crop. International Journal of Recycling of Organic Waste in Agriculture, 3(1), 47-57.

19. Gustavsson J, Cederberg C.,Sonesson U. 2011. Global food losses and food waste: Extent, causes and prevention. Rome: Food and Agriculture Organization of the United Nations (FAO), 1-29.

20. Hussain N., Das S., Goswami L., Das P., Sahariah B. 2018. Intensification of vermitechnology for kitchen vegetable waste and paddy straw employing earthwormconsortium: Assessment of maturity time, microbial structure, and economic benefit. Journal of Cleaner Production, 182, 414-426.

21. Kadam D., Pathade G. 2014. Effect of tendu (Diospyros melanoxylon RoxB.) leaf vermicompost on growth and yield of French bean (Phaseolus vulgaris L.). International Journal of Recycling of Organic Waste in Agriculture, 3, 44-50.

22. Kashem M. A., Sarker A., Hossain I., Islam M. S. 2015. Comparison of the vermicompost and inorganic fertilizers on vegetative growth and fruit production of tomato (Solanum lycopersicum L.). Open Journal of Soil Science, 5, 53-58.

23. Kaushik P., Garg V. K. 2003. Vermicomposting of mixed soil textile mill sludge and cow dung with the epigeic earthworm Eisenia foetida. Bioresource Technology, 90, 311-316.

24. Kołodziej M., Kostecka J. 1994. Method for vermicompost evaluation directly in the earthworm's bed. Metoda oceny wermikompostów bezpośrednio w siedlisku hodowlanym dżdżownic. Zeszyty Naukowe AR w Krakowie. Sesja Naukowa, 41, 95-98. (in Polish).

25. Kostecka J. 1994. Guide for earthworm breeders. Poradnik hodowcy dżdżownic. Akademia Rolnicza w Krakowie. Filia w Rzeszowie, 1-40.

26. Kostecka J. 2000. Badania nad wermikompostowaniem odpadów organicznych. Zeszyty Naukowe AR w Krakowie. Rozprawy, 268, 1-88.

27. Kostecka J., Błażej J. 2000. Growing plants on vermicompost as a way to produce high quality foods. Bull. of the Polish Acad. of Scien. Biol. Scien. 48(1), 1-10. [http://repozytorium.ur.edu.pl/handle/item/3113].

28. Kostecka J., Garczyńska M., Pączka G. 2018. Food waste in the organic recycling system and a sustainable development. Problems of Sustainable Development, 13(2), 157-164.

29. Kostecka J., Kaniuczak J., Nowak M. 1999. Wybrane cechy wermikompostów z organicznych odpadów domowych. Folia Univ. Agric. Stet. 200. ser. Agricultura, 77, 173-177.

30. Kostecka J., Pączka G. 2006. Possible use of earthworm Eisenia fetida (Sav.) biomass for breeding aquarium fish. European Journal of Soil Biology, 42, 231-233

31. Kostecka J., Pączka G. 2011. Kitchen wastes as a source of nitrogen and other macroelements according to technology of vermiculture. Ecological Chemistry and Engineering, A.18(12), 1683-1689.

32. Kostecka J., Pączka G., Garczyńska M., Podolak-Machowska A., Dunin-Mugler C, Szura R. 2014. Wykorzystanie wermikompostowania do zagospodarowania odpadów organicznych w gospodarstwach domowych. Inżynieria i Ochrona Środowiska, 17(1), 21-33.

33. Krutwagen B., Kortman J., Verbist K. 2008. Inventory of Existing Studies Applying Life Cycle Thinking to Biowaste Management. Office of Official Publications of the European Communities, Luxembourg. [Available at: http://eplca.ire. ec.europa.eu/nploads/Waste-InYentorv-of-existing-stiidies-applving- life-cycł e-thinking-to-biowaste-management.pdf (accessed 2 August 2018).

34. Loh T.C., Lee Y.C., Liang J.B., Tan D. 2005.Vermicomposting of cattle and goat manures by Eisenia foetida and their growth and reproduction performance. Bioresource Technology, 96(1), 111-114.

35. Monier V, Mudgal S, Escalon V. 2010. Final report – Preparatory study on food waste across EU 27; European Commission [DG ENV – Directorate C]. Paris: BIO Intelligence Service, 1-205.

36. Padmavathiamma P. K., Li L. Y., Kumari U. R. 2008. An experimental study of vermin-biovaste composting for agricultural soil improvement. Bioresource Technology, 99, 1672-1681.

37. Paradelo R., Moldes A. B., Barral M. T. 2009. Properties of slate mining wastes incubated with grape marc compost under laboratory conditions. Waste Management, 29, 579-584.

38. Parvaresh A., Movahedian H., Hamidian L. 2004. Vermistabilization of Multicipal Wastewater Sludge with Eisenia fetida. Iranian Journal of Environmental Health, Science and Engineering, 1, 43-50.

39. Pączka G., Kostecka J. 2012. Trends in organic waste vermicomposting. In: Kostecka J., Kaniuczak J. (Eds.) Internetowa Promocja Nauki. Nauka dla gospodarki 3/2012. Practical Applications of Environmental Research, 22, 267-281. http://www2.ur.edu.pl/wbr_monografie/Practical_Applications.pdf

40. Pączka G., Kostecka J. 2013. The influence of vermicompost from kitchen waste on the yield-enhancing characteristic of peas *Pisum sativum* L. Var. Saccharatum Ser. Bajka variety. Journal of Ecological Engineering, 14, 49-53.

41. Sangwan P., Kaushik C. P., Garg V. K. 2008. Feasibility of utylization of horse dung spiked filter cake in vermicomposters using exotic earthworm *Eisenia foetida*. Bioresource Technology, 99, 2442-2448.

42. Sapek A. 2013. Dissipation of fertiliser compo-

nents from food treated as waste. Water-Environment-Rural Areas, 1(41), 129-142.

43. Sherman R. 2003. Raising earthworms successfully. North Carolina Cooperative Extension Service. North Carolina State University, Raleigh, NC, 1-26.

44. SinghR., Sharma R.R., Kumar S., Gupta R.K., Patil R.T. 2008. Vermicompost substitution influences growth, physiological disorders, fruit yield and quality of strawberry (Fragaria x ananassa Duch.). Bioresource Technology, 99, 8507-8511.

45. Song X., Liu M., Wu D., Griffiths B.S., Jiao J., Li H., Hu F. 2015. Interaction matters: Synergy between vermicompost and PGPR agents improves soil quality, crop quality and crop yield in the field. Applied Soil Ecology, 89, 25-34.

46. The Fertilisers and Fertilisation Act (Journal of Laws of the Republic of Poland, 28 June 2018, item 1259) [Avaiable at https://www.infor.pl/akt-prawny/DZU.2018.124.0001259,ustawa-o-nawozach-i-nawozeniu.html, accessed 2 October 2018]

47. The National Waste Management Plan [KPGO 2022]. [Avaiable at https://www.mos.gov.pl/komunikaty/szczegoly/news/krajowy-plan-gospodarki-odpadami-2022/, accessed 2 October 2018].

48. The Regulation of the Minister of Economy of 16 July 2015 on landfill waste disposal (Journal of Laws of the Republic of Poland [Dz. U. RP], item 1277). [Avaiable at http://prawo.sejm.gov.pl/isap.nsf/DocDetails.xsp?id=WDU20150001277, accessed 2 October 2018]

49. The Regulation of the Minister of the Environment of 20 January 2015 on the process of R10 recovery (Journal of Laws of the Republic of Poland [Dz. U. RP], item 132).

50. Thematic Strategy on Soil Protection [COM (2002) 179], [COM (2006) 231], [COM (2006) 232]. [Available at: https://www.eea.europa.eu/policy-documents/soil-thematic-strategy-com-2006-231, accessed 2 October 2018]

51. Tripathi G., Bhardwaj P. 2004. Decomposition of kitchen waste amended with cow manure using an epigeic species (*Eisenia fetida*) and an anecic species (*Lampito mauritii*). Bioresource Technology, 92(2), 215-218.

9

Preparing and using Cellulose Granules as Biodegradable and Long-Lasting Carriers for Artificial Fertilizers

Tobiasz Gabryś[1], Beata Fryczkowska[1*]

[1] Institute of Textile Engineering and Polymer Materials, University of Bielsko-Biala, ul. Willowa 2, 43-309 Bielsko-Biala, Poland

[*] Corresponding author's e-mail: bfryczkowska@ath.bielsko.pl

ABSTRACT

The paper presents the results of research on the preparation and use of cellulose granules as carriers of nutrients in the cultivation of plants. The granules were prepared from a cellulose solution in 1-ethyl-3-methylimidazole acetate followed by coagulation in water and primary alcohols: methanol, ethanol, 1-propanol, 1-butanol, 1-pentanol, 1-hexanol and 1-octanol. Modifications of granules were also carried out by hydrophobization at elevated temperature and by encapsulation in a polylactide solution. As a result of the research, cellulose granules were obtained, which were characterized by different porosity, depending on the type of coagulant used. The morphology of granules surface and cross-sections was examined by means of scanning electron microscopy (SEM). The cellulose granules exhibited good sorption/desorption properties which were investigated by conductometry and UV-Vis spectroscopy. The longest desorption time of NH_4NO_3 was characteristic of granules obtained as a result of thermal hydrophobization of the surface, which were used in the cultivation of the spider plant. As a result of the research, cellulose granules were obtained which may find potential application in crop production, as long-acting, non-dusting and fully biodegradable fertilizers.

Keywords: cellulose, ionic liquid, granule, long-acting fertilizers

INTRODUCTION

Contemporary agriculture and gardening develop very dynamically. Continuous economic development and population growth force the farmers to obtain the highest possible yields. Intensification of agriculture has led to the soils impoverishment in terms of macro- and micronutrients. Macronutrients play a structural role and in large quantities are taken from the soil. The micronutrients include about 30 elements. The microcomponents used as fertilizer preparations include eight of them: iron, manganese, zinc, copper, boron, molybdenum, nickel and chlorine. They take part in biochemical processes, mainly as an element necessary for the proper functioning of enzymes. They also show activity stimulating the effectiveness of macronutrients.

One of the methods guaranteeing high and predictable yields is the intensification of agriculture. It is based on increasing agricultural production by increasing expenditures and material resources per unit of space or by increasing labour inputs. It leads to an increase in the specialization of production, especially to the separation of plant and animal production. This is related, among others, to an increase in the use of mineral fertilizers due to limited possibilities of natural fertilization [Wang et al. 2017]. Methods of fertilization used currently are based mainly on the use of artificial fertilizers with an extended period of release. They are widely used in gardening, nursery and agriculture. Their main advantage is the fact that at the beginning of the growing season they gradually release their minerals and macro- and micronutrients for up to several months allowing the plant to freely extract nutrients from the substrate. Micronutrient fertilizers are produced in liquid or finely crystalline form. Liquid fertilizers are characterized by effective and fast action and high efficiency. Trace element chelates have gained the topmost importance. These compounds are resistant to external factors, are characterized by high durability, provide a high level of bioavailability of micronutrients and there is a low probability of their phytotoxicity. Fertilizer chelates are well soluble in water, and thanks to slow dissociation, ions are released gradually, which enables optimum performance [Klem-Marciniak et al. 2015].

Substances selected for fertilizing purposes allow for permanent stabilization of the micronutrient cation in a wide pH range and in the presence of other fertilizing components. The compounds belonging to the aminopolycarboxylic group (APCAs) are particularly durable. The fertilizer industry uses such ligands as nitrilo triacetic acid (NTA), ethylene diamine tetraacetic acid (EDTA), diethylene triamine pentaacetic acid (DTPA), iminodisuccinic acid (IDHA), hydroxy-2-ethylenediaminetriacetic acid (HEEDTA) and others [Klem-Marciniak et al. 2015].

EDTA is used in most conventional artificial fertilizers [Bloem et al. 2017]. The disadvantage of this compound is very low biodegradability [Allard et al. 1996), because it can persist in soil for up to 15 years [Alvarez et al. 1996, Meers et al. 2005]. Another disadvantage of EDTA is the ability to absorb heavy metals from soil, which creates a huge threat to the environment [Luo et al. 2017].

Improper use of mineral fertilizers may lead to undesirable effects, which undoubtedly include: dusting of mineral compounds, penetration of fertilizers into ground and underground water, and crop over-fertilization [Gagliardi & Pettigrove 2013]. Excess of micronutrients in the soil solution may lead to eutrophication of waters, remobilization of heavy metals from benthic and river sediments, and consequently to their introduction into the food chain.

An example of fertilizers that solve these problems appear to be the cellulose granules described in this paper. Cellulose is the most widespread, inexpensive, biodegradable polymer derived from renewable sources, which is used both in unprocessed form and as derivatives. Ionic liquids (IL), also referred to as "green solvents," can be used to dissolve cellulose [Gathergood et al. 2004, Earle & Seddon, 2000, Novoselovet al. 2007, Weerachanchai et al. 2014]. Cellulose solutions in IL can be used to obtain fibers, nanofibers, gels and aerogels, flocs, membranes and granules [Wendler et al. 2012].

The literature reports a method for obtaining cellulose granules that Suzuki et al. obtained from microcrystalline cellulose dissolved in 1-butyl-3-methylimidazolium chloride [Suzuki et al. 2014]. Another ionic liquid, 1-allyl-3-methylimidazolium chloride, was used by Voon et al. to prepare porous cellulose granules [Voon et al. 2015, Voon et al. 2017b]. Luo et al. obtained composite granules of cellulose, chitosan and magnetite, using the NaOH/urea/ H_2O to prepare the mixture [Luo et al. 2015]. The same solvent system was used by other researchers to obtain granules that were coagulated in HNO_3 [Trygg et al. 2013, Yildir et al. 2013]. A similar system of solvents: NaOH/thiourea/urea was used to obtain smaller particles: cellulose micro- and nano-granules [Voon et al. 2017a].

In this paper, research on obtaining biodegradable, long-acting carriers of artificial fertilizers in the form of cellulose granules was undertaken. The cellulose was dissolved in 1-ethyl-3-methylimidazoline acetate (EMIMAc) to obtain a 5% solution from which drops were formed, which were then coagulated in water and selected primary alcohols. Modifications of cellulose granules, consisting in hydrophobization at elevated temperature and coating with a polylactide solution, were also carried out, in order to prolong the NH_4NO_3 desorption process. As a result of the study, granules characterized by good sorption / desorption properties relative to the mineral salt were obtained. The hydrophobization of cellulose granules has positively influenced the slowdown of the NH_4NO_3 release. The obtained granules can be successfully used in plant production as non-dusty, biodegradable fertilizers with a prolonged release.

MATERIALS AND TEST METHODS

Reagents

Cellulose (long fibers), 1-ethyl-3-methylimidazolium acetate with (97%), 1-octanol (99%) were purchased from Sigma-Aldrich. Polylactide (3050) was purchased from Nature Works LLC. Methanol (99.8%), ethanol (96%), 1-propanol (99.5%), 1-butanol (99.5%), 1-pentanol (98.5%), 1-hexanol (98%), NH_4NO_3, $CHCl_3$ were purchased from Avantor Performance Materials Poland S.A.

Formation of cellulose granules

A 5% solution of cellulose (CEL) in the ionic liquid 1-ethyl-3-methylimidazolium acetate (EMIMAc) was prepared. At the beginning adequate amounts of cellulose and ionic liquid were weighed. The whole was then mixed thoroughly, then heated in a laboratory microwave oven at 3 intervals of 5 seconds, keeping the temperature of the mixture below approx. 40 °C. The obtained cellulose solutions were allowed to deaerate for 24 hours The process to obtain cellulose solution was already described in our previous paper [Fryczkowska et al. 2017a].

The cellulose granules were formed using an infusion pump (Fig. 1) fitted with a 20 cm³ syringe. The syringe filled with the CEL solution was placed in a holding fixture of the KdScientific KDS-100 infusion pump, which, while working, extruded drops through a nozzle with a diameter of 2 mm into the crystallizer with the appropriate coagulant. The cellulose granules were coagulated using: distilled water, methanol, ethanol, 1-propanol, 1-butanol, 1-pentanol, 1-hexanol and 1-octanol. Under the influence of the coagulant, EMIMAc was eluted and cellulose coagulated. The granules obtained in this way were transferred together with the coagulant into a conical flask, equipped with a magnetic stirrer and stirred for one week. The granules were then filtered and air dried for 72 hours (Table 1).

Preparation of long-acting fertilizers based on cellulose granules

The cellulose granules were obtained in two series. In the first series of 2 g of granules: W, M, E, P, B, Pe, H, O were introduced into 100 cm³ of 1% w/w aqueous solution of NH_4NO_3 and were left for 10 minutes. In the second one, a 25% w/w solution of NH_4NO_3 was used. The impregnated granules were filtered and air-dried.

Physical modification of cellulose granules

In order to slow down the release of NH_4NO_3 from cellulose granules, which eventually were to be used as long-acting fertilizers, modification of their surface was carried out. For this purpose, cellulose granules that were coagulated in methanol were selected

At the beginning, two samples of dried cellulose granules (M) were prepared. One portion of granules (M_a) was added to 1% w/w aqueous

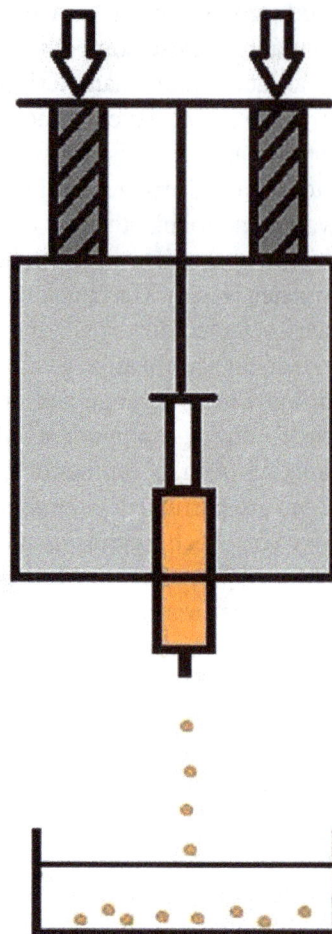

Fig. 1. Cellulose granules forming sequence (diagram made by yourself)

solution of NH_4NO_3. A second portion of granules (M_b) was added to 25% w/w of aqueous solution NH_4NO_3. The granules were then left for 24 hours, filtered and air-dried (Table 2). The amount of solution and the weight of the granules were the same as abowe.

To modify the granules, the simplest method was used, consisting in hydrophobization by heating. Samples of granules M, M_a and M_b were put into a laboratory dryer at 65 °C and heated for 4 hours. The samples thus obtained were marked: M1, M_a1 and M_b1 (Table 2) and were used for further research.

The second method consisted in coating the granules with a biodegradable polymer solution. Polylactide (PLA) was selected for the study, from which 1% w/w polymer solution in $CHCl_3$ was prepared. Then, subsequently, granules M, M_a and M_b were put into the PLA/$CHCl_3$ solution, then they were quickly removed and placed on the glass plate until the solvent evaporated. As a result of this process, cellulose granules M2,

M_a2 and M_b2 were obtained (Table 2), which were used for further research.

Bulk density study

To study the bulk density cellulose granules were placed in a graduated cylinder with a capacity of 5 cm³ and their volume was precisely measured, then they were weighed on the Sartorius CP224S-0CE analytical balance. The bulk density of cellulose granules (d) (g/cm³) was calculated according to formula (1).

$$d = \frac{W}{V} \qquad (1)$$

where: W – mass of granules [g];
V – volume of granules [cm³]

The test results for unmodified and modified granules are summarized in Table 1 and Table 2, respectively.

Study of water absorption

The water absorption study was performed as follows. 0.1 g of cellulose granules was weighed on the analytical balance, then placed in glass vials, followed by adding 10 cm³ of distilled water into each and left for 10 minutes. Then, the granules were filtered off, the remaining water removed with a filter paper, and they were re-weighed. The water absorption (U) was calculated from the formula (2).

$$U = \frac{W_w - W_d}{W_d} \times 100\% \qquad (2)$$

where: W_w – mass of wet granules [g];
W_d – mass of dry granules [g].

Salt sorption and desorption studies on cellulose granules

The study of sorption properties of cellulose granules were carried out in a solution of NH_4NO_3 at a concentration of 1% w/w. 0.25 g of cellulose granules were weighed on an analytical balance and put in a beaker, then 40 cm³ of previously prepared NH_4NO_3 solution was added and left for 24 hours. Next, the spectrum of the solution from over the granules ($\lambda = 302$ nm) was made on HACH DR/4000U UV-Vis spectrophotometer and the concentration of the salt in the solution was determined, and the results were summarized

in Table 3. Finally, samples of cellulose granules were filtered and allowed to dry.

Salt desorption studies (NH_4NO_3), on the other hand, were carried out in such a way that 40 cm³ of distilled water was added to the dried cellulose granules (0.25 g) and left for 24 hours. Next, the spectrum of the solution from over the granules was made and the concentration of the salt in the tested solution was determined.

Investigation of the kinetics of NH_4NO_3 salt ion release from cellulose granules

The cellulose granules were tested in terms of kinetics of release of the salt contained in them (NH_4NO_3) into the aqueous solution. The tests were carried out in parallel on granules soaked with 1% w/w and 25% w/w solution of the salt. Initially, samples of granules weighing 0.5 g were prepared and put into beakers with distilled water. For granules M_a, M_a1, M_a2, 80 cm³ of distilled water was used, and 800 cm³ for granules M_b, M_b1, M_b2. Subsequently, Elmetron CC-315 microprocessor conductometer was used to measure the conductivity of the solution every 5 minutes. The tests were carried out for 30 minutes.

Scanning electron microscope

Observations of the granules surface morphology and their cross sections were carried out using JSM – 5500 LV JEOL scanning electron microscope (SEM). All samples were coated with a layer of gold in JEOL JFC 1200 vacuum coater at 3×10^{-5} Tr.

Using granules in the cultivation of spider plant (Chlorophytum comosum)

In order to check the possibility of using cellulose granules in the cultivation of plants, the experiment was carried out using spider plant (*Chlorophytum comosum*), which is relatively easy to grow. Plants were divided into two groups. The first one were non-fertilized plants and the second one-plants fertilized with cellulose granules soaked in NH_4NO_3. Hydrophobized granules (M_b1) were selected for the study. The method of fertilization consisted in scattering a portion of 0.5 g of cellulose granules on the surface of the flower bed filling the pots with plants. Re-fertilization was carried out after 1.5 months, during which the total distribution of granules took place. In total, 4 doses, i.e. 2g of cellulose granules per plant in a pot with a diameter of 12 cm, were used.

During the six months of the study, the manner of NH_4NO_3 release to the external environment, the time of granules action and the time of their decomposition, and above all the effect of the use of granules on plant development were observed. At the end, the aboveground parts of the plants were cut down and the green mass was weighed.

RESULTS AND DISCUSSION

Physicochemical properties

The cellulose granules (Fig. 2), which were obtained as a result of coagulation in water and primary alcohols, were subject to physicochemical properties testing including the check of: bulk density, water absorption and sorption and desorption properties for NH_4NO_3.

First, the bulk mass was determined (Tab. 1). Analysing the results of the research it was concluded that granules M are characterized by the highest bulk density of 0.69 g/cm³. The obtained result confirms the observations, which showed that the fastest cellulose granules coagulation occurred in methanol. In the case of cellulose granules coagulated in other primary alcohols, their bulk densities were: 0.66 (granules E); 0.64 (granules P); 0.59 (granules B); 0.54 (granules Pe); 0.53 g/cm³ (granules H). The decrease in bulk density is closely related to the process of granules coagulation, which depends on the polarity of the coagulant. In the case of primary alcohols, their polarity drops from methanol towards 1-hexanol. The lowest bulk density value was 0.07 g/cm³ which characterized granules O. The obtained result is due to the slow process of cellulose coagulation in 1-octanol and is closely related to the low polarity of this alcohol. On the other hand, the bulk density of granules W, which were co-

agulated in water, is 0.55 g/cm³ and is close to the average value typical for higher alcohols.

The results of the water absorbency tests (Table 1) clearly indicate that membrane O is the most porous of all granules obtained. Water absorption for granules O is 442%. Whereas the water absorption for granules W is minor and amounts to ~ 31%, which proves their compact structure. Analysing the sorption properties of granules M, it can be noticed that water absorption is ~ 43%, which may indicate the presence of fine pores. In contrast, granules coagulated in other alcohols have a water absorption capacity of ~75 (E); 67 (P); 27 (B); 25 (Pe); 26% for granules H.

Water absorption is influenced by porosity. The tests show that granules B, Pe, H, characterized by similar bulk density values, absorb the least amount of water. Such a result may indicate the presence of closed pores in the structure of granules, preventing rapid sorption of water in the whole mass of the granule. The results obtained confirm our previous research [Fryczkowska et al. 2017b]. Analysing the results, it can be concluded that cellulose granules do not have good water sorption properties, with the exception of granules O.

Interesting results were obtained during studying salts sorption capacity of the granules (NH_4NO_3) from 1% aqueous solution (Table 1). It turned out, the granules only capture 20% of the salt from the solution. Also this time granules O show the best properties. Granules O adsorbed ~25%, while the remaining granules sorbed 17–20% of salt. The low values of NH_4NO_3 sorption may be due to the low concentration of salt used in the test.

The aim of the research was to obtain granules that could potentially be used as fertilizer carriers. Therefore, in addition to the studies of sorption properties, tests of the reverse process seemed to be useful. The obtained NH_4NO_3 desorption results (Table 1) showed that all obtained granules easily and effectively desorbed the salt contained in them. The NH_4NO_3 desorption ef-

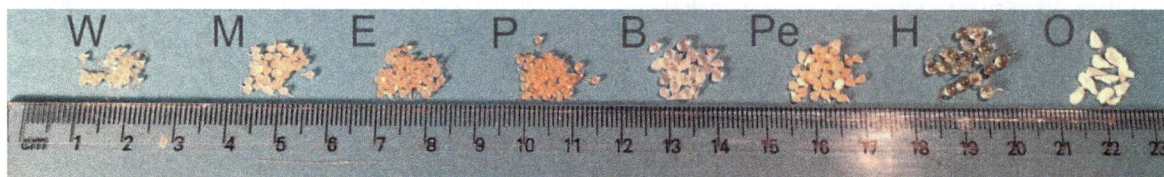

Fig. 2. Made by yourself pictures of cellulose granules obtained during coagulation of the cellulose solution in water (W) and primary alcohols (M, E, P, B, Pe, H, O).

Table 1. Physicochemical properties of cellulose granules

Type of coagulant	Designation of cellulose granules	d [g/cm³]	U [%]	Concentration of NH_4NO_3 adsorbed on granules [%]	Concentration of NH_4NO_3 released from the granules [%]
water	W	0.55	31.27	17.55	17.30
methanol	M	0.69	42.55	20.04	19.77
ethanol	E	0.66	74.47	18.88	18.45
1-propanol	P	0.64	66.67	18.62	18.55
1-butanol	B	0.59	26.87	17.44	17.39
1-pentanol	Pe	0.54	25.36	17.04	16.97
1-hexanol	H	0.53	25.57	17.22	17.19
1-octanol	O	0.07	442.37	25.42	25.38

Table 2. Comparison of physicochemical properties of cellulose granules coagulated in methanol (M), soaked with NH_4NO_3 and subject to physicochemical modification

Type of granules	NH_4NO_3 concentration [%]	Designation of cellulose granules	d [g/cm³]	U [%]	Concentration of NH_4NO_3 adsorbed on granules [%]	Concentration of NH_4NO_3 released from the granules [%]
unmodified	0	M	0.69	42.55	-	-
	1	M_a	0.70	39.23	20.04	19.77
	25	M_b	0.71	41.55	25.82	24.28
hydrophobized	0	M1	0.59	7.10	-	-
	1	M_a1	0.60	7.81	20.33	12.97
	25	M_b1	0.62	8.60	25.90	14.99
PLA-coated	0	M2	0.70	9.03	-	-
	1	M_a2	0.71	9.89	20.26	19.42
	25	M_b2	0.73	11.00	25.23	25.01

ficiency is similar for all granules tested and is close to ~ 100%.

The test results obtained for cellulose granules, formed in various solvents, led us to the preparation of granules with slow release of salt. Granules M, which were modified by two methods, were selected for the tests. One modification method consisted in hydrophobization, carried out by heating the granules at elevated temperature. The second technique consisted in coating the granules with a layer of a biodegradable polymer, which was PLA.

Interesting results were obtained for the water absorption of the initial granules M (Table 2) subject to modification. It turned out that granules M1 absorb 6 times less, and granules M2 4.5 times less water than unmodified granules M. Analysing the water absorption results for M_a and M_b granules (containing NH_4NO_3), it was observed that they are characterized by a small decrease in water sorption in relation to the water absorption of granules M. For hydrophobized granules (M_a1 and M_b1) and coated granules (M_a2 and M_b2), the water absorption is slightly increased, which may be due to the presence of salt, released from the inside of the granules through osmosis.

Low values of sorption (~ 20% and ~ 36%) were recorded for granules soaked with 1% and 25% solutions of NH_4NO_3, respectively. The obtained results may stem from low porosity of the granules, which determines the amount of adsorbed salt. Studies of NH_4NO_3 adsorption/ desorption (Table 2) demonstrated that granules M_a and M_b as well as granules M_a2 and M_b2 adsorb and desorb similar amounts of salt within 24 hours. However, in the case of granules M_a1 and M_b1, it is observed that desorption of salt is lower by half.

Desorption kinetics

An important feature that demonstrates the possibility of using granules as long-acting fertilizers is the release kinetics of the salt used in the experiment (NH_4NO_3). After preliminary tests, W, M, E, P, B, Pe, H, O granules were selected for testing and soaked with a 25% salt solution. In Figure 3, we can observe that all obtained cellulose granules undergo almost total desorption of salt as soon as after 5 minutes of testing. Over the next 25 minutes, ionic conductivity increases, but to a small extent. The highest values of ionic conductivity were obtained for granules O, which is easy to explain, as these granules contained the highest amount of salt (Table 1), and the process of desorption of salt from granules O is similar as in all other types of cellulose granules. Water, penetrating the inside of granules, causes rapid osmotic desorption of mineral compounds to the external environment, which is not a beneficial phenomenon in the context of the study. The conducted research prompted the authors to modify cellulose granules in order to slow down the desorption process.

Cellulose granules M, characterized by high bulk density and good sorption of salt, were selected for physical modification. Two granule modification techniques have been proposed and carried out: hydrophobization by heating at elevated temperature and coating with PLA solution.

Studies on the release kinetics of NH_4NO_3 salt ions from modified cellulose granules (Fig. 4) show that this process is the slowest in the heated granules (M_b1). In Figure 4 we observe that the conductivity of the solution containing granules M_b1 during the first minutes slowly increases from ~0.7 mS to ~0.8 mS, then the increase speeds up to reach 1.35 mS after 30 minutes. The observed phenomenon may result from the properties of granules, which were subject to hydrophobization. It is known that when the cellulose is heated at an elevated temperature, it is crosslinked, resulting in increased water resistance. The analysis

of the test results indicates that a similar phenomenon occurred during the hydrophobization of cellulose granules M_b1. It can, therefore, be proposed that when these granules come in contact with water, it slowly permeates them, causing the granules to become softened from the inside with slow desorption of salt.

When analysing the desorption curves of the granules (Fig. 4), their similarity can be observed. Both types of cellulose granules release ions faster and in greater quantities than M_b1. The ionic conductivity of solutions after 5 minutes of testing is ~ 1 mS and 1.35 mS, for M_b2 and M_b granules, respectively. At the end of the process, the desorption of ions from the granules stabilizes and amounts to ~ 3 mS. Thus, the encapsulation of cellulose granules did not bring the intended effect of slowing down the release of NH_4NO_3.

Morphology of the surface and structure of granules

Scanning electron microscopy (SEM) allowed to observe of the tested cellulose granules (Fig. 5 and 6). The microphotographs in Figure 5 show the structure of the outer surface – the skin layer, and cross-section of cellulose granules coagulated in water and in individual primary alcohols. Analysing SEM images of the surface of subsequent granules, it was observed that granules W have a compact surface with no clear pores (W-1). For granules coagulated in alcohols (Fig. 5–2), it is observed that the surface begins to wrinkle, there are corrugations, which

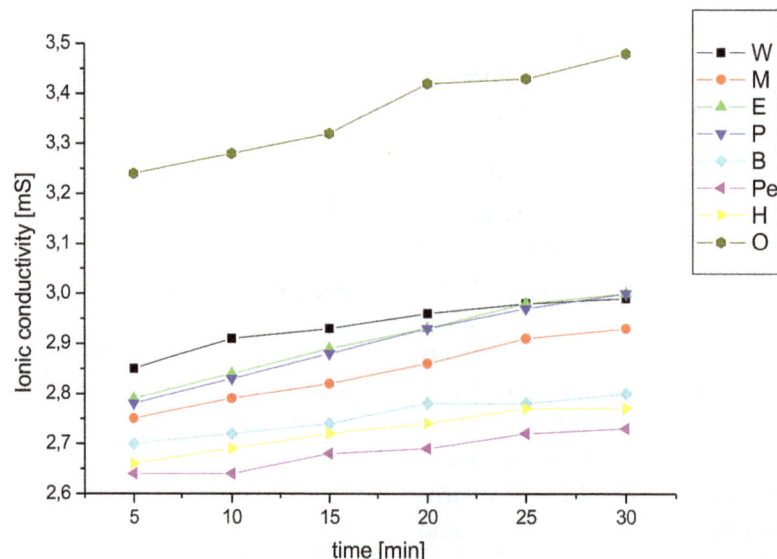

Fig. 3. The kinetics of NH_4NO_3 salt desorption for cellulose granules W, M, E, P, B, Pe, H, O.

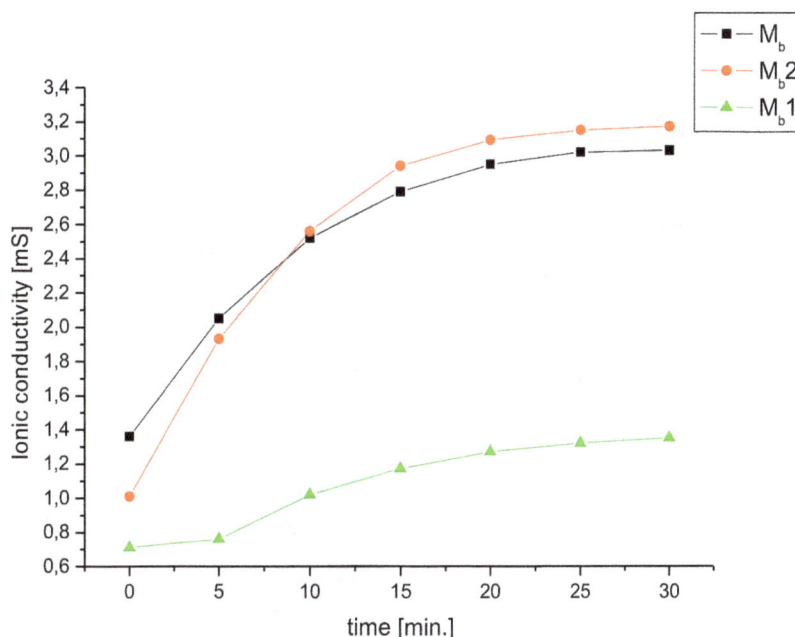

Fig. 4. The kinetics of desorption of NH_4NO_3 salts for M_b, M_b1, M_b2 cellulose granules

are increasing in number and getting deeper (in the direction from methanol to higher alcohols). O granules, which are much larger than their predecessors (Fig. 2), are characterized by most developed surface and the wrinkles make them white in the visible light.

Analysing the cross sections of cellulose granules (Fig. 5–4) it was observed that the sample preparations in some cases revealed their structure. The cross-sections of granules E and P (Fig. 5. E-4; P-4) show fine pores. In the pictures of successive granules B and Pe, a highly wrinkled surface was revealed, originating from the pores disrupted during sample preparation. The microphotograph in Figure 5. Pe-4 uncovers the layered structure of the granule Pe, which is built of large chambers near the surface (in the skin layer) . While, SEM images of granules O (Fig. 5 O-4) have shown that inside they are made of numerous large chambers and of a layered skin on the outside.

The morphology of the surface and cross-sections of cellulose granules confirmed our assumptions, based on previous studies on the production of cellulose membranes [Fryczkowska et al. 2017c].

SEM microscopy (Fig. 6) allowed to observe changes in the appearance of the surface structure and cross-sections of modified cellulose granules. M_b, M_b1, M_b2 granules were selected for the tests. In the general view of the cellulose granules (Fig. 6–1, 2) it can be observed that salt crystals

(granule M_b) appear on the surface, and are then coated with a polymer film (granule M_b) and are not present on the surface of granule M_b1. A different appearance of the M_b1 granule may indicate that during hydrophobization, salt crystals detached from its outer surface.

Analysing the cross-sections (Fig 6–3, 4), it can be seen that the sample preparation allowed uncovering the corrugations on the surface of the cellulose granules that look different than the initial granule M (Fig 5, M-3). In the cross-sectional image of granule M_b, a coating of salt crystals can be observed, which disappear in the cross-section of granule M_b1. However, on the cross-section of the M_b2 granule, it can be observed that a second layer, similar to skin layer, is clearly visible at the edge of the granule. The outer layer of the M_b2 granule is formed by NH_4NO_3 crystals, coated with a PLA layer.

The use of granules in cultivation

Cellulose granules (M) prepared in the experiment, after being soaked with NH_4NO_3 solution, were physically modified in the heating process and served as fertilizers for growing plants. Spider plant was selected for the study and fertilized with cellulose pellets for 6 months. Of all the types of granules, M_b-1, characterized by a long release of NH_4NO_3, were selected. The result of the experiment was very satisfactory (Fig. 7). Green mass increase of the plants fertilized with

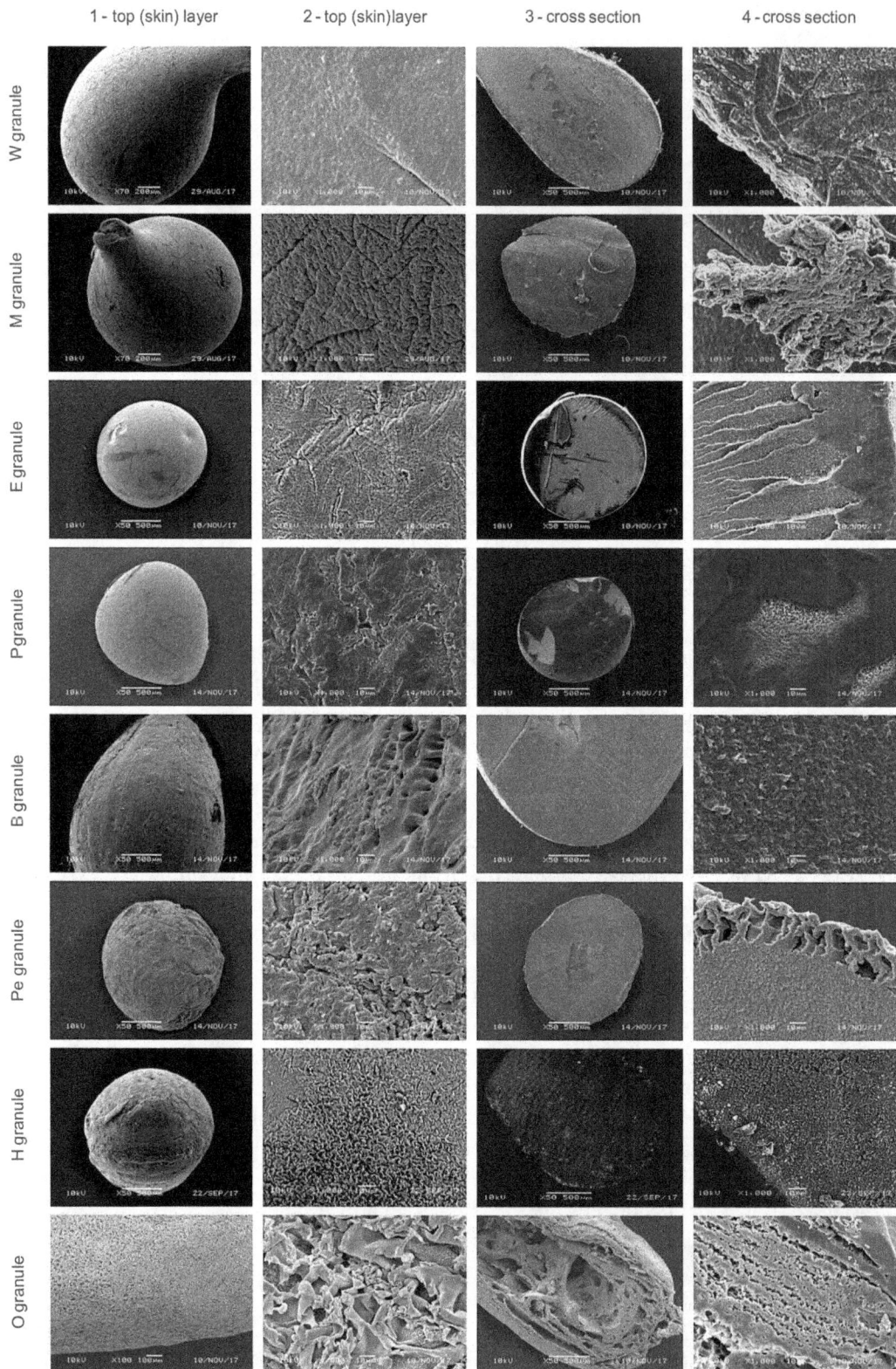

Fig. 5. Self-made SEM images of the surface and cross-sections of cellulose granules

CEL granules was 34% higher than plants culti-vated under the same environmental conditions, but not fertilized. The fertilized plants were char-acterized by full leaf colouring, high growth force and a strong root system, and tolerated periodic water shortages better.

The analysis of the fertilizer dose showed that NH_4NO_3 in the amount of 0.148 kg/10 m^2 of surface was used throughout the growing pe-riod. For comparison, the recommended dose of the popular fertilizer – Azofoska – is 0.16–0.25 kg/10 m^2.

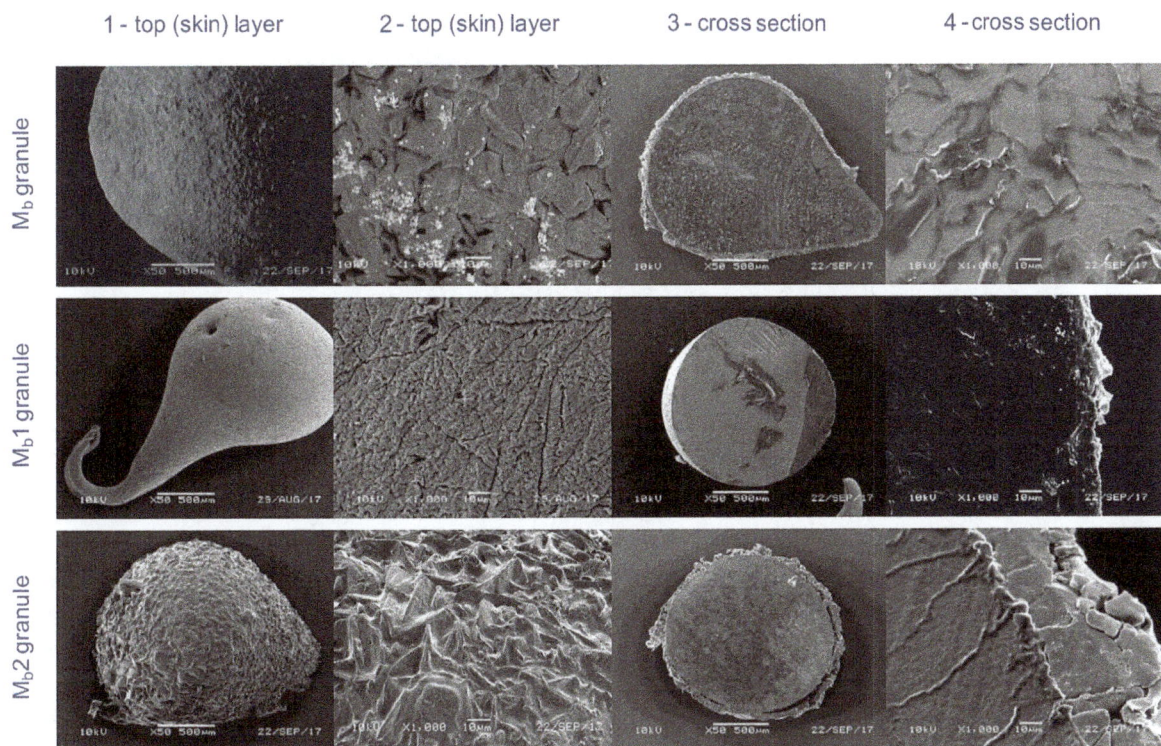

Fig. 6. Self-made SEM images of the surface and cross-sections of cellulose granules after physical modification (M$_a$ - heated granules, M$_b$ – PLA-coated granules)

Fig. 7. Plant (*Chlorophytum comosum*) development throughout 6 months of cultivation. On the right-hand side, the plant after the use of granules; on the left-hand side, the plant without their use (photos taken by hand).

Fig. 8. Covering of the M_b-1 cellulose granule with salt (photos taken by hand).

During the experiment, it was observed that the M_b-1 cellulose granules, which were evenly distributed on the cultivation substrate, in the first stage were covered with the salt coating (Fig. 8). After some time, this coating disappeared and the granule decomposed and disappeared in the substrate.

CONCLUSIONS

This paper has undertaken study on obtaining fully biodegradable, long-acting carriers of artificial fertilizers in the form of cellulose granules, which could be an alternative to ETDA chelates. The granules were prepared from a 5% cellulose solution in EMIMAc, coagulated in water and selected primary alcohols: methanol, ethanol, 1-propanol, 1-butanol, 1-pentanol, 1-hexanol and 1-octanol. The method of forming cellulose granules affected their internal and surface structure, which was confirmed by SEM images. The cellulose granules obtained in the experiment were characterized by a bulk density of 0.53–0.69 g/cm^3, water absorption at the level of ~ 26–75% and the ability to adsorb ~ 20% w/w of NH_4NO_3 in relation to the cellulose weight. Physicochemical properties of granules O differed from the others: bulk density – 0.07 g/cm^3; water absorption – 442.37. Analysis of the salt desorption process from cellulose granules showed that removal of NH_4NO_3 takes place immediately, regardless of the coagulant used.

The purpose of the study was obtaining cellulose granules with a prolonged duration of action, therefore two methods of modification were used: hydrophobization at elevated temperature and coating with 1% w/w PLA/$CHCl_3$ solution. As a result of the study, beads characterized by good sorption/desorption properties relative to

the mineral salt were obtained. The hydrophobization of cellulose granules had a beneficial effect on the slowdown of the NH_4NO_3 release process, which is why they were used in the cultivation of the spider plant. During the six-month study on plants, full degradation of fertilizer granules was observed, resulting in a 34% higher increase in green mass as compared to non-fertilized plants. The fertilized plants were characterized by full leaf colouring, high growth force and a strong root system, as well as better toleration to periodic water shortages.

The cellulose granules modified by hydrophobization obtained in the experiment can be successfully used in plant production as modern, non-dusty, biodegradable and environmentally safe fertilizers with extended duration of action.

REFERENCES

1. Allard A. S., Renberg L., Neilson A. H. 1996. Absence of $^{14}CO_2$ evolution from ^{14}C-labelled EDTA and DTPA and the sediment/water partition ratio. Chemosphere, 33(4), 577–583.

2. Alvarez J. M., Obrador A., Rico M. I. 1996. Effects of chelated zinc, soluble and coated fertilizers, on soil zinc status and zinc nutrition of maize. Communications in Soil Science and Plant Analysis, 27(1–2), 7–19.

3. Bloem E., Haneklaus S., Haensch R., Schnug E. 2017. EDTA application on agricultural soils affects microelement uptake of plants. Science of the Total Environment, 577, 166–173.

4. Earle M. J., Seddon K. R. 2000. Ionic liquids. Green solvents for the future. Pure and Applied Chemistry, 72(7), 1391–1398.

5. Fryczkowska B., Wiechniak K. 2017a. Preparation and properties of cellulose membranes with graphene oxide addition. Polish Journal of Chemical

Technology, 19(4), 41–49.

6. Fryczkowska B., Wyszomirski M., Puzoń M. 2017b. Obtaining and Application of New Cellulose- and Graphene Oxide-Based Adsorbents for Treatment of Industrial Waste Containing Heavy Metals. Journal of Ecological Engineering, 18(6), 43–52.

7. Fryczkowska B., Kowalska M., Biniaś D., Ślusarczyk C., Janicki J., Sarna E., Wyszomirski M. 2017c. Properties and structure of cellulosic membranes obtained from solutions in ionic liquids coagulated in primary alcohols. Autex Research Journal, in press.

8. Gagliardi B., Pettigrove V. 2013. Removal of intensive agriculture from the landscape improves aquatic ecosystem health. Agriculture, Ecosystems and Environment, 176, 1–8.

9. Gathergood N., Garcia M. T., Scammells P. J. 2004. Biodegradable ionic liquids: Part I. Concept, preliminary targets and evaluation. Green Chemistry, 6(3), 166.

10. Klem-Marciniak E., Huculak-Mączka M., Hoffman K., Hoffmann J. 2015. Effect of reaction time to receive fertilizer chelates, 9(2), 15–17.

11. Luo J., Cai L., Qi S., Wu J., Gu X. W. S. 2017. Improvement effects of cytokinin on EDTA assisted phytoremediation and the associated environmental risks. Chemosphere, 185, 386–393.

12. Luo X., Zeng J., Liu S., Zhang L. 2015. An effective and recyclable adsorbent for the removal of heavy metal ions from aqueous system: Magnetic chitosan/cellulose microspheres. Bioresource Technology, 194, 403–406.

13. Meers E., Ruttens A., Hopgood M. J., Samson D., Tack F. M. G. 2005. Comparison of EDTA and EDDS as potential soil amendments for enhanced phytoextraction of heavy metals. Chemosphere, 58(8), 1011–1022.

14. Novoselov N. P., Sashina E. S., Kuz'mina O. G., Troshenkova S. V. 2007. Ionic liquids and their use for the dissolution of natural polymers. Russian Journal of General Chemistry, 77(8), 1395–1405.

15. Suzuki T., Kono K., Shimomura K., & Minami H. 2014. Preparation of cellulose particles using an ionic liquid. Journal of Colloid and Interface Science, 418, 126–131.

16. Trygg J., Fardim P., Gericke M., Mäkilä E., Salonen J. 2013. Physicochemical design of the morphology and ultrastructure of cellulose beads. Carbohydrate Polymers, 93(1), 291–299.

17. Voon L. K., Pang S. C., Chin S. F. 2015. Highly porous cellulose beads of controllable sizes derived from regenerated cellulose of printed paper wastes. Materials Letters, 164, 264–266.

18. Voon L. K., Pan, S. C., Chin, S. F. 2017a. Optimizing Delivery Characteristics of Curcumin as a Model Drug via Tailoring Mean Diameter Ranges of Cellulose Beads, 2017.

19. Voon L. K., Pang S. C., Chin S. F. 2017b. Porous Cellulose Beads Fabricated from Regenerated Cellulose as Potential Drug Delivery Carriers, 2017.

20. Wang Y., Fang N., Tong L., Shi, Z. 2017. Source identification and budget evaluation of eroded organic carbon in an intensive agricultural catchment. Agriculture, Ecosystems and Environment, 247(2017), 290–297

21. Weerachanchai P., Wong Y., Lim K. H. Tan T. T. Y., Lee, . M. 2014. Determination of solubility parameters of ionic liquids and ionic liquid/solvent mixtures from intrinsic viscosity. ChemPhysChem, 15(16), 3580–3591.

22. Wendler F., Todi L. N., Meister F. 2012. Thermostability of imidazolium ionic liquids as direct solvents for cellulose. Thermochimica Acta, 528, 76–84.

23. Yildir E., Kolakovic R., Genina N., Trygg J., Gericke M., Hanski, L., Ehlers H., Rantanen J., Tenho M., Vuorela P, Fardim P., Sandler N. 2013. Tailored beads made of dissolved cellulose – Investigation of their drug release properties. International Journal of Pharmaceutics, 456(2), 417–423.

10

Pretreatment of Stabilized Landfill Leachate using Ozone

Anna Kwarciak-Kozłowska[1]

[1] Institute of Environmental Engineering, Czestochowa University of Technology, Czestochowa, Poland, e-mail: akwarciak@is.pcz.czest.pl

ABSTRACT

The paper presents the possibility of using the ozonation process in landfill leachate pretreatment. The study was conducted in three stages. In the first stage, the landfill leachate was subjected only to the ozone, with the dose varying from 10 mg/dm^3 to 40 mg/dm^3. As part of this stage of research, the effect of changes in the pH of wastewater undergoing the process of ozonation on the efficiency of TOC removal was examined. For all the tested pH values (pH = 3.5, pH = 7 pH = 8.5 pH = 10), the TOC removal rate constant (k_{Rowo}) during ozonization was determined. In the second stage of the study, the ozonation process was facilitated by UV radiation. Additionally in this stage, the rate of generation of OH• radicals was accelerated by the addition of hydrogen peroxide to the reactor. The COD: H$_2$O$_2$ ratio by weight was 1:2.5, 1:5 and 1:10 and 1:20. In the last stage of the study, we attempted to assist the ozonation process using ultrasonic field. The employed vibration amplitude amounted to 25μm and sonication time equalled 300 seconds. It was found that the ozonation process is the most effective at alkaline pH (8.5). The TOC removal efficiency was 37% (346 mg/dm^3) after 60 minutes of ozonation. The best results of pollutants oxidation measured as COD and TOC removal were observed when the dose of ozone was 20 mg/dm^3. The combination of sonication and ozonation has resulted in a reduction of COD and TOC values by 370 mg/dm^3 and 126 mg/dm^3, respectively, in comparison to the ozonation process alone. It was found that the most effective process in landfill leachate treatment is the combination of ozonation with hydrogen peroxide addition (COD:H$_2$O$_2$=1:10). The COD, TOC and BOD values were 65%, 62% and 36% lower, respectively, in comparison to ozonation process conducted alone.

Keywords: landfill leachate, ozonation, hydrogen peroxide, ultrasonic field, advanced oxidation process

INTRODUCTION

The term 'leachate' means the water generated in landfills due to the seepage of rainwater through the bed. They elute organic and mineral compounds resulting from the biological and physicochemical changes from the landfills waste. Although many methods may be applied, the most appropriate leachate treatment choice will depend on its features, technical applicability, cost effectiveness, and other factors related to the quality requirements of the effluents. The leachate produced in mature landfills contains significant amounts of humic acids having a very stable structure which is difficult to breach by bacteria. Therefore, the purification efficiency by the biological method of the activated sludge is not very high [Kurniwana and Wai-hung 2006; Kwarciak-Kozłowska and Sławik-Dembiczak 2016]. Advanced oxidation processes (AOP) represent relatively new methods while their mechanism is based on the generation of free hydroxyl radicals. AOP encompasses ozonation which is a treatment based on the high oxidant power of ozone that can be used to decompose large organic molecules into smaller and less complex ones occurring at normal pressure and temperature being, hence, industrially interesting [Amaral-Silva et al. 2016]. Ozone belongs to very strong oxidants and therefore starts reactions with many organic and inorganic compounds. It is selective in direct reactions with organic compounds; Conversely, hydroxyl radicals are usually non-selective with respect to these compounds [Zarzycki 2002]. Ozonation alters the molecular structure of refractory organic compounds present in the leachate, turning them into compounds that are easily assimilated biologically. The ozone processes can be made more effective for example at high pH (O_3/OH^-) and by the addition of hydrogen peroxide (O_3/H_2O_2). These systems (Peroxone process) favour the production of hydroxyl radicals (•OH), which are highly reactive species. OH^- and H_2O_2 initiate a series of radical reactions that enhance the ozone decomposition to yield •OH [Bila et al. 2005; Tizaoui et al. 2007, Catalkaya and Kargi 2007; Hansen et al. 2016; Liu et al. 2015; Zarzycki 2002]. Furthermore, the effect of the ultrasonic wave is connected with the stimulation of the cavitation phenomenon [Suslick 1989]. The energy released during the implosion of the cavity bubbles is able to initiate a thermal decomposition of water particles using the hydroxyl radicals [Barbusiński 2013; Cesaro et al. 2013; Xu et al. 2005].

As a result, this research was carried out in order to investigate the treatment of landfill leachate using ozone-bases processes (i.e. O_3 alone, O_3/H_2O_2 and O_3/US). This study investigated the effects of different operating parameters such as, H_2O_2 concentration (COD: H_2O_2 was 1:2.5, 1:5 and 1:10 and 1:20), dose of ozone (10–40 mg/dm^3) and sonication process (ampitude 25 μm, sonication time 300).

MATERIALS AND METHODS

The leachate was derived from the regional municipal waste landfill in Silesian Province (Poland). The leachate was characterised by the pH of 7.89 and light brown colour. The COD value was 2885 mg/dm^3 on average and BOD was under 190 mg/dm^3. The BOD/COD ratio was at a low level (0.06) that indicated low biodegradability. The amount of TOC was 550 mg/dm^3.

The process of landfill leachate ozonation was performed in a cylindrical glass reactor with the diameter of 10 cm and active volume of 4 dm^3. The positive pressure system for supplying of air-ozone mixture to the reaction chamber was used. Fine-bubble aeration was performed using the diffuser located at the reactor bottom. Ozone was generated from oxygen using the OZOMATIC LAB 802 ozonizer, with the range of air-ozone mixture production ranging from 0.1 to 1 dm^3/min. The gas flow rates were adjusted using the RO6 RP T96193 rotameter (max 45 dm^3/h). The measurement of the ozone content at the out-

let from the reaction chamber was conducted using the Dreschel washing bottle with the absorption solution connected with the valve located at its top [Kwarciak-Kozłowska et al. 2016].

The sonication of the landfill leachate was performed using the Sonics vibro cell ultrasonic disintegrator which generated the ultrasonic wave with vibration frequency 40 kHz. The power adjustment in the device was possible through the digital adjustment of the percentage of maximal amplitude of the sonotrode (max 123 μm) [Kwarciak-Kozłowska and Sławik-Dembiczak 2016].

The examinations were performed at three stages. At the first stage, the leachate was exposed to the effect of ozone, with the doses changing in the range from 10 mg/dm^3 to 40 mg/dm^3. The ozone efficiency was controlled by the adjustment of the generator efficiency (from 5% to 100%) to the specific gas flow rate (from 10 dm^3/h to 40 dm^3/h). At this stage, the effect of pH of the liquors subjected to ozonation on the degree and rate of TOC removal was determined.

At the second stage of the examinations, the rate of generation of OH• radicals in the ozonation process was accelerated by adding hydrogen peroxide to the reaction chamber. Its amount was increased in the geometrical progress by q=2, at initial amount of 1:2.5 (ChZT:H_2O_2).

At the last stage, the ozonation process was combined with the ultrasonic field. The ultrasonic field parameters such as sonication time (300 sec) and vibration amplitude (25 μm) used at this stage were the most advantageous values determined from the previous examinations concerning demonstration of the effect of the ultrasonic field on the change of the biodegradability of landfill leachate [Kwarciak-Kozłowska and Sławik-Dembiczak 2016]. The efficiency of the unit processes used in the study was controlled based on the changes in the levels of COD, TOC and BOD. The study design and methodology of determination are presented in Figure 1.

RESULTS AND DISCUSSION

Effect of pH of landfill leachate and ozone doze on the degree of ozonation of contaminants (stages 1)

The first step of the experiment examined the effect of the initial pH value of the landfill leachate subjected to ozonation on the TOC removal

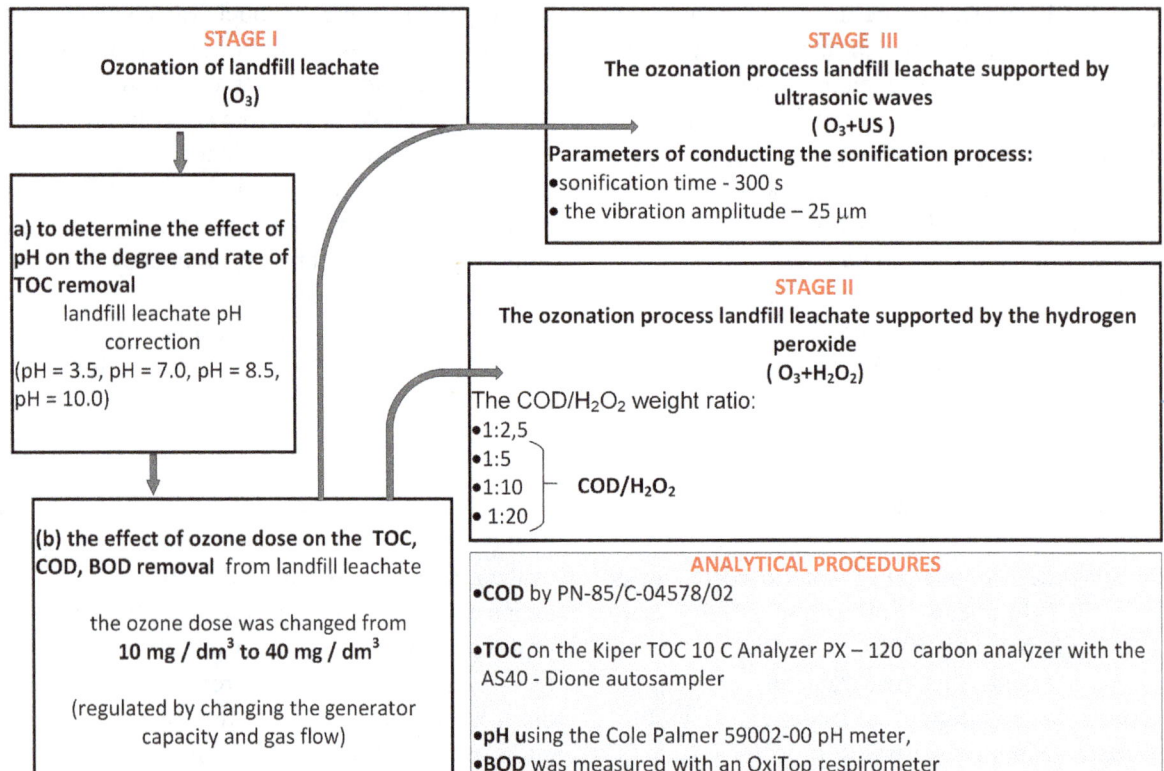

Figure 1. The scheme of treatment the landfill leachate in ozonation process

rate. This stage was performed at a constant ozone dose of 10 mg/dm³. The ozonation was performed for the initially acidified liquor (pH 3.5) and the TOC removal rate ranged from 7% (5 min of the process) to 13% (60 min of the process).

After 60 minutes of the process, the TOC value reduced from 550 mg/dm³ to 451 mg/dm³. It was found that the landfill leachate ozonation process occurs most effectively when their pH is initially at the level of 8.5. After 60 minutes, the TOC removal rate reached 37% (346 mg/dm³). Further alkalization of raw leachate to pH=10 resulted in a slight decline in the removal rate to 25% (412 mg/dm³). Similar effects were observed when the landfill leachate reaction was neutral. The TOC removal rate reached 28% (396 mg/dm³). These changes are illustrated in Fig. 2a. At this stage, the changes in the TOC removal rate were also analysed. For a specific time range (from 5 min to 30 min), the TOC degradation process was the first order reaction which then was transformed into a quasi-static reaction (Fig. 2b). Assuming that the decomposition of organic compounds determined as TOC in the ozonation process was the first order reaction, a reaction rate constant was determined based on the Langmuir-Hinshelwood formula [Kudlek et al. 2017].

The analysis of the results obtained for constant TOC degradation rates revealed that the oxidation of organic compounds contained in landfill leachate occurs much faster in the first minutes of the process. The value of k_{rTOC} for wastewater with pH=3.5, pH=7.0, pH=8.5 and pH=10 in the 5th minute of the process was 0.014 min⁻¹, 0.027 min⁻¹, 0.054 min⁻¹ and 0.023 min⁻¹. In the 60th minute of the process, the value of these coefficients were substantially reduced and amounted to 0.0039 min⁻¹ (pH=3.5), 0.0066 min⁻¹ (pH=7.0), 0.0092 min⁻¹ (pH=8.5) and 0.0057 min⁻¹ (pH=10).

The second step of this stage of the examinations was to evaluate the effect of the contaminant removal from the landfill leachate on the amount of the ozone supplied to the reaction chamber. The ozone dose was changed from 10 mg/dm³ to 40 mg/dm³. The COD and TOC of the raw landfill leachate were 2885 mg/dm³ and 550 mg/dm³. The worst effects of ozonation of contaminants determined as COD and TOC were observed when the lowest ozone dose was used. The 60-minute period of ozonation contributed to the removal of contaminants denoted as COD and TOC by 37% (220 mg/dm³) and 42% (1673 mg/dm³). The double ozone dose (20 mg/dm³) resulted in the increase in the removal rates for TOC (by 19%)

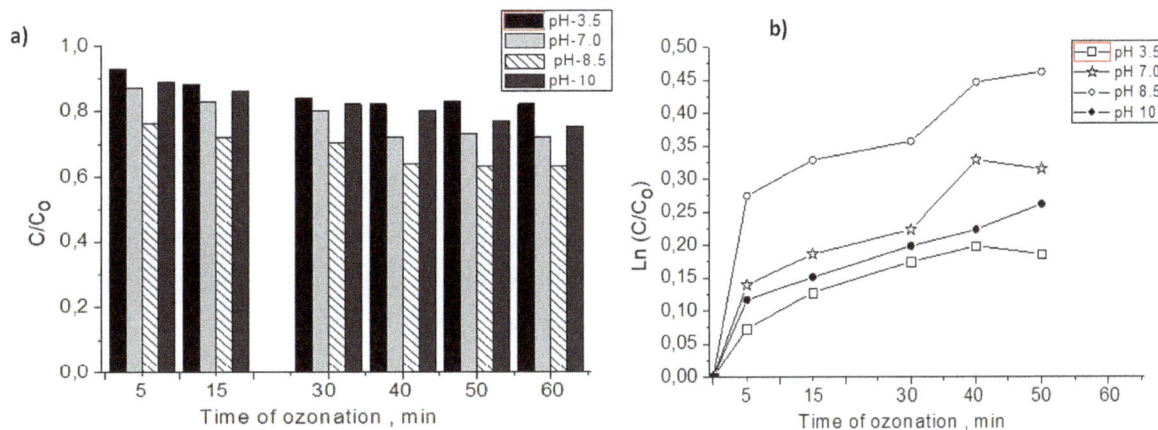

Figure 2. Effect of pH on the ozonation and removal of TOC- C/Co (a) and lnC/Co (b)

and COD (by 18%). Further increase of the ozone dose resulted in the reduction in the effectiveness of contamination oxidation. The rate of COD removal from landfill leachate after an hour process of ozonation at the ozone dose of 30 mg/dm^3 was 54%, whereas the COD value for treated leachate was 1327 mg/dm^3. With this ozone dose, the rate of TOC removal was 50%, whereas its content in the treated landfill leachate reduced to the level of 275 mg/dm^3. It was found that for all the used doses of ozone, the elongation of the time of the

process to over 45 minutes did not lead to an increase in the oxidation of the contaminants contained in the landfill leachate. These changes are illustrated in Figures 3 and 4.

After determination of the most advantageous ozone dose (20 mg/dm^3) the BOD was evaluated during a 60 minute process of landfill leachate oxidation. After the 5th minute of the ozonation process, the value of BOD reduced from the level of 190 mg/dm^3 to 135 mg/dm^3. Extension of the ozonation time to 15 minutes resulted in

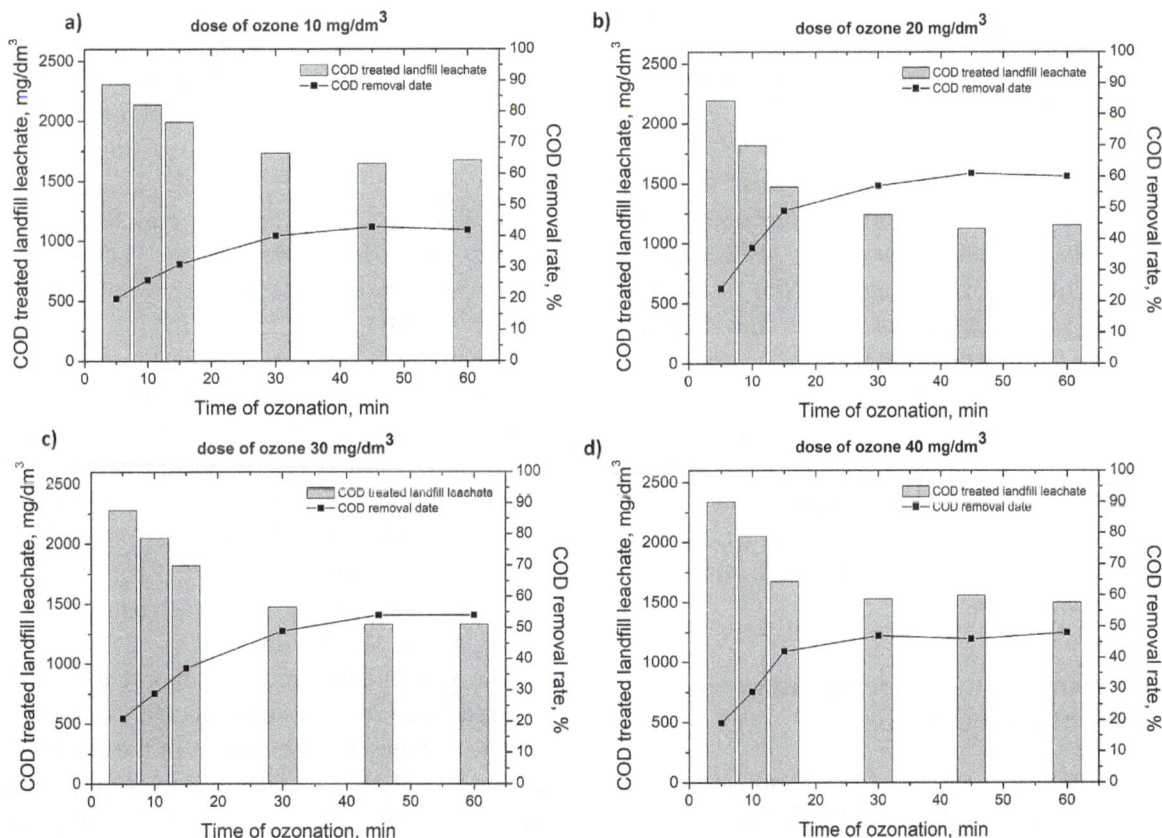

Figure 3. The effect of the ozone dose on the COD removal rate

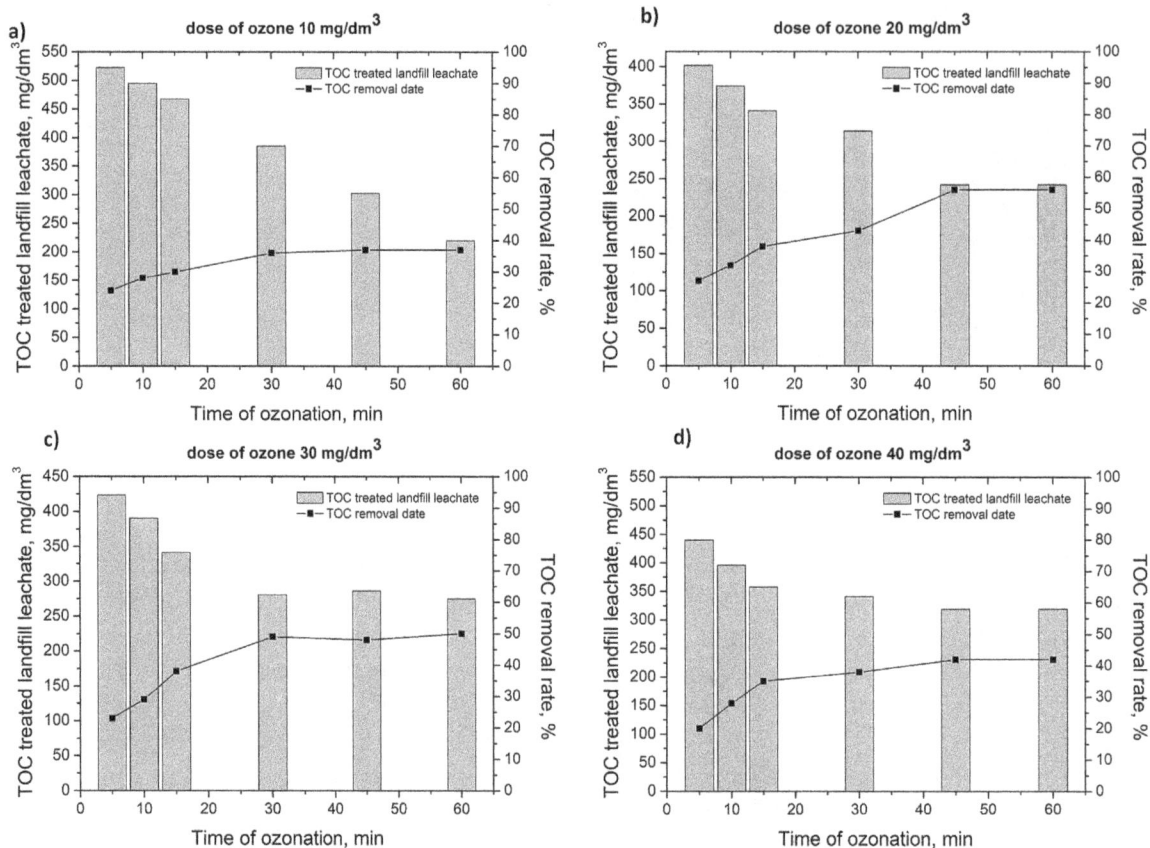

Figure 4. The effect of the ozone dose on the TOC removal rate

the increase in its removal rate by another 30%, i.e. to the level of 59% (78 mg/dm³). After 30 minutes and 60 minutes of ozonation, the value of the BOD did not change and was at a similar level of 42 mg/dm³.

Leszczyński et al. [2016] conducted research on the use of the ozonization process for stabilized landfill leachate treatment. The average values of its main parameters were: pH 8.32, COD 870 mg/dm³, BOD 90 mg/dm³, NH_4^+ 136.2 mg/dm³, UV254 absorbance 0.312 and turbidity 14 NTU. The ozone dosages were used in the range of 115.5 to 808.5 mg/dm³ of the leachate. The maximum COD, colour and UV254 absorbance adjustment were 37.3%; 81.6% and 59.2%, respectively, by applying a high ozone dose of 808.5 mg/dm³. After oxidation, the ratio of BOD/COD was increased from 0.1 up to 0.23 [Leszczyński et al. 2016].

Treatment of landfill leachate using the system of $O_3+H_2O_2$ (stages 2)

At two next stages of the examinations, the attempts were made to intensify the ozonation process by its combination with another unit pro-

cess. In the first step, the ozonation process was supported by the effect of hydrogen peroxide. Four doses were used in the experiment, with the mass ratio of H_2O_2 to COD of raw landfill leachate of 2.5:1, 5:1, 10:1 and 20:1. The ozone dose was maintained at the level of 20 mg/dm³.

It was found that during the process that combines ozonation with hydrogen peroxide (system $O_3+H_2O_2$) at the lowest dose (COD: H_2O_2=1:2.5), the effectiveness of oxidation of the contaminants denoted as COD and TOC was the lowest. After 5 minutes of the process, the COD removal rate was 46% and its extension to 60 minutes caused an increase in the efficiency of oxidation of contaminants to 72% (COD of the treated landfill leachate: 807 mg/dm³). With this dose of the hydrogen peroxide, the TOC removal rate following an hour of ozonation was 70% and its value in the raw landfill leachate was reduced from 550 mg/dm³ to 165 mg/dm³.

The most effective process supported by H_2O_2 occurred when the ratio of COD:H_2O_2 was 1:10. After 45 minutes and 60 minutes of the combined process ($O_3+H_2O_2$), the COD value of the treated landfill leachate was similar, with 375 mg/dm³ (87%) and 403 mg/dm³ (86%). The degree of

TOC removal after 60 minutes of the treatment process reached the level of 83%, and its value in the treated leachate was 93 mg/dm^3.

Figure 5 illustrates the changes in TOC (a) and COD (b) of the treated landfill leachate depending on the dose of hydrogen peroxide.

The research confirms that the support of ozone treatment with H_2O_2 results in an increase in the efficiency of stabilized leachates treatment [Amaral-Silva et al. 2016; Tizaoui et al. 2007]. In the Amaral-Silva studies, the highest organic load removal and biodegradability improvement was observed with the O_3/H_2O_2 process using 4 g/dm^3 H_2O_2. This system was able to eliminate 45% of the COD, 89% of the leachate BOD/COD ratio from 0.05 to 0.29 permitting the treated wastewater for the local sewage collector [Amaral-Silva et al. 2016]. Tizaoui et al. 2007 found that the ozone efficiency was almost doubled when combined with hydrogen peroxide at 2 g/dm^3 and higher. H_2O_2 gave lower performances. Enhancement in the leachate biodegradability from about 0.1 to about 0.7 was achieved by the O_3/H_2O_2 system [Tizaoui et al.2007].

Treatment of landfill leachate using the system of O_3+US (stage 3)

The last stage of the experiment was aimed at examining the effect of ultrasonic waves (US) on the change in the efficiency of treatment of landfill leachate subjected to ozonation. It was assumed that the free radicals of H•, HO•, HOO•, ozone, and hydrogen peroxide generated through the effect of ultrasonic waves should contribute to the increased degree of oxidation and/or destabilization of the organic substance compared to the independently performed ozonation process.

The examination performed by the authors previously demonstrated that the most effective sonication process occurs at the vibration amplitude of 25 μm and reaction time of 300 seconds. The process of liquor sonication substantially affected the generation and release of aliphatic compounds. Furthermore, a substantial number of compounds with the -OH functional group that can originate from alcohols and carboxyl acids was observed. [Kwarciak-Kozłowska and Sławik-Dembiczak 2016]. The value of acoustic energy supplied to the landfill leachate sample for such parameters of the ultrasonic field was on average 32459J. The changes caused in liquors are connected with e.g. cavitation. It was found that with these parameters of the ultrasonic field, the intensity of the ultrasound wave was higher than the theoretical threshold of cavitation (1.0 W/cm^2) and was on average 1.34 W/cm^2 [Kwarciak-Kozłowska and Sławik-Dembiczak 2016].

Figure 6 compares the value of organic contaminants determined as COD and TOC in the independent process of ozonation with the process where ozonation effect was intensified by ultrasound waves.

It was found that it is more advantageous to conduct the process of landfill leachate treatment in the combined system (system O_3+US), since the level of COD leachate treated after each analysed time of the process was on average by 370 mg/dm^3 lower, compared to the liquors which were only subjected to ozonation. After 60 minutes of ozonation and combination of ozonation with sonication, COD in the treated leachate was 1154 mg/dm^3 and 865 mg/dm^3. In the case of TOC, the landfill leachate treated using the integrated system was characterized by lower value of the index by 126 mg/dm^3, on average. Furthermore, the time of the process of leachate

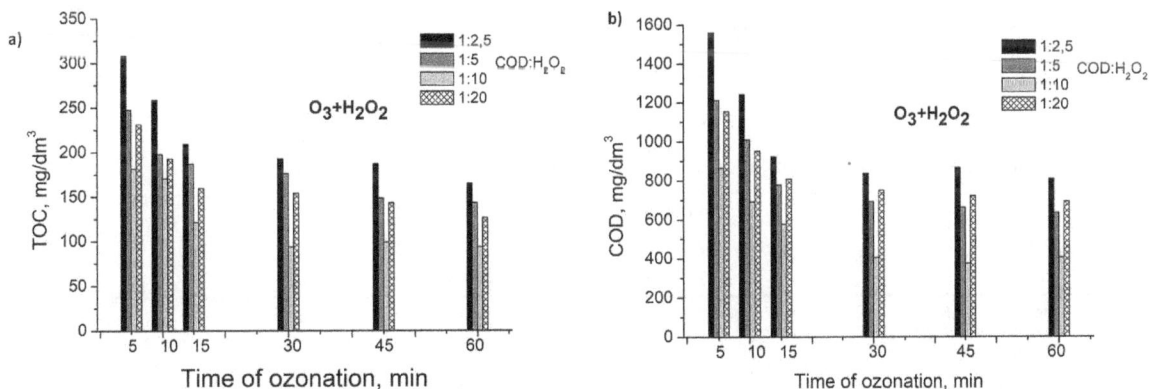

Figure 5. Changes in TOC (a) and COD (b) of the treated leachate depending on the dose of hydrogen peroxide

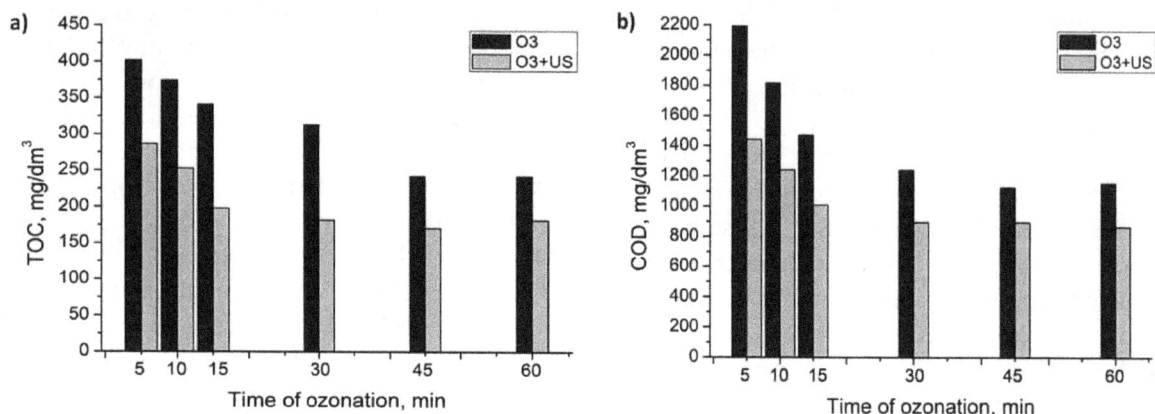

Figure 6. Changes in TOC (a) and COD (b) of the treated leachate in O_3 and O_3+US system

treatment using the O_3+US method for over 15 minutes (TOC) and 30 minutes (COD) did not impact on the improvement in oxidation of these contaminants.

The last step was the comparison of the results obtained in individual systems of landfill leachate treatment. The values in the tables represent the results obtained in each of the systems under most advantageous conditions (Table 1). It was found that the landfill leachate can be treated in the arrangement that combined ozonation with the addition of hydrogen peroxide (COD:H_2O_2=1:10). The value of COD, TOC and BOD in such landfill leachate was lower compared to independent ozonation process by 65%, 62% and 36 %, respectively. This landfill leachate is planned to be additionally treated using the low-pressure membrane processes (e.g. ultrafiltration).

CONCLUSIONS

The results obtained in the study led to the following conclusions:
1. The landfill leachate ozonation process occurs most effectively if the pH is 8.5. After 60 minutes of ozonation, the TOC removal rate reached 37% (346 mg/dm^3);

2. It was found that the ozonation of leachate is most effective if its dose is 20 mg/dm^3;
3. The most advantageous system of landfill leachate treatment was the the arrangement that combined ozone with hydrogen peroxide (COD:H_2O_2=1:10). The value of COD, TOC and BOD in such leachate was lower compared to independent ozonation process by 65%, 62% and 36%, respectively.
4. Inclusion of the sonication process in ozonation of landfill leachate resulted in a reduction in COD and TOC by 370 mg/dm^3 and 126 mg/dm^3 compared to ozonation alone.

Acknowledgements

The study has been funded by BS/PB–401–301/11

REFERENCES

1. Amaral-Silva N., Martinsa R.C., Castro-Silva S., Quinta-Ferreira R.M. 2016.Ozonation and perozonation on the biodegradability improvement of a landfill leachate, Journal of Environmental Chemical Engineering 4, 527–533.
2. Barbusiński K. 2013. Advanced oxidation in the treatment of selected industrial wastewater, Pub-

Table 1. Comparison of the quality of landfill leachate treated in the examined combination systems

Indicator	Raw Landfill leachate	Purification system applied (characteristics of treated landfill leachate – after 60 min)			Permissible Standards
		O_3	O_3+H_2O_2	O_3+US	
COD, mg/dm³	2885	1154	404	865	125
BOD, mg/dm³	190	42	27	32	25
TOC, mg/dm³	550	242	93	181	30

* Regulation of the Ministry of Environmental Protection, Natural Resources and Forestry of 18 November 2014 on the classification of water and conditions the sewage discharged to waters and soil should satisfy.

lisher of the Silesian University of Technology, Gliwice (in Polish).

3. Bila D.M., Montalvai A.F., Silva A.C., Dezotti M. 2005. Ozonation of a landfill leachate: evaluation of toxicity removal and biodegradability improvement, Journal of Hazardous Materials B117, 235–242.

4. Catalkaya, E.C., Kargi, F. 2007. Color, TOC and AOX removals from pulp mill effluent by advanced oxidation processes: A comparative study, Journal of Hazardous Materials 139, 244–253.

5. Cesaro, A., Naddeo, V., Belgiorno, V. 2013. Wastewater treatment by combination of advanced oxidation processes and conventional biological systems. Journal of Bioremediation & Biodegradation, 4, 208.

6. Hansen, K.M.S., Spiliotopoulou, A., Chhetri, R.K., Escolà Casas, M., Bester, K., Andersen, H.R. 2016. Ozonation for source treatment of pharmaceuticals in hospital wastewater – Ozone lifetime and required ozone dose, Chemical Engineering Journal 290, 507–514.

7. Kudlek E., Dudziak M., Bohdziewicz J., Kamińska G. 2017. The role of pH in the decomposition of organic micropollutants during the heterogeneous photocatalysis proces, E3S Web of Conferences 17, 00047.

8. Kurniwana T., Wai-hung L. 2006. Physico-chemical treatment for removal of recalcitrant contaminansts from landfill leachate, Journal of Hazardous Materials 129, 1–3, 80–100.

9. Kwarciak-Kozłowska A., Krzywicka A., Gałwa-Widera M. 2016. The use of ozonation process in coke wastewater treatment, Rocznik Ochrona Środowiska, 18, 61–73.

10. Kwarciak-Kozłowska A. Sławik-Dembiczak L. 2016. Characterization of the organic fraction of pretreated leachate from old landfill after sonication exposure, Inżynieria i Ochrona Środowiska 19(4), 561–575.

11. Leszczyński J., Tałałaj I., Walery M., Biedka P. 2016. Landfill leachates pretreatment by ozonation, Ecological Engineering, 143–146 (in Polish).

12. Liu, Y., Jiang, J., Ma, J., Yang, Y., Luo, C., Huangfu, X., Guo, Z. 2015. Role of the propagation reactions on the hydroxyl radical formation in ozonation and peroxone (ozone/hydrogen peroxide) processes. Water Research 68, 750–758.

13. Regulation of the Minister of Environment of 18 November 2014 on the conditions to be met when discharging sewage into water or soil, and on the substances particularly harmful to the aquatic environment (in Polish).

14. Suslick, K.S. 1989. The chemical effects of ultrasound. Scientific American 2, 80–86

15. Tizaoui Ch., Bouselmi L., Mansouri L., Ghrabi A. 2007. Landfill leachate treatment with ozone and ozone/hydrogen peroxide systems, Journal of Hazardous Materials 140, 316–324.

16. Zarzycki R. 2002. Advanced oxidation techniques in environmental protection, Łódź, PA Publishing (in Polish).

11

Spatial and Temporal Variability of Moisture Condition in Soil-Plant Environment using Spectral Data and Gis Tools

Henryk Grzywna[1*], Paweł B. Dąbek[1], Beata Olszewska[1]

[1] Wrocław University of Environmental and Life Sciences, Institute of Environmental Protection and Development, pl. Grunwaldzki 24, 50-363 Wrocław, Poland

* Corresponding author's e-mail: henryk.grzywna@upwr.edu.pl

ABSTRACT

The studies on agricultural droughts require long-term atmospheric, hydrological and meteorological data. On the other hand, today, the possibilities of using spectral data in environmental studies are indicated. The development of remote sensing techniques, increasing the spectral and spatial resolution of data allows using remote sensing data in the study of water content in the environment. The paper presents the results of the analysis of moisture content of soil-plant environment in the lowland areas of river valley using the spectral data from Sentinel-2. The analyses were conducted between February and November 2016. The spectral data were used to calculate the Normalize Differential Vegetation Index (NDVI) which provided the information about the moisture content of the soil-plant environment. The analyses were performed only on grasslands, on 22 objects located in the research area in the Oder river valley between Malczyce and Brzeg Dolny, Poland. The NDVI values were correlated with the hydrological and meteorological parameters. The analyses showed spatial and temporal variability of the moisture conditions in the soil-plant environment showed by the NDVI variability and existence some relationships between the climatic and spectral indices characterizing the moisture content in the environment.

Keywords: drought, soil moisture, NDVI, Sentinel–2, satellite data, remote sensing

INTRODUCTION

The human has no influence on the occurrence of drought and can only affect the social, economic, technical or political factors that present the scale of the phenomenon in a given region [Jarząbek et al. 2013, Łabędzki and Bąk 2013]. Drought is a climatic phenomenon defined as a noticeable lack of water. It causes damage to nature and the economy, as well as a nuisance to the population and even a threat to human life [Nichol and Abbas 2015, Hisdal and Tallaksen 2000]. Precipitation is the main source of water supply for plants and their amount and distribution are regarded as a factor determining their size and conditions [Dzieżyc et al. 1987]. The dynamics of changes in soil moisture significantly affect the condition of plants and crops, the moisture deficit reduces the quantity and quality of yields, which results in, i.a. the losses in the economy and agriculture [Dubicki et al. 2002]. In a surface layer of a soil, where the scale and dynamics of moisture changeability is the largest, the impact of precipitation variability is visible the most [Lauzon et al. 2004]. The soil properties that affect the water holding capacity are particularly important from the agricultural point of view [Biniak-Pieróg 2008, Jankowiak and Bieńkowski 2011]. Soil retention has a fundamental role in the water balance of the agricultural areas [Nyc 1994, Pływaczyk and Olszewska 2014].

Agricultural drought is a state in which the soil moisture is insufficient to meet the water needs of plants [Łabędzki and Bąk 2015a, Nyc 1994]. The studies pertaining to agricultural droughts are focused mainly on using the methods based on the long-term atmospheric data, temperature, precipitation and evaporation measurements without considering the changes in soil properties and parameters. Such analyses characterize local habitats and they are only representative for small areas [Martínez-Fernández et al. 2016, Torres et al.

2013]. The data comes from a wide measurement network, only in limited number of cases [Jackson et al. 2004]. Remote sensing measurements and analyzes using GIS technology have only enabled obtaining the current information about soil moisture in large areas, mainly in upper (active) soil layer [Łabędzki and Bąk 2015].

Currently, remote sensing techniques are rapidly developing. New devices, enabling to use cameras installed on aircraft boards and unmanned aerial vehicles or using satellites have appeared. The quick and significant increase in a satellite remote sensing in the spectral and spatial resolution are observed – from NOAA (National Oceanic and Atmospheric Administration) satellite with spatial resolution ca. 1000 m, through Landsat satellite series with spatial resolution 100 m to 20 m, to one of the latest Sentinel-2 missions with spatial resolution 10 and 20 m [Drusch et al. 2012]. In an aircraft-based remote sensing using hyperspectral device, it is possible to obtain the images with the spectral resolution of 1 nm. Depending on the height from which the image will be taken, the spatial resolution can reach even 1 cm [Mohd et al. 2018].

The use of remote sensing techniques in the analyses of droughts is based mainly on employing spectral indicators for determining the condition of the vegetation (vegetation health) [Nicolai-Shaw et al. 2017, Dąbrowska-Zielińska et al. 2011]. Connecting the reflection of light with the biophysical characteristics of plants enabled to use the spectral indicators for assessing biomass production, forecasting of yields, and susceptibility of plant organisms to the influence of stressors such as water retention in the soil. It is possible to use plants as indicators in the study of soil-plant environment moisture [Dąbrowska-Zielińska et al. 2011], and thus in the assessment or even prediction of drought. Normalized Difference Vegetation Index (NDVI) [Pettorelli et al. 2005, Huete et al. 2002] is one of the most recognized and described spectral indices. This index is calculated based on the difference between reflectance of sunlight in the near–infrared and red bands normalized by the sum of both of them [Wójtowicz et al. 2005]. Its basic function is determining the amount of chlorophyll in the plants cells. According to this parameter, it is possible to determine the condition of the plant and its susceptibility to external stressors [Carlson and Ripley 1997]. NDVI can be used as a substitute indicator for assessing the water content in the environment be-

cause of the correlation between the vegetation condition and availability of water and the hydrological regime [Taylor et al. 2003, Pettorelli et al. 2005]. The high value of the indicator is exhibited by dense vegetation growth in the favorable conditions for the environment. The NDVI value more than 0.5 identifies the plants in good condition [Carlson and Ripley 1997].

The aim of the study is to prove the possibility of using NDVI in the analyses of the relation between remote sensing indicator and meteorological and hydrological parameters. On the basis of the spectral images from Sentinel–2 NDVI values, the relations were calculated. The values of the indicator provided the information about the current state of the plants, their condition and reaction to the soil-plant environmental conditions and its variability. The correlation values will help to answer the question, if it is possible to use NDVI for monitoring agricultural and soil drought in lowland areas in the river valleys.

MATERIALS AND METHODS

The study area is located on the left bank of the Oder River valley from city of Brzeg Dolny to Malczyce, in Lower Silesian voivodeship, in Poland (N: 51° 13' 53.51", E: 16° 40' 47.02". PL–92: 338057, 338057) (Figure 1) [Olszewska et al. 2012]. Since 1993, the studies of soil moisture measurements and the analyses of ground water tables and water levels in the rivers have been conducted in the area. The area of analyses is mainly under the agricultural use, there are only scattered buildings. The research area is 7.73 km². This paper presents the analyses which were carried out from February to November 2016.

The spectral data for the analyses were obtained from the Sentinel-2 satellite of European Space Agency (ESA). The set comprises the two satellites: Sentinel-2A and Sentinel-2B which move simultaneously. The main instrument of Sentinel-2 satellite is a single multi-spectral instrument (MSI) with 13 spectral channels, in the spectral range of 413–2210 nm with a spatial range from 10 to 60 m [Drusch et al. 2012]. ESA for scientific purposes provides data of varying degrees of processing, available on the Sentinels Scientific Data Hub website. For the presented study, the data referred to as 2A were obtained. These include the data pertaining to the reflection in the top of atmosphere and they are after

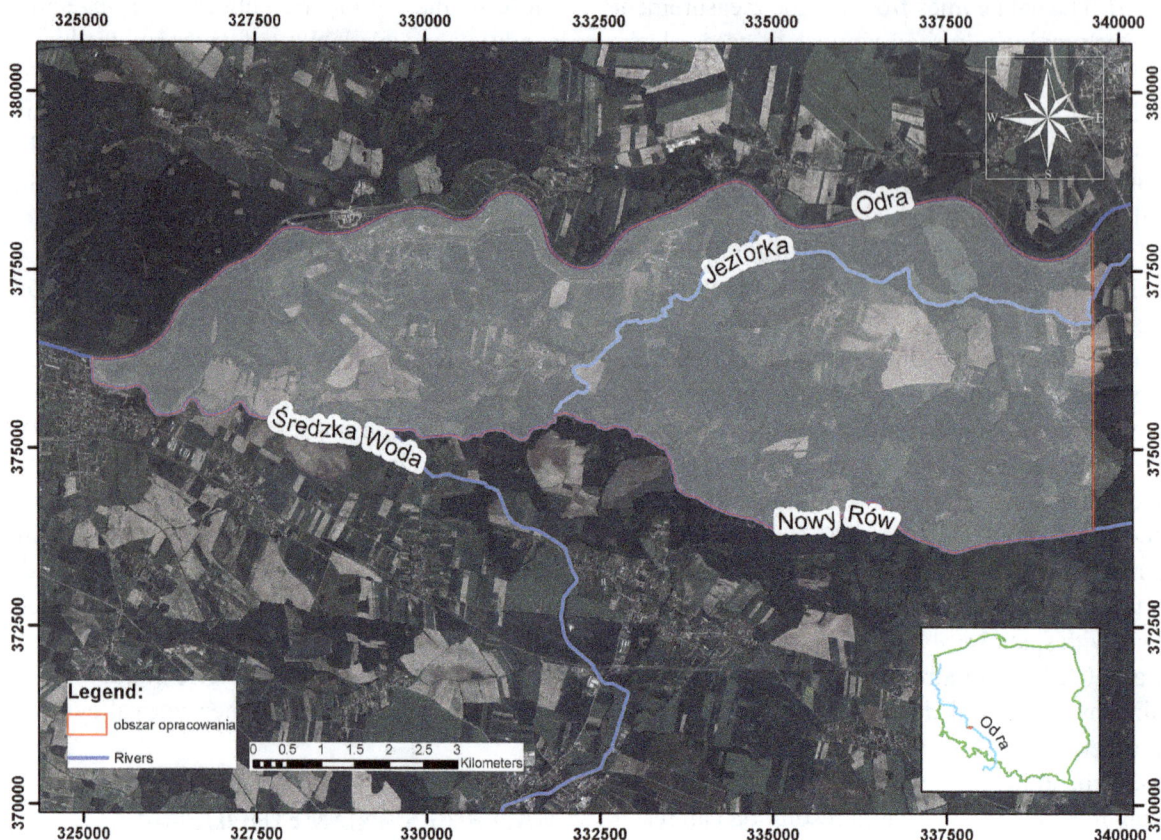

Figure 1. Location of the study area in the left bank of the Odra River valley.

necessarily radiometric and geometric corrections. In order to obtain the surface reflection, MSI bands were scaled to surface reflectance using Dark Object Subtraction (DOS) methods in Semi-Automatic Classification Plugin for QGIS [Congedo 2016]. The spectral images prepared in this way were used to determine the relevant spectral indicators.

Eight spectra images with a cloud cover less than 5% for the research area were selected from Sentinel-2 mission: Feb 6, March 17, March 27, May 6, Jun 25, Aug 4, Sep 13 and Nov 22.

Daily sum of precipitation [mm], average daily temperature [°C] for Wrocław–Strachowice station, water levels [cm] in the Odra from the gauge stations of Brzeg Dolny were obtained from the Institute of Meteorology and Water Management – National Research Institute. This iconforms to the act of the re-use of public sector information, which implements the directive 2013/37/UE into Polish law. The Climatic Water Balance was calculated by means of the Jaworski method [Jaworski 2004], with regard to the evapotranspiration rate according to the model of Doroszewski and Górski [Doroszewski and Górski 1995]. For the basic analysis, the average temperature (T.mean), the precipitation sum (P.sum), the average water level (W.mean) and the climatic water balance (CWB) for periods of 5, 10, 20 and 30 days before the satellite image was taken, were calculated (Table 1). The adopted meteorological and hydrological parameters were the same for the entire research area. This was related to the fact that the individual research objects are close to each other and to the reference station.

The basic analyses were conducted in two consecutive phases. The first phase included selection of grasslands from other types of the land cover on the research area and calculation the NDVI for grasslands for every object (Figure 2). In the second phase, the statistical analyses of correlation between data describing moisture condition in soil-plan environment on the research area were conducted. Correlation was done between the in situ (meteorological and hydrological) data and spectra data describing state and condition of plants (indicator of environmental moisture) [Dąbrowska-Zielińska et al. 2011].

The analyses were performed on grassland. The condition of the grassland depends only on the availability of water resources, not on the growth stages during the vegetation season. It

Table 1. Specification of meteorological and hydrological data (T.mean – average temperature [°C]; P.sum – precipitation sum [mm]; W.mean – average water level [cm]; CWB – climatic water balance [mm]; 5, 10, 20, 30 – the length of the previous period [days]).

Date	T.mean 5 [°C]	T.mean 10 [°C]	T.mean 20 [°C]	T.mean 30 [°C]	P.sum 5 [mm]	P.sum 10 [mm]	P.sum 20 [mm]	P.sum 30 [mm]	W.mean 10 [cm]	W.mean 20 [cm]	CWB 5 [mm]	CWB 10 [mm]	CWB 20 [mm]	CWB 30 [mm]
Feb 6, 2016	6.4	6.9	2.8	2.2	5.8	12.9	26.7	41.4	115.0	89.0	7.0	14.7	30.8	46.8
March 17, 2016	3.6	3.7	3.3	3.6	6.6	21.5	49.0	71.8	251.0	246.0	-14.0	2.6	30.4	53,6
March 27, 2016	6.3	5.8	4.8	4.2	4.5	5.1	26.6	54.1	214.0	232.0	-23.7	-23.3	0.4	27.1
May 6, 2016	12.3	10.1	9.4	9.8	21.2	23.5	33.3	67.6	176.0	204.0	-37.7	-36.5	-36.9	-5.6
June 25, 2016	22.2	20.4	19.4	19.4	4.4	37.4	41.4	46.7	80.0	89.0	-91.5	-60.0	-74.1	-88.7
Aug 4, 2016	21.1	21.7	20.6	20.4	7.4	15.7	30.2	108.1	108.0	119.0	-72.4	-71.8	-68.6	-10.4
Sep 13, 2016	22.1	20.5	20.3	19.7	0.0	15.7	15.7	35.2	63.0	71.0	-63.3	-50.8	-77.2	-81.4
Nov 22, 2016	8.4	5.0	4.6	6.0	0.0	9.2	31.0	35.3	-	-	15.2	24.4	35.0	40.7

is the opposite situation than in the case of cereals, for which the spectrum of the reflected light changes along with changes of phonological phase [Martyniak et al. 2007]. The analyses performed by the Institute of Environmental Protection and Development of Wrocław University of Environmental and Life Sciences showed that the water management of plants in a river valley between Brzeg Dolny and Malczyce based mainly on the use of water from precipitation and this is related to the fact of lowering the groundwater table [Pływaczyk and Olszewska 2014, Głuchowska and Pływaczyk 2008]. The ground water table is so deep that it reaches the levels which are not available for the root systems of most plants [Pływaczyk and Olszewska 2014, Chalfen et al. 2014].

Corine Land Cover 2012 was used for identification of grassland from other type of land cover. The result of the identification of the grassland areas was verified using the available material coming from the documentation from an older research conducted in the area. 22 objects were chosen for next analyses, which constituted compact complexes with the area greater than 150 m². A grid with mesh size 10 m was used for analyses, which is consistent with the spatial resolution of Sentinel-2 data (Figure 2).

For each of the 22 objects, based on the NDVI value for every grid cell for 8 available scenes the median, minimum and maximum NDVI values were calculated used the formula [Kycko and Zagajewski 2013]:

$$NDVI = \frac{(NIR - VIS)}{(NIR + VIS)} \quad (1)$$

where: NIR – near infrared, Sentinel-2 band no. 8, VIS – visible spectrum, Sentinel-2 band no. 4 [Drusch et al. 2012].

The statistical analyses were performed in the Statistica 13.1 software. In order to find out if the dependent variables (NDVI) and independent variables (meteorological and hydrological parameters) have a normal distribution the Shapiro-Wilk test were performed. The correlation between dependent and independent variables, describing the moisture conditions of the soil-plant environment, were sought next. The Pearson linear correlation coefficient was used to show the linear relationship of independent and dependent variable having a normal distribution. For the variable pairs where at least one did not have a normal distribution, the non-parametric Spearman correlation coefficient was used. The confidence interval of 95% was used.

RESULTS

For the objects having continuity of data in the analyzed period, the changes of the NDVI values for 8 measurement campaigns during the research period are presented in the chart (Figure 3). The objects are numbered from 1 to 23; due to technical reasons the object no. 2 was removed at the stage of data preparation (Figure 2).

The increase of NDVI values was observed in the following periods:
- Feb 6 – Jun 25 for objects no.: 1, 3, 8÷9,
- Feb 6 – May 6 for objects no.: 5÷7, 10, 14÷15, 18÷23,
- Feb 6 – Aug 4 for objects no.: 11÷13, 16÷17,
- Feb 6 – Sep 13 for objects no.: 4.

For the objects no. 1, 4÷11, 16÷23 NDVI values exceeds 0.5 during the period of 6th Feb to 13th Sep which was 77% of objects. On this basis, it can be concluded that in this period the vegetation in the research area was in good condition at

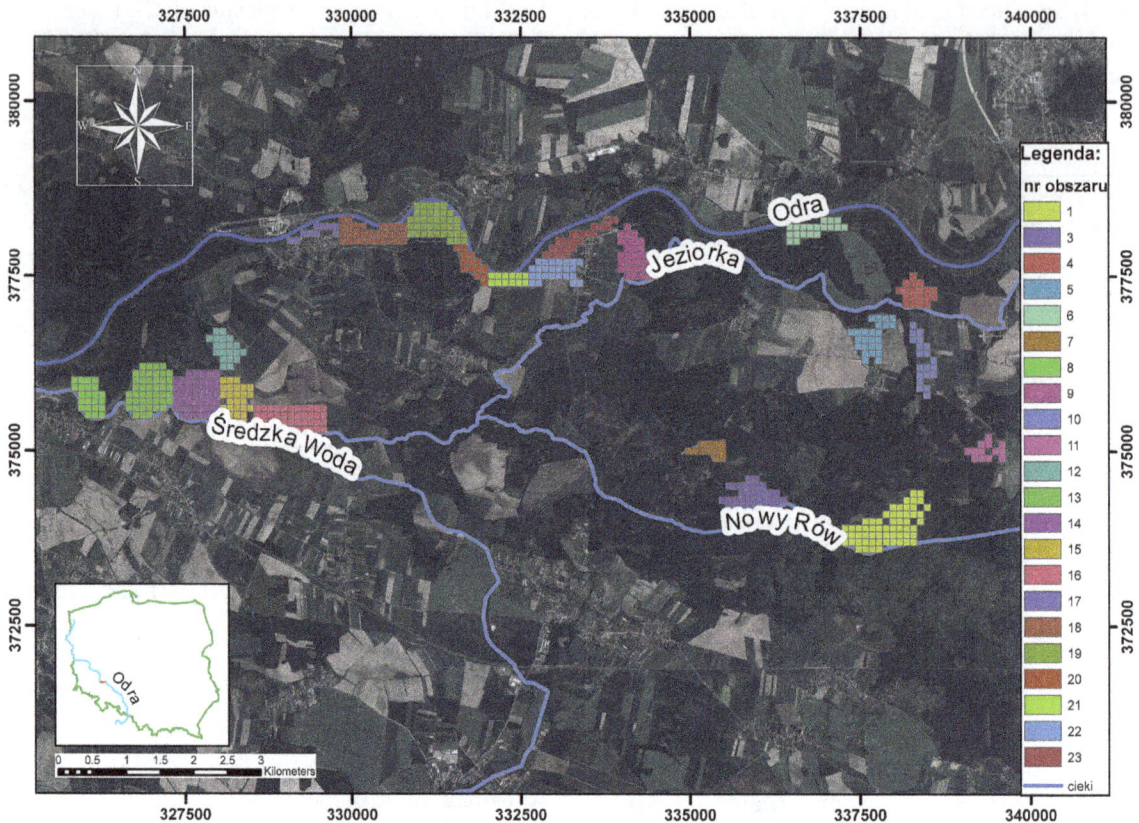

Figure 2. Location of the 22 objects in the research area, grid 10 m.

the culminating point of vegetation in the end of this periods [Grzywna et al. 2018]. The median of NDVI had never increased above 0.5 for the object no. 15. For the object no. 14, the NDVI value was slightly higher than 0.5 (0.52) on the 6th May. For object no. 12, the value of NDVI was higher than 0.5 for only one campaign on 4th Aug. The NDVI value was greater than 0.5 for objects no.: 4, 6, 8, 11, 16, in the period from 6th Feb to 27th March. Most often, the NDVI value was higher than 0.5 for the objects 4, 6 ÷ 7, 9, 11 (6 times) and 7 ÷ 8, 22 ÷ 23 (5 times). However, for the objects 22 and 23, this occurred in the period from 6th May to 22nd Nov, whereas on other objects, in the period 17th Mar to 22nd Nov (Figure 3).

The smallest NDVI values were most often observed on the object no. 15 (3 times) and no. 12 (3 times) among the 8 campaigns on 22 objects. In each event, when the minimum value of NDVI was noted for object no. 12, the next lowest value was observed on the object no. 15.

The NDVI value was greater than 0.5 on each objects in the following dates: 6th Feb on none of the objects, 17th March on the objects no.: 4, 9, 11 (3 objects, 13.6%). 27th March on the objects no.: 4, 6, 8, 16 (4 objects, 18.2%), 6th May on the objects no.:1, 4÷11, 14, 16÷23 (18 objects, 81.8%),

25th Jun on the objects no.: 1, 3, 6÷9, 11, 17÷23 (14 objects, 63.6%). 4th Aug on objects no.: 1, 3÷13, 16÷20, 22, 23 (19 objects 86.4%), 13th Sep on the objects no.: 1, 3÷11, 13, 17 ÷23 (18 objects, 81.8%), 22nd Nov on the objects no.: 4, 6÷7, 9, 11, 13, 21÷ 23 (9 objects, 40.1%) (Table 2).

Normal distribution had the following independent variables: T.mean 5, P.sum 10, P.sum 20, P.sum 30, W.mean 10, W.mean 20, CWB 5, CWB 10, CWB 30 and distribution of the NDVI values (dependent variables) in the analyzed period for objects: 1, 5, 7÷8, 10÷11, 13÷14, 16÷23. Other variables did not have normal distribution.

In general, no correlation was found between the distribution of the NDVI values and independent variables in terms of the sum of precipitation and average water levels. Only between the NDVI values and P.sum 10 for about half of the objects was correlation found at the medium level ($0.50 < R < 0.70$).

For the analyzed objects in the research periods, no correlation was found between the NDVI values and precipitation and river water level. However, there is a noticeable difference in the case of the correlation between the NDVI values and average temperature or CWB. The thermal parameter had a positive correlation with NDVI.

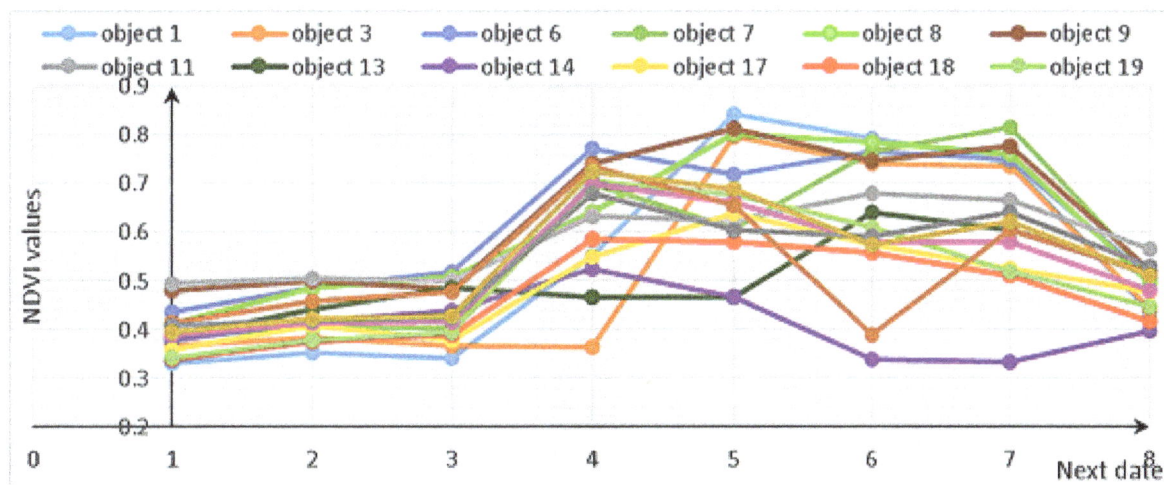

Figure 3. The NDVI values for subsequent measurement campaigns: 1 – Feb 6, 2 – Mar 17, 3 – Mar 27, 4 – May 6, 5 – Jun 25, 6 – Aug 4, 7 – Sep 13, 8 – Nov 22.

Table 2. Median value of NDVI indicator for 22 research objects in the following 8 campaigns

Campaign [date]	Object number																						
	1	3	4	5	6	7	8	9	10	11	12	13	14	15	16	17	18	19	20	21	22	23
Feb 6	0.33	0.37	0.47	0.36	0.44	0.39	0.41	0.48	0.4	0.49	0.28	0.38	0.38	0.28	0.48	0.36	0.34	0.34	0.39	0.41	0.41	0.39
March 17	0.35	0.38	0.53	0.42	0.49	0.41	0.49	0.5	0.44	0.5	0.29	0.44	0.42	0.29	0.48	0.41	0.38	0.38	0.41	0.46	0.42	0.42
March 27	0.34	0.37	0.53	0.43	0.52	0.4	0.51	0.48	0.44	0.5	0.25	0.49	0.44	0.28	0.51	0.38	0.39	0.39	0.42	0.48	0.42	0.43
May 6	0.55	0.36	0.78	0.68	0.77	0.7	0.64	0.74	0.67	0.63	0.36	0.47	0.52	0.4	0.58	0.55	0.59	0.71	0.7	0.74	0.68	0.73
Jun 25	0.84	0.8	x	x	0.72	0.61	0.8	0.81	x	0.62	x	0.47	0.47	x	x	0.64	0.58	0.68	0.66	0.65	0.61	0.69
Aug 4	0.79	0.74	0.8	0.62	0.77	0.77	0.79	0.75	0.65	0.68	0.69	0.64	0.34	0.34	0.6	0.57	0.56	0.61	0.58	0.39	0.59	0.58
Sep 13	0.74	0.74	0.82	0.57	0.75	0.82	0.76	0.78	0.58	0.67	0.41	0.61	0.33	0.29	0.32	0.52	0.51	0.52	0.58	0.61	0.64	0.62
Nov 22	0.44	0.44	0.54	0.42	0.51	0.51	0.49	0.52	0.48	0.57	0.43	0.52	0.4	0.34	0.34	0.48	0.42	0.45	0.48	0.52	0.53	0.51

This correlation stated that as the average temperature rises, the condition of plants represented by the increase in the NDVI values, increased proportionally. In the case of correlation to CWB, this was a reverse–proportional relationship – a negative correlation. As the value of CWB increased, the NDVI values decreased – the condition of plants decreased.

The correlations between the NDVI values and CWB 5, and CWB 30 were very high for many objects. The high and very high correlation (R>0.70) for more than half of the objects were recorded for the average temperature T.mean 5, T.mean 20 and T.mean 30, most objects showed the correlation for the T.mean 30 (over 75%). Less than half of the objects showed a high and very high correlation between the NDVI values and T.mean 10 (Table 3).

The high and very high correlation for more than half of the objects was recorded for CWB 5, CWB 10, CWB 30. Less than half of the objects showed a high and very high correlation for CWB 20 (Table 3).

High or very high correlation between the NDVI values and all or almost all temperature and CWB indicators was exhibited by objects no.: 1, 4÷9, 11. Objects no.: 3, 12÷16 and 21 did not show high or very high correlations or no other. Objects no.: 10, 17÷20, 22÷23 showed a high or very high correlation to a part of the temperature value and CWB (Table 3).

DISCUSSION AND CONCLUSION

The analyses conducted by other researchers show high relevance of using the spectral data for assessing the plants condition and biomass production [Gao 1996, Martínez-Fernández et al. 2016]. NDVI shows the condition of plants based on the chlorophyll amount in the cells [Pettorelli et al. 2005]. Groeneveld and Baugh [2007] proved that NDVI can be used as an indicator of moisture content of the environment in which plants live and thus can be helpful in assessing the drought [Dąbrowska-Zielińska et al. 2011]. Bar-

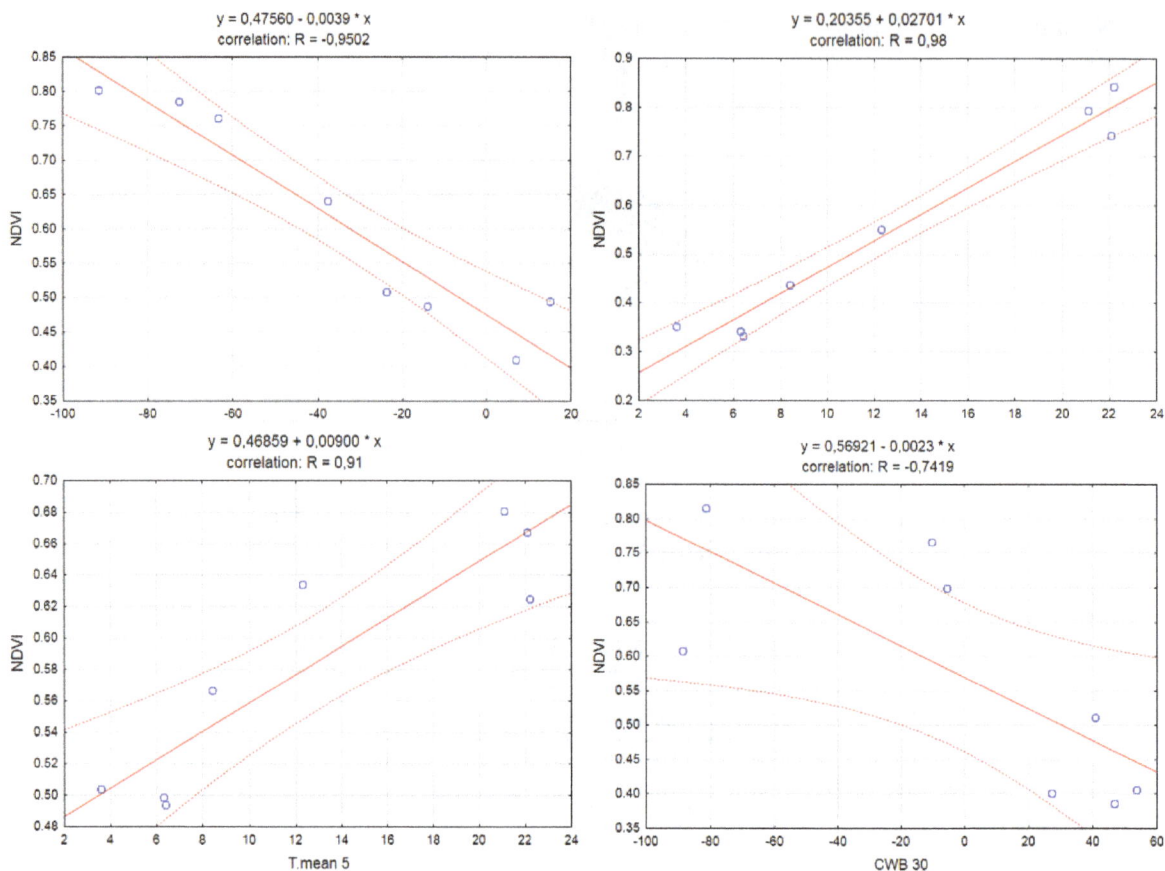

Figure 4. Selected of high or very high correlations (R>0.70) between the NDVI values with selected independent variables along with regression equations (diagram A – object no. 1, diagram B – object no. 11, diagram C – object no. 8, diagram D – object no. 7).

bosa showed the high correlation (0.84) between NDVI and precipitation which, in his opinion, is caused by the limited availability of precipitation [Barbosa et al. 2009]. Researchers conducted by authors like Plessis [1999] Schmidt and Gitelson [2000] showed close correlation between NDVI and precipitation, especially in dry and sub-humid environments, where precipitation is a factor limiting plant growth and their condition. However, the relationship between NDVI and precipitation is dependent on the characteristics of the type of vegetation and soil [Al-Bakri and Suleiman 2004]. It has been proven that NDVI can give a misrepresentation of the state of vegetation if the impact of greening – natural changes in the color of plants during the growing season, will not be removed [Wang et al. 2012]. Many factors, such as plant diseases, the amount of minerals or other stressors, can affect the NDVI values.

The changes in the index value (NDVI) do not necessarily have to be related to soil moisture [Nichol 2015]. This have been also confirmed by the studies conducted throughout the whole research area Brzeg Dolny – Malczyce in 2016

[Grzywna et al. 2018]. Similarly as for the whole research area, on particular objects, no high correlation was found between the NDVI values and precipitation or water levels.

It cannot be clearly determined which environmental parameter describing the moisture content in the environment has the highest impact on the NDVI value, representing the state of the plants.

The lack of correlation between the NDVI value and independent variables of precipitation and average water levels enables to state that it is impossible to use the NDVI in analysis of moisture content in soil-plane environment to forecast drought in agriculture.

High or very high correlation between NDVI, and temperature and CWB is a natural thing. This is related to the changes in the thermal conditions during the growing season.

There was no relationship between the location of the object on the research area and the values of the correlation between the NDVI and meteorological or hydrological parameters.

Table 3. Values of correlation coefficients between the NDVI values and meteorological as well as hydrological parameters. x – no correlation (R<0.50); 0.50<R<0.70 – average correlation, R>0.70 – high and very high correlation.

Variable	Object number																					
	1	3	4	5	6	7	8	9	10	11	12	13	14	15	16	17	18	19	20	21	22	23
T.mean 5	0.98	0.64	0.89	0.74	0.64	0.87	0.96	0.9	0.78	0.91	0.71	0.68	x	0.64	x	0.86	0.83	0.74	0.76	x	0.78	0.75
T.mean 10	0.71	x	0.71	0.68	0.71	0.74	0.79	0.69	0.64	0.76	x	0.55	x	x	x	0.67	0.62	0.62	0.6	x	0.62	0.6
T.mean 20	0.86	0.62	0.93	0.89	0.86	0.88	0.93	0.83	0.86	0.9	0.61	0.79	x	0.57	x	0.81	0.76	0.76	0.74	x	0.76	0.74
T.mean 30	0.9	0.67	0.96	0.86	0.83	0.93	0.9	0.88	0.89	0.95	0.79	0.81	x	0.71	x	0.86	0.79	0.79	0.76	x	0.79	0.76
P.sum 5	x	x	x	x	x	x	x	x	x	x	x	x	x	x	0.81	x	x	x	x	x	x	x
P.sum 10	0.58	x	x	0.55	x	x	0.56	0.63	0.51	x	x	x	x	0.61	x	0.69	0.62	0.66	0.65	0.57	x	0.64
P.sum 20	x	x	x	x	x	x	x	x	x	x	x	x	x	x	x	x	x	x	x	x	x	x
P.sum 30	x	x	x	x	x	x	x	x	0.5	x	x	x	x	0.8	x	x	x	x	x	x	x	x
W.mean 10	-0.75	-0.57	-0.66	x	x	-0.63	-0.65	-0.71	x	-0.65	-0.66	x	x	x	x	-0.57	x	x	x	x	-0.54	x
W.mean 20	-0.62	x	x	x	x	x	x	x	x	x	x	x	0.51	x	x	x	x	x	x	x	x	x
CWB 5	-0.91	-0.6	-0.75	-0.8	-0.71	-0.71	-0.95	-0.81	-0.78	-0.75	x	-0.52	x	x	x	-0.81	-0.82	-0.75	-0.73	x	-0.65	-0.7
CWB 10	-0.86	-0.57	-0.75	-0.82	-0.76	-0.77	-0.93	-0.76	-0.8	-0.79	x	-0.62	x	x	x	-0.76	-0.83	-0.75	-0.72	x	-0.67	-0.68
CWB 20	-0.76	-0.52	-0.79	-0.75	-0.71	-0.76	-0.86	-0.81	-0.64	-0.71	x	x	x	x	x	-0.69	-0.64	-0.64	-0.69	x	-0.71	-0.69
CWB 30	-0.89	-0.62	-0.93	-0.7	-0.74	-0.74	-0.89	-0.88	-0.68	-0.76	-0.5	x	x	x	x	-0.79	-0.77	-0.69	-0.75	-0.58	-0.74	-0.77

REFERENCES

1. Al-Bakri J.T., Suleiman A.S. 2004. NDVI response to rainfall in different ecological zones in Jordan. International Journal of Remote Sensing, 25(19), 3897–3912.

2. Barbosa H.A.T.V., Lakshmi K., Aydin E.G. 2009. Using the Satellite- Derived NDVI- OLR Feedbacks over West Sahel Africa to Assess Land- Atmosphere Responses to Environmental Change. In AIP Conference Proceedings. pp. 357–360.

3. Biniak-Pieróg M. 2008. Wpływ elementów agrometeorologicznych na zmienność zasobów wodnych gleby w półroczu zimowym. Współczesne Problemy Inżynierii Środowiska, VII.

4. Carlson T.C., Ripley D. a. 1997. On the relationship between NDVI, fractional vegetation cover, and leaf area index. Remote Sensing of Environment, 62, 241–252.

5. Chalfen M., Lyczko W., Plywaczyk L. 2014. The prognosis of influence of the Oder River waters dammed by Malczyce barrage on left bank areas. Journal of Water and Land Development, 21(1), 19–27.

6. Congedo L. 2016. Semi-Automatic Classification Plugin Documentation. Release 6.0.1.1.

7. Dąbrowska-Zielińska K., Ciołkosz A., Malińska A., Bartold M. 2011. Monitoring of agricultural drought in Poland using data derived from environmental satellite images. Geoinformation Issues, 3(1), 87–97.

8. Doroszewski A., Marcinkowska I. 1995. Llimatyczny bilans wodny sezonów wegetacyjnych 1921–1993 w Puławach. Środowisko Przyrodnicze Lubelszczyzny. Gleby i Klimat Lubelszczyzny, cz.

II – Klimat (J. Kołodziej, R. Turski, red.). Materiały z Konferencji Naukowej Lublin, 25 kwietnia 1994, 193–197.

9. Drusch M., Del Bello U., Carlier S., Colin O., Fernandez V., Gascon F., Hoersch B., Isola C., Laberinti P., Martimort P., Meygret A., Spoto F., Sy O., Marchese F., Bargellini P. 2012. Sentinel-2: ESA's Optical High-Resolution Mission for GMES Operational Services. Remote Sensing of Environment, 120, 25–36.

10. Dubicki A. i in. 2002. Zasoby wodne w dorzeczu górnej i środkowej Odry w warunkach suszy. IMGW, Warszawa, s. Atlasy i monografie.

11. Dzieżyc J, Nowak L., Panek K. Średnie regionalne niedobory opadów i potrzeby deszczowania roślin uprawnych na glebach lekkich i średnich. Zesz. Probl. Post. Nauk Roln., 314, 1987b, s. 35–47.

12. Gao B.C. 1996. NDWI – A normalized difference water index for remote sensing of vegetation liquid water from space. Remote Sensing of Environment, 58(3), 257–266.

13. Głuchowska B., Pływaczyk L. 2008. Wody gruntowe w dolinie Odry poniżej stopnia wodnego w Brzegu Dolnym. Współczesne Problemy Inżynierii Środowiska. V.

14. Groeneveld D.P., Baugh W.M. 2007. Correcting satellite data to detect vegetation signal for eco-hydrologic analyses. Journal of Hydrology, 344(1–2), 135–145.

15. Grzywna H., Dąbek P.B., Olszewska B. 2018. Analysis of moisture conditions in the lowland areas using high resolution spectral data from the Sentinel-2 satellite and the GIS tools. In Proc. 10th Conference on Interdisciplinary Problems in Environmental Protection and Engineering EKO-DOK 2018.

16. Hisdal H., Tallaksen L.. 2000. Technical Report No. 6 Drought Event Definition.

17. Huete A., Didan K., Miura H., Rodriguez E.P., Gao X., Ferreira L.F. 2002. Overview of the radiometric and biopyhsical performance of the MODIS vegetation indices. Remote Sensing of Environment, 83, 195–213.

18. Jackson T.J., Chen D., Cosh M., Li F., Anderson M., Walthall C., Doriaswamy P., Hunt E.R. 2004. Vegetation water content mapping using Landsat data derived normalized difference water index for corn and soybeans. Remote Sensing of Environment, 92(4), 475–482.

19. Jankowiak J., Bieńkowski J. 2011. Kształtowanie I Wykorzystanie Zasobów Wodnych W Rolnictwie. (5), 39–48.

20. Jarząbek A., Sarna S., Karpiarz M. 2013. Ochrona przed suszą w planowaniu gospodarowania wodami.

21. Kycko M., Zagajewski B. 2013. Wpływ geometrii źródło promieniowania-roślina-detektor na wartość teledetekcyjnych wskaźników roślinności Assessment of geometry of radiation source-plant-detector on value of the remote sensing indices. Teledetekcja Środowiska, 49, 15–26.

22. Łabędzki L., Bąk B. 2015.a. Assessment of Soil Moisture on Permanent Grassland in Upper Noteć Valley Based on Soil Moisture Index. Ecological Engineering, 43, 153–159.

23. Łabędzki L., Bąk B. 2015.b. Method of Indicator-Based Assessment and Classification of Soil Moisture on Permanent. Infrastructure and Ecology of Rural Areas, (III), 515–531.

24. Łabędzki L., Bąk B. 2013. Monitoring i prognozowanie przebiegu i skutków deficytu wody na obszarach wiejskich. Infrastruktura i ekologia terenów wiejskich, (2), 65–76.

25. Lauzon N., Anctil F., Petrinovic J. 2004. Characterization of soil moisture conditions at temporal scales from a few days to annual. Hydrological Processes, 18(17), 3235–3254.

26. Martínez-Fernández J., González-Zamora A., Sánchez N., Gumuzzio A., Herrero-Jiménez C.M. 2016. Satellite soil moisture for agricultural drought monitoring: Assessment of the SMOS derived Soil Water Deficit Index. Remote Sensing of Environment, 177, 277–286.

27. Martyniak L., Dabrowska-Zielinska K., Szymczyk R., Gruszczynska M. 2007. Validation of satellite-derived soil-vegetation indices for prognosis of spring cereals yield reduction under drought conditions – Case study from central-western Poland. Advances in Space Research, 39(1), 67–72.

28. Mohd Asaari M.S., Mishra P., Mertens S., Dhondt S., Inzé D., Wuyts N., Scheunders P. 2018. Close-range hyperspectral image analysis for the early detection of stress responses in individual plants in a high-throughput phenotyping platform. ISPRS Journal of Photogrammetry and Remote Sensing, 138, 121–138.

29. Nichol J.E., Abbas S. 2015. Integration of remote sensing datasets for local scale assessment and prediction of drought. Science of the Total Environment, 505, 503–507.

30. Nicolai-Shaw N., Zscheischler J., Hirschi M., Gudmundsson L., Seneviratne S.I. 2017. A drought event composite analysis using satellite remote-sensing based soil moisture. Remote Sensing of Environment, 203, 216–225.

31. Nyc K. 1994. Rola retencji gruntowej w bilansowaniu zasobów wodnych. Zeszyty Naukowe Akademii Rolniczej we Wrocławiu, nr 248, Konferencje V: 247–251.

32. Olszewska B., Pływaczyk L., Łyczko W. 2012. Oddziaływanie spiętrzenia Odry stopniem wodnym w Brzegu dolnym na przepływach w cieku Jeziorka w latach 1971–2020. Water-Environment-Rural Areas, 12(3), 161–170.

33. Pettorelli N., Vik J.O., Mysterud A., Gaillard J.M., Tucker C.J., Stenseth N.C. 2005. Using the satellite-derived NDVI to assess ecological responses to environmental change. Trends in Ecology and Evolution, 20(9), 503–510.

34. Du Plessis W.P. 1999. Linear regression relationships between NDVI, vegetation and rainfall in Etosha National Park, Namibia. Journal of Arid Environments, 42(4), 235–260.

35. Pływaczyk L., Olszewska B. 2014. Changes in the moisture content of the middle fen soils in the Odra river valley in the region of Brzeg Dolny in the vegetation periods 2004–2009. Journal of Ecological Engineering, 15(4), 61–68.

36. Schmidt H., Gitelson A. 2000. Temporal and spatial vegetation cover changes in Israeli transition zone: AVHRR-based assessment of rainfall impact. International Journal of Remote Sensing, 21(5), 997–1010.

37. Torres G.M., Lollato R.P., Ochsner T.E. 2013. Comparison of drought probability assessments based on atmospheric water deficit and soil water deficit. Agronomy Journal, 105(2), 428–436.

38. Wang D., Morton D., Masek J., Wu A., Nagol J., Xiong X., Levy R., Vermote E., Wolfe R. 2012. Impact of sensor degradation on the MODIS NDVI time series. Remote Sensing of Environment, 119, 55–61.

39. Wang J., Rich P.M., Price K.P. 2003. Temporal responses of NDVI to precipitation and temperature in the central Great Plains , USA. International Journal of Remote Sensing, 24(11), 2345–2364.

40. Wójtowicz A., Wójtowicz M., Piekarczyk J. 2005. Zastosowanie teledetekcji do monitorowania i oceny produktywności plantacji rzepaku. Oilseed Crops, XXVI, 269–276.

12

The Aerophytic Diatom Assemblages Developed on Mosses Covering the Bark of *Populus alba* L.

Mateusz Rybak[1], Teresa Noga[2*], Robert Zubel[3]

[1] Department of Agroecology, Faculty of Biology and Agriculture, University of Rzeszów, Ćwiklińskiej 1A, 35–601 Rzeszów, Poland

[2] Department of Soil Studies, Environmental Chemistry and Hydrology, Faculty of Biology and Agriculture, University of Rzeszów, Zelwerowicza 8B, 35–601 Rzeszów, Poland

[3] Department of Botany and Mycology, Maria Curie-Skłodowska University, ul. Akademicka 19, 20-033 Lublin, Poland

* Corresponding author's e-mail: teresa.noga@interia.pl

ABSTRACT

The study was conducted in an old, historical park, in the northern part of Stalowa Wola city (south-eastern Poland). The aim of the study was to investigate the diversity of moss-inhabiting diatoms of the white poplar (*Populus alba* L.) bark. During the study, a total of 47 diatom taxa were found, three out of which were considered as dominant. Three other species are mentioned in the Red List of the Algae in Poland: *Achnanthes coarctata* (Brébisson) Grunow, *Luticola acidoclinata* Lange-Bertalot and *Stauroneis thermicola* (Petersen) Lund. For three species: *Luticola sparsipunctata* Levkov, Metzeltin & Pavlov, *L. vanheurckii* Van de Vijver & Levkov and *Hantzschia subrupestris* Lange-Bertalot, this is the first report from Poland.

Keywords: diatoms, tree bark, mosses, diversity, ecology

INTRODUCTION

Many algae species can survive and develop outside of the aquatic habitats – on a trees bark, on leaves, on wet walls and rocks, or on wet to semi-wet soils. They occur in continuously or only periodically wet habitats, using water in the form of atmospheric precipitation, dew, fog and moisture contained in the air. The aerophytic algae develop most abundantly in a warm and humid climate (Podbielkowski 1996).

The arboreal algae find a specific and favourable microclimate on the bark of trees that provide shade, nutrients and shelter against wind. In Europe, the trees bark is usually covered with green algae (mostly *Pleurococcus vulgaris* Meneghini, *Protococcus viridis* C. Agardh and *Chlorella vulgaris* Beyerinck), rarely with blue-green algae and diatoms (the latter presumably due to the absence of silica in the environment). The arboreal algae have wide ecological amplitude. They are known for their nitrogenous compounds toler-

ance, so they can live even on the trees growing in cities (Kawecka, Eloranta 1994).

The mosses growing on the tree bark use it only as a substrate, while all nutrients are taken by mosses from an atmospheric precipitation or from the air humidity. The diversity of epiphytic bryophytes is affected by both the chemical (pH) and physical (including structure of surface and water absorption capacity) properties of the bark. In numerous ecosystems, bryophytes form an expansive cover that accumulates water and organic particles, providing a favourable environment for many organisms, from bacteria, algae and fungi to small animals (Lindo, Gonzales 2010, Plášek 2013, Glime 2017).

The studies on the moss-inhabiting diatoms (bryophytic diatoms) are most often related to peatland areas, mainly including the mosses from the genus *Sphagnum* spp. (Nováková, Poulíčková 2004, Poulíčková et al. 2004). The diatoms developing among the typical terrestrial bryophytes are very rarely studied, mainly on the mosses growing on rocks (Round 1957) and on terrestrial mosses occurring in the polar regions (e.g. Hickman, Vitt 1973, Van de Vijver, Beyens 1997, Gremmen et al. 2007). Until now, the algological studies of tree barks have focused mainly on the green and blue-green algae in tropical zones (Foerester 1971, Mrozińska 1990, Thompson, Wujek 1997, Salleh, Milow 1999, Neustupa 2003, 2005, Neustupa, Škaloud 2008, 2010, Kharkongor, Ramanujam 2014, Štifterová, Neustupa 2015), while diatoms were noted rarely, usually mentioned only in the lists of species (Lakatos et al. 2004, Geissler et al. 2006, Neustupa, Škaloud 2010, Kharkongor, Ramanujam 2014, Štifterová, Neustupa 2015, Qin et al. 2016). In Poland, no studies on the diatom assemblages developing among mosses on the tree barks have been conducted so far.

The aim of the study was to investigate the diversity of diatom assemblages among the mosses growing on the bark of white poplar (*Populus alba* L.) in relation to chemical parameters (pH, conductivity, anions and cations) measured in the filtrates obtained from the tree bark.

METHODS

The conducted research concerns the mosses growing on white poplar tree trunk (*Populus alba*) in an old park, in the northern part of the Stalowa Wola city. The tree from which the research material was taken (50°36'01.1"N 22°01'53.3"E) grew in an exposed place (about 30 m from the nearest tree), which resulted in the trunk being directly illuminated. Mosses covered the base of the trunk in almost 100%, while on the trunk up to the height of 150 cm above the ground level, they formed small clumps (less than 5% of the bark surface).

The samples were collected in March 2016 and in March, June and August 2017. Small pieces of bark (approx. 8×8 cm) covered by mosses were collected from two heights: 20 cm and 150 cm above the ground level. The collected material was placed in paper envelopes. In August 2017, the material was collected only at the base of the trunk, because at the height of 150 cm, mosses were absent. During the each sampling, a piece of bark from height of 150 cm was collected for chemical analyses.

The filtrates for chemical analyses were prepared according to Schmidt et al. (2001). The filtrates were obtained by soaking pieces of bark in deionized water (in weight ratio 1:10) for 24 hours. For the preparation of filtrates, the intact pieces of bark were used to obtain a solution similar to that forming on the surface of the bark that is a source of water and nutrients for epiphytic organisms.

The pH and electrolytic conductivity were measured using a MARTINI pH56 pH meter and MARTINI EC59 conductivity meter. The ions content was determined using a Thermo scientific DIONEX ICS–5000+DC device in the Departmental Laboratory of Analysis of Environmental Health and Materials of Agricultural Origin at the University of Rzeszów.

In order to obtain cleaned diatom material, a modified Qin et al. (2016) method was used:
1. For the purpose of separating mosses and diatoms from the bark surface, a part of the collected material was placed in beakers and 50 ml of 30% hydrogen peroxide (H_2O_2) was added, and left at room temperature for 48 hours.
2. In the next step, the bark fragments were rinsed with deionized water, and the formed solution was collected into the same beaker, in which the bark was digested.
3. In order to obtain clean diatom valves, the solution was centrifuged to remove the excess of hydrogen peroxide and again digested in the mixture of sulfuric acid and potassium dichromate, until organic matter was completely dissolved.
4. In the last step the chromic mixture was removed by centrifugation with distilled water (at 2 500 rpm).

For LM (light microscope) slides, pure diatom valves were mounted in Pleurax resin (refractive index 1.75). Diatoms were identified under a Carl Zeiss Axio Imager.A2 light microscope (LM) with a Zeiss AxioCam ICc 5 camera at 1000 × magnification. For Scanning Electron Microscope (SEM), the observations samples were coated at Turbo-Pumped Sputter Coater Quorum Q 150OT ES with a 20 nm layer of gold and observed under a Hitachi SU 8010 microscope.

The diatoms identification was conducted in accordance with the following references: Krammer and Lange-Bertalot (1986, 1988, 1991a,b), Krammer (2000), Lange-Bertalot et al. (2003), Hofmann et al. (2011) and Levkov et al. (2013). The nomenclature of mosses is according to Ochyra et al. (2003). The habitat and substrate preferences of mosses were determined according to Dierβen (2001), Smith (2004) and Ellenberg, Leuschner (2010).

The species composition of the samples was determined by counting specimens on randomly selected fields of view under light microscope according to a modified method described by Lakatos et al. (2004). There were about 300 valves. In the samples with low number of valves on slide, all specimens were counted. The species with a share more than 50 valves were defined as dominant.

The categories of endangered for diatom taxa were distinguished using the Red List of the Algae in Poland (Siemińska et al. 2006).

The statistical analyses with graphical interpretation were performed using the Canoco software (version 5.03). In order to analyse the differentiation of individual samples collected at different heights above ground level and in different seasons, the Principal Component Analysis (PCA) was used (Ter Braak, Šmilauer 2012).

RESULTS

The chemical analysis of filtrates obtained from soaking the bark showed that the solutions had a slightly acidic to neutral pH (5.3–6.9), while the conductivity values ranged from 43 to 427 µS cm⁻¹. The values of ions were significantly higher in the early spring periods compared to the summer season, when the concentration of nutrients decreased even below the limit of quantification (Table 1).

The examined tree was abundantly covered with mosses at the trunk base, while at a height of 150 cm, the mosses were less numerous. The base of the trunk was overgrown by typical terrestrial species: *Amblystegium serpens* Schimp, *Brachythecium salebrosum* (Web. Mohr) Schimp., *Herzogiella seligeri* (Brid.) Z. Iwats., *Rosulabryum capillare* (Hedw.) and epiphytic: *Hypnum cupressiforme* Hedw. and *Platygyrium repens* (Brid.) Schimp., J.R. Spence. The mosses formed a uniform, thick layer of about 5 cm, completely surrounding the base of the trunk. At the height of 150 cm, there were only single, small clumps of two species *Platygyrium repens* (Brid.) Schimp. and *Orthotrichum speciosum* Nees.

During the research carried out in 2016–2017, 47 diatom taxa from 23 genera were observed (Table 2). The most numerous genera were *Luticola* (9 taxa), followed by *Hantzschia* (4 taxa). The selected diatom taxa are presented in Figures 1 and 2.

Diatoms were numerous only at the base of the trunk. At the height of 150 cm above the ground level, only single valves were observed (Table 3). The PCA analysis also split the samples collected at different heights above the ground into two groups. The samples taken from a height of 20 cm were more varied, compared to those collected from a height of 150 cm, which formed a more compact group (Fig. 3).

Among all the observed diatoms, the dominant ones were: *Hantzschia amphioxys*, *Luticola acidoclinata* and *Pinnularia borealis*. In early spring (both in 2016 and 2017), only one species prevailed in the diatom assemblage – *Luticola acidoclinata* in 2016 and *Hantzschia amphioxys* in 2017. In the summer (both in June and August), all three species dominated the diatom flora (Table 3).

In the studied assemblages, three diatom species from Red List of the Algae in Poland were recorded (Siemińska et al. 2006). All of them are classified in the R category (rare): *Luticola acidoclinata* (which is also the dominant species), *Achnanthes coarctata* and *Stauroneis thermicola* (both observed as single specimens). In addition, three species have not been reported

Table 1. Values of chemical parameters measured in filtrates obtained from poplar bark in years 2016–2017

Date	03.2016	03.2017	06.2017	08.2017
pH	6.8	6.9	5.8	5.3
Conductivity [µS cm⁻¹]	397	427	145	43
Cl⁻ [mg l⁻¹]	12.55	14.43	1.19	0.33
NO₂⁻ [mg l⁻¹]	1.14	2.91	<0.001	<0.001
NO₃⁻ [mg l⁻¹]	3.63	1.37	<0.001	<0.001
PO₄³⁻ [mg l⁻¹]	3.93	3.29	<0.001	<0.001
SO₄²⁻ [mg l⁻¹]	23.45	42.06	13.35	0.18
Na⁺ [mg l⁻¹]	8.80	8.44	0.40	0.21
NH₄⁺ [mg l⁻¹]	0.43	2.23	1.10	0.08
K⁺ [mg l⁻¹]	25.03	61.69	35.22	10.88
Mg²⁺ [mg l⁻¹]	23.24	20.34	1.73	0.64
Ca²⁺ [mg l⁻¹]	23.14	20.04	4.94	1.91

Table 2. List of diatom taxa found in epiphytic mosses overgrowing the bark of white poplar, at different highs above soil level in years: 2016–2017

Date	03.2016	03.2017	06.2017	08.2017
Achnanthes coarctata (Brébisson) Grunow			+ ○	
Aulacoseira distans (Ehrenberg) Simonsen	+			
Aulacoseira granulata (Ehrenberg) Simonsen	+	○	+	
Aulacoseira muzzanensis (Meister) Krammer			○	+
Cocconeis placentula var. *lineata* (Ehrenberg) Van Heurck	+			
Cocconeis pseudolineata (Geitler) Lange-Bertalot	○			
Cyclostephanos dubius (Hustedt) Round		+	+	+
Epithemia adnata (Kützing) Brébisson	+			
Eunotia bilunaris (Ehrenberg) Schaarschmidt	+	+		
Fallacia insociabilis (Krasske) D.G. Mann		+		
Frustulia saxonica Rabenhorst	+			
Halamphora montana (Krasske) Levkov	+ ○			+
Hantzschia abundans Lange-Bertalot	+	+ ○	+	+
Hantzschia amphioxys (Ehrenberg) Grunow	+ ○	+ ○	+ ○	+
Hantzschia calcifuga Reichardt & Lange-Bertalot	+	+	+	+
Hantzschia subrupestris Lange-Bertalot	+ ○			
Humidophila contenta (Grunow) Lowe, et al.	+	+	+ ○	+
Humidophila gallica (W. Smith) Lowe, et al.	+			
Luticola acidoclinata Lange-Bertalot	+	+	+ ○	+
Luticola sparsipunctata Levkov, Metzeltin & A. Pavlov	+	+	+ ○	+
Luticola vanheurckii Van de Vijver & Levkov		○		
Luticola ventricosa (Kützing) D.G. Mann, Morphotyp I		+		
Luticola ventricosa (Kützing) D.G. Mann, Morphotyp II	+	+ ○		+
Luticola cf. *nivalis* (Ehrenberg) D.G. Mann	+	+ ○		
Luticola sp. 1	+	+		+
Luticola sp. 2	+ ○	+ ○	+	+
Luticola sp. 3	+	+	+	+
Mayamaea atomus (Kützing) Lange-Bertalot	+		+	+
Mayamaea excelsa (Krasske) Lange-Bertalot			+	
Mayamaea permitis (Hustedt) K. Bruder & Medlin			+	
Meridion circulare (Greville) C. Agardh		○		
Muelleria gibbula (Cleve) Spaulding & Stoermer			○	
Navicula wiesneri Lange-Bertalot	+	○	○	
Nitzschia pusilla Grunow			+	+
Nitzschia solgensis Cleve-Euler			○	
Orthoseira dendroteres (Ehrenberg) Genkal & Kulikovskiy	+		+ ○	+
Pantocsekiella sp.	+	○	○	
Pinnularia borealis Ehrenberg	+ ○	+ ○	+ ○	+
Pinnularia issealana Krammer	○			
Pinnularia obscura Krasske	+	+ ○	+	+
Sellaphora atomoides (Grunow) Wetzel & Van de Vijver	+	+		+
Sellaphora saugerresii (Desmazières) Wetzel & D.G. Mann		+		
Stauroneis borrichii (Petersen) Lund				+
Stauroneis lundii Hustedt				+
Stauroneis thermicola (Petersen) Lund		+		+
Surirella minuta Brébisson	+			+
Surirella terricola Lange-Bertalot & Alles	+ ○			

+   20 cm above soil level
○   150 cm above soil level

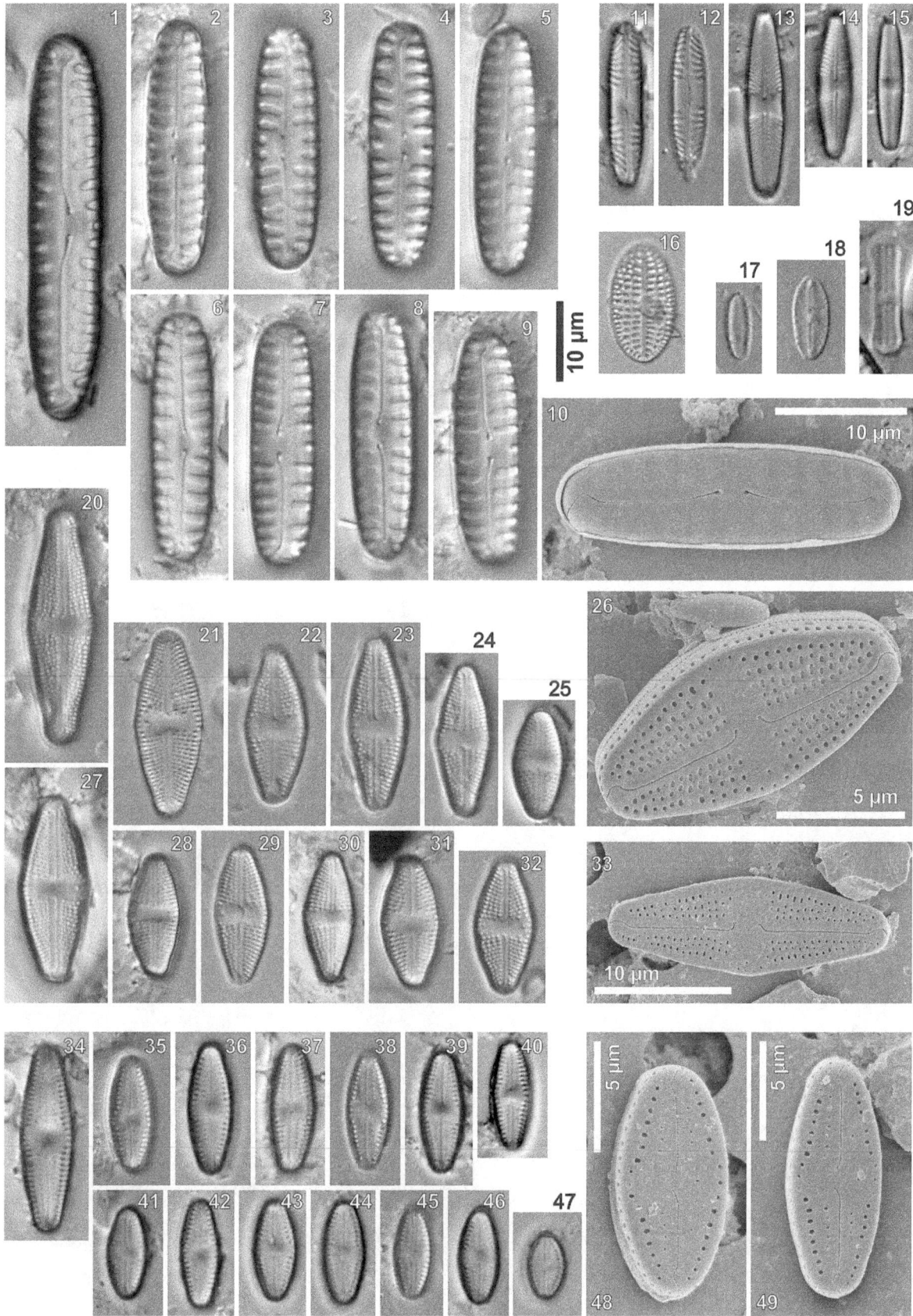

Fig. 1. Selected diatom taxa: 1–10 – *Pinnularia borealis*, 11–12 – *P. obscura*, 13 – *Stauroneis borrichii*, 14 – *S. lundii*, 15 – *Sellaphora atomoides*, 16 – *Cocconeis pseudolineata*, 17 – *Humidophila gallica*, 18 – *Mayamaea atomus*, 19 – *Humidophila contenta*, 20–33 – *Luticola acidoclinata*, 34–49 – *L. sparsipunctata*

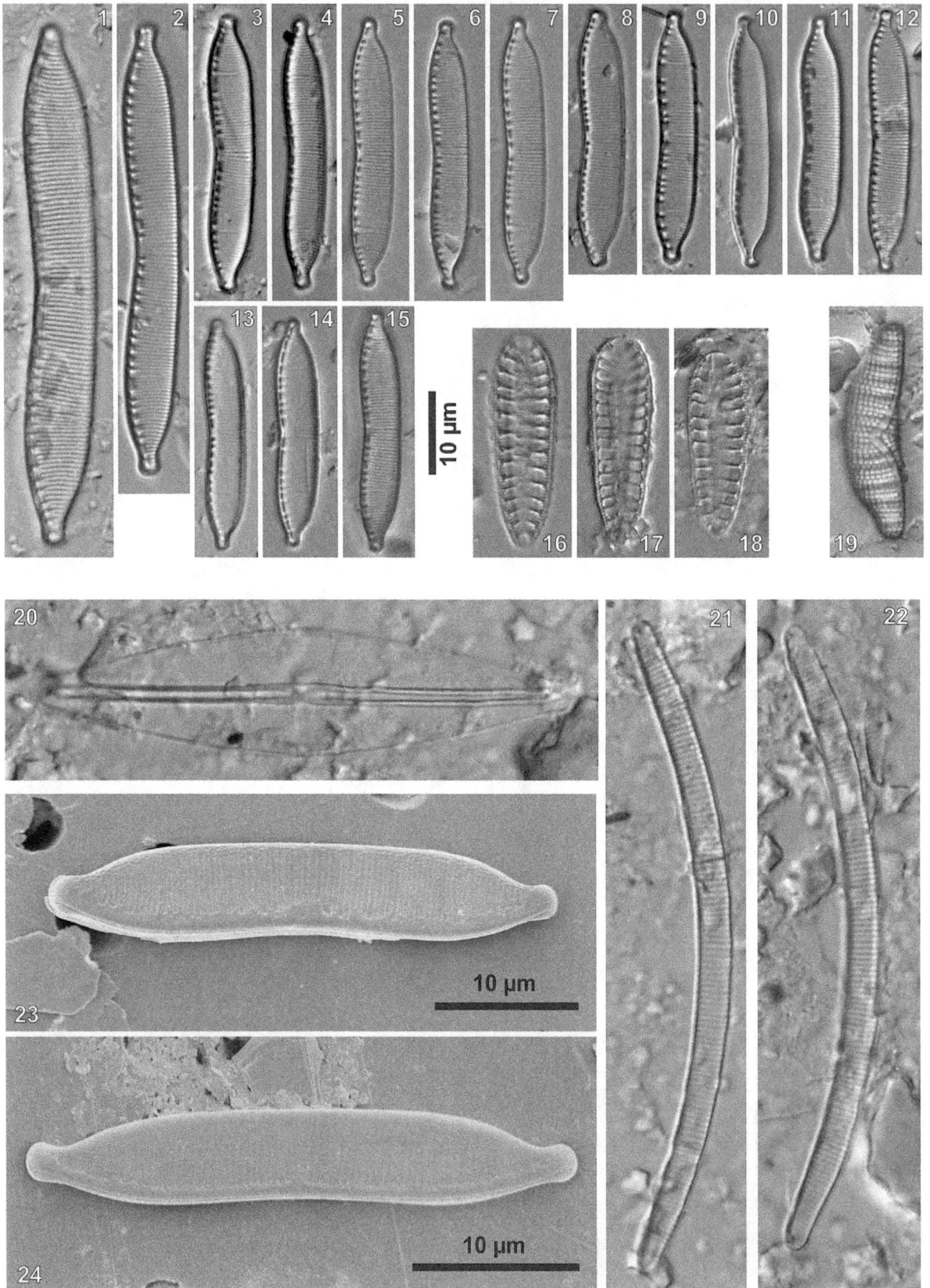

Fig. 2. Selected diatom taxa: 1 – *Hantzschia subrupestris*, 2 – *H. calcifuga,* 3–15 and 23–24 – *H. amphioxys*, 16–18 – *Surirella minuta*, 19 – *Epithemia adnata*, 20 – *Frustulia saxonica*, 21–22 – *Eunotia bilunaris*

Table 3. The number of diatoms frustules with the number of taxa and dominants found at different heights above ground level in particular research seasons. for the dominants, the total number of frustules counted in the sample was given

Date	03.2016		03.2017		06.2017		08.2017
High above soil level	20 cm	150 cm	20 cm	150 cm	20 cm	150 cm	20 cm
Number of all frustules counted in sample	224	22	292	31	247	21	293
Total number of taxa	29	6	20	12	19	10	23
Number of dominant taxa in sample	1	0	1	0	3	0	3
Hantzschia amphioxys	43	9	144	13	54	3	103
Luticola acidoclinata	97	0	22	0	79	10	70
Pinnularia borealis	43	7	39	1	78	1	68

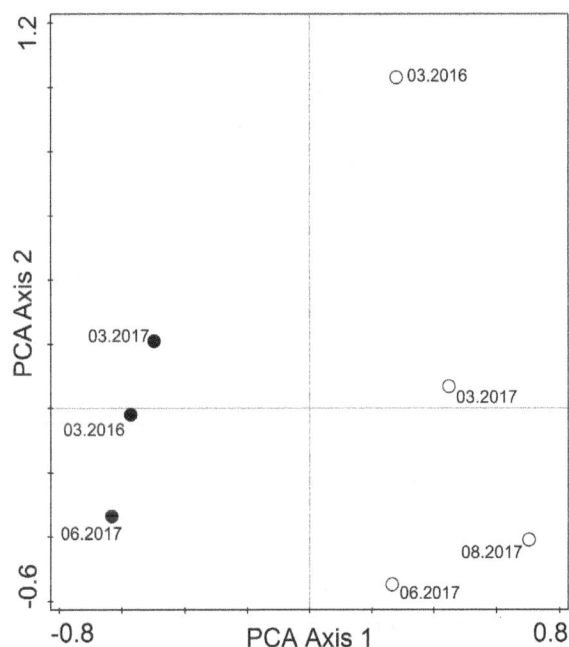

Fig 3. PCA ordination of sampling sites determined by relative community composition. Clusters represent sites of similarity based on height of sampling on three bark (black points – 150 cm, white points – 20 cm)

from Poland: *Hantzschia subrupestris, Luticola sparsipunctata* and *L. vanheurckii*.

DISCUSSION

The values of chemical parameters measured in the filtrates of the white poplar bark showed large fluctuations throughout the year. The electrolytic conductivity, the ions (especially nutrients) and pH decreased significantly in summer in comparison to the early spring in 2016 and 2017. All chemical parameters had the lowest values in August. This is probably the result of nutrients leaching by rainwater (from retained dust and decay of the bark).

The bark of trees, due to its porosity, absorbs rainwater and in this way undergoes acidification. The degree of acidification depends on the bark structure, which is different depending on the tree species (Zimny 2006). The pH characterizing the bark of the *Acer platanoides* L., *Fraxinus excelsior* L. and *Tilia cordata* Mill. in clean areas is usually subneutral (4.9–7.5), while in the *Betula pendula* Roth, *Picea abies* (L.) H. Karst. and *Pinus sylvestris* L. is generally acidic (Wirth 1995). Štifterová and Neustupa (2015) measured the pH of the bark in KCl and obtained the highest values (about 5.5) for the maple (*A. platanoides, A. pseudoplatanus* L.), while Marmor and Randlane (2007) measured the bark pH of the *Tilia cordata* (also in KCL), which resulted in the pH range from 4.1 to 5.5.

Epicolous bryophytes also have specific preferences related to the chemical properties of a bark (Bates 2009). Typical epiphytic species inhabit the bark of deciduous trees, while the acidic bark of conifers is inhabited only at the base by forest floor species (Rydin 2009).

Only two species of mosses were found during the study, which can be considered as epiphytic. Both have overgrown bark at a height of 150 cm above soil level. *Platygyrium repens* (Brid.) Schimp. is a moss occurring mainly on the bark of trees and especially on rotting wood, while *Orthotrichum speciosum* Nees. is a typical epiphytic species associated with the bark of deciduous trees (Dierßen 2001, Ellenberg, Leuschner 2010).

The moss-inhabiting diatom assemblages were characterized by a high proportion of aerophytic species, which are able to grow outside the typical aquatic habitats. Many species from the genera: *Hantzschia, Luticola* and *Mayamaea,* as well as species: *Muelleria gibbula, Pinnularia borealis, P. obscura* or *Stauroneis thermicola* are usually noted in soils and very often dominated the diatom flora (Stanek-Tarkowska, Noga 2012a,b, Levkov et al. 2013, Stanek-Tarkowska et al. 2013, 2015, 2016, Barragán et al. 2017).

Hantzschia amphioxys and *Pinnularia borealis* form the largest populations in terrestrial environments and have wide ecological preferences. They are typical aerophytic and soil species, which can also develop besides aquatic environment, i.e. on moist walls, in wet rock crevices and among mosses (Bąk et al. 2012, Hofmann et al. 2011, Lange-Bertalot et al. 2017). *Pinnularia borealis* var. *borealis* is a cosmopolitan species that can be often widely distributed by winds (Krammer 2000). In the studied material, *H. amphioxys* formed the most numerous populations and dominated the assemblages at the base of the trunk in almost every season, while *P. borealis* dominated the diatom flora only in the summer. In Podkarpacie region both species were noted in many rivers and streams, but always as single specimens (Noga et al. 2014), while on soils they are numerous and often dominant species (Stanek-Tarkowska, Noga 2012a,b, Stanek-Tarkowska et al. 2013, 2015, 2016).

The third dominant species – *Luticola acidoclinata* occurs in oligotrophic waters with circumneutral to slightly acidic. Formerly, it was reported as *Luticola mutica* var. *intermedia* (Hustedt) Hustedt (Hofmann et al. 2011, Bąk et al. 2012, Lange-Bertalot et al. 2017). Levkov et al. (2013) frequently reported populations of *L. acidoclinata* from oligotrophic, slightly acidic waters (springs, small rivers and peat bogs) and subaerial habitats, usually as epiphytic on mosses. *Luticola acidoclinata* is noted in the Red list of Algae in Poland as a rare species (R category) (Siemińska et al. 2006). In the Podkarpacie region, the species is found always as single specimens in the upper sections of small rivers and streams (Noga et al. 2014). The conducted study showed that in the aerophytical environment (among the mosses growing on the tree bark), *L. acidoclinata* developed in large numbers and dominated the diatom assemblages in three seasons (it was the only one dominant in March 2016). This species is also often observed on soils in southern Poland, especially on meadows, pastures, fallow lands, etc., which are overgrown with varying degrees by mosses (Poradowska, personal communication). All the information mentioned above suggests that *L. acidoclinata* is a rare species only in waters, but has the optimum occurrence in aerophytical environments, especially among mosses.

The remaining species from the R (rare) category were noted individually. Both *Achnanthes coarctata* and *Stauroneis thermicola* are aerophytic species that do not form numerous populations in aquatic environments. They are found in periodically dry habitats, especially on wet rocks and mosses (Hofmann et al. 2011, Bąk et al. 2012, Lange-Bertalot et al. 2017). In the Podkarpacie region, both species were reported individually in different types of waters (Noga et al. 2014), while *Stauroneis thermicola* developed numerously also in soils, where it was often one of the main dominant species. Therefore, it was considered as the soil species (Stanek-Tarkowska et al. 2013).

In the studied material, three species that have not previously been reported from Poland, were found. Among them, only *Luticola sparsipunctata* occurred in all of the studied seasons, but not as the dominant species. The other two species *Hantzschia subrupestris* and *Luticola vanheurckii* were found as single specimens only in the spring seasons.

So far, *Luticola sparsipunctata* is known only from type locality in the Czech Republic and Austria (Levkov et al. 2013) and from caves in Hawaii (Miscoe et al. 2016). It was described as a new species in 2013 (Levkov et al. 2013), therefore its ecology is still poorly known.

Luticola vanheurckii is also known only from the type locality in Belgium (Levkov et al. 2013). It was firstly observed in Poland in a small puddle in the city of Stalowa Wola – a few kilometres from the presented site (Noga, Rybak – in press). In the mosses collected from white poplar, only one valve was found.

In the spring of 2016, few *Hantzschia subrupestris* frustules were noted both at the base of the trunk and at a height of 150 cm. So far, *H. subrupestris* is known only from a few localities, mainly from Europe (Lange-Bertalot 1993, 1996, Lange-Bertalot et al. 2003, Denys, Oosterlynck 2015, Veen et al. 2015), less often from North America (Bahls 2009).

Epicolous mossy sites, as habitat diatoms, have so far been poorly studied. Both rare and well-known diatom species can develop in this habitat. Usually, this kind of habitat is dominated by the aerophytic diatoms, which are resistant to lack of water and changing nutrient levels. These species also often occur in soils and on wet rocks. For this reason, the studies on diatom assemblages from arboreal mosses can extend the knowledge about the ecology, distribution and adaptability of many diatom species.

Acknowledgement

The author wish to thank Prof. Andrzej Massalski for language correction of the manuscript.

REFERENCES

1. Bahls L.L. 2009. A checklist of diatoms from inland waters of the Northwestern United States. Proceedings of the Academy of Natural Sciences of Philadelphia, 158(1), 1–35.

2. Barragán C., Wetzel C.E., Ector L. 2017. A standard method for the routine sampling of terrestrial diatom communities for soil quality assessment. J. Appl. Phycol., https://doi.org/10.1007/s10811–017–1336–7.

3. Bates J.W. 2009. Mineral nutrition and substratum ecology. In: B. Goffinet B., A.J. Shaw (Eds). Bryophyte Biology 2nd edition, Cambridge University Press, Cambridge, pp. 299–357.

4. Bąk M., Witkowski A., Żelazna-Wieczorek J., Wojtal A.Z., Szczepocka E., Szulc A. & Szulc B. 2012. The guide to identyfication diatoms in phytobenthos for the purpose of assessing the ecological status of surface waters in Poland. Biblioteka Monitoringu Środowiska. pp. 1–452. Warszawa, (in Polish).

5. Denys L., Oosterlynck P. 2015. Diatom assemblages of non–living substrates in petrifying Cratoneurion springs from lower Belgium. Fottea, 15(2), 123–138.

6. Dierβen K. 2001. Distribution, ecological amplitude and phytosociological characterization of European bryophytes. Bryophytorum Bibliotheca, 56, 1–289.

7. Ellenberg H., Leuschner C. 2010. Vegetation Mitteleuropas mit den Alpen. Ulmer, Stuttgart, pp 1357.

8. Foerester J.W. 1971. The ecology of an elfin forest in Puerto Rico. 14. The algae of Pico del Oeste. Arnold Arboretum, 52, 86–109.

9. Geissler U., Kusber W-H., Jahn R. 2006. The diatom flora of Berlin (Germany): A spotlight on some documented taxa as a case study on historical biodiversity. In: A. Witkowski (Ed.) Eighteenth International Diatom Symposium 2004 Międzyzdroje, Poland, pp. 91–105, Biopress Limited, Bristol. 3.

10. Glime J. M. 2017. Chapter 1 – The Fauna: A Place to Call Home. Bryophyte Ecology Volume 2: Bryological Interaction, 1, 1–16.

11. Gremmen N., Van De Vijver B., Frenot Y., Lebouvier M. 2007. Distribution of moss-inhabiting diatoms along an altitudinal gradient at sub-Antarctic Îles Kerguelen. Antarctic Science, 19(1), 17–24.

12. Hickman M., Vitt D.H. 1973. The aerial epiphytic diatom flora of moss species from subantatctic Campbell Island. Nova Hedwigia, 24, 443–458.

13. Hofmann G., Werum M., Lange-Bertalot H. 2011. Diatomeen im Süßwasser – Benthos von Mitteleuropa. Bestimmungsflora Kieselalgen für die ökologische Praxis. Über 700 der häugfisten Arten und ihre Ökologie. [In:] H. Lange-Bertalot, (Ed.): 908 pp. A.R.G. Gantner Verlag K.G., Ruggell.

14. Kawecka B., Eloranta P.V. 1994. Basics of the ecology of freshwater algae and terrestrial environments. PWN, Warszawa, 256 pp. (in Polish).

15. Kharkongor D., Ramanujam P. 2014. Diversity and species composition of subaerial algal communities on forested areas of Meghalaya, India. Hindawi Publishing Corporation International Journal of Biodiversity, Vol. 2014, Article ID 456202, 10 pages, http://dx.doi.org/10.1155/2014/456202.

16. Krammer K. 2000. The genus *Pinnularia*. [In:] H. Lange-Bertalot (Ed.): Diatoms of Europe. Vol. 1., 703 pp. A.R.G. Gantner Verlag K.G.

17. Krammer K., Lange-Bertalot H. 1986. Bacillariophyceae. 1. Naviculaceae. In: H. Ettl, J. Gerloff, H. Heyning, D. Mollenhauer (Eds): Süsswasserflora von Mitteleuropa 2(1), 876 pp. G. Fischer Verlag, Stuttgart – New York.

18. Krammer K., Lange-Bertalot H. 1988. Bacillariophyceae. 2. Bacillariaceae, Epithemiaceae, Surirellaceae. In: H. Ettl, J. Gerloff, H. Heyning, D. Mollenhauer (Eds): Süsswasserflora von Mitteleuropa 2(2), 596 pp. G. Fischer Verlag, Stuttgart – New York.

19. Krammer K., Lange-Bertalot H. 1991a. Bacillariophyceae. 3. Centrales, Fragilariaceae, Eunotiaceae. In: H. Ettl, J. Gerloff, H. Heyning, D. Mollenhauer (Eds): Süsswasserflora von Mitteleuropa 2(3), 576 pp. G. Fischer Verlag, Stuttgart – Jena.

20. Krammer K., Lange-Bertalot H. 1991b. Bacillariophyceae. 4. Achnanthaceae, Kritische Ergänzungen zu *Navicula* (Lineolate) und *Gomphonema*, Gesamtliteraturverzeichnis. In: Ettl, H., Gerloff, J., Heyning, H., Mollenhauer, D. (Eds): Süsswasserflora von Mitteleuropa 2(4), G. Fischer Verlag, Stuttgart – Jena, 437 pp.

21. Lakatos M., Lange-Bertalot H., Büdel B. 2004. Diatoms living inside the thallus of the green algal lichen *Coenogonium linkii* in neotropical lowland rain forests. Journal of Phycology, 40, 70–73.

22. Lange-Bertalot H. 1993. 85 neue Taxa und über 100 weitere neu definierte Taxa ergänzend zur Süsswasserflora von Mitteleuropa, Vol. 2/1–4. Bibliotheca Diatomologica, 27, 1–164.

23. Lange-Bertalot H. 1996. Rote liste der limnischen Kieselalgen (Bacillariophyceae) Deutschlands. Schriftenreihe für Vegetationskunde, 28, 633–677.

24. Lange-Bertalot H., Cavacini P., Tagliaventi N., Alfinito S. 2003. Diatoms of Sardinia: Rare and 76 new species in rock pools and other ephemeral wa-

ters. Iconographia Diatomologica, 12, 1–438.

25. Lange-Bertalot H., Hofmann, Werum M., Cantonati M. 2017. Freshwater benthic diatoms of Central Europe: over 800 common species used in ecological assessments. English edition with updated taxonomy and added species. In: M. Cantonati et al. (Eds). Koeltz Botanical Books, Schmitten-Oberreifenberg, 942 pp.

26. Levkov Z., Metzeltin D., Pavlov A. 2013. *Luticola* and *Luticolopsis*. In: H. Lange-Bertalot, (Ed.), Diatoms of Europe. Vol. 7. Köningstein/Germany: Koeltz Scientific Books, 698 pp.

27. Lindo Z., Gonzales A. 2010. The Bryosphere: An Integral and Influential Component of the Earth's Biosphere. Ecosystems, 13, 612–627.

28. Marmor L., Randlane T. 2007. Effects of road traffic on bark pH and epiphytic lichens in Tallinn. Folia Cryptog. Estonica, Fasc., 43, 23–37.

29. Miscoe, L.H., Johansen, J.R., Kociolek, J.P., Lowe, R.L. 2016. The diatom flora and cyanobacteria from caves on Kauai, Hawaii. I. Investigation of the cave diatom flora of Kauai, Hawaii: an emphasis on taxonomy and distribution. Bibliotheca Phycologica, 123, 3–74.

30. Mrozińska T. 1990. Aerophitic algae from North Korea. Algological Studies, 58, 28–47.

31. Neustupa J. 2003. The genus *Phycopeltis* (Trentepohliales, Chlorophyta) from tropical Southeast Asia. Nova Hedwigia, 76, 487–505.

32. Neustupa J. 2005. Investigations on genus *Phycopeltis* (Trentepohliaceae, Chlorophyta) from South-East Asia, including the description of two new species. Cryptogamia Algologica, 26, 229–242.

33. Neustupa J., Škaloud P. 2008. Diversity of subaerial algae and cyanobacteria on tree bark in tropical mountain habits. Biologia, 63(6), 806–812.

34. Neustupa J., Škaloud P. 2010. Diversity of subaerial algae and cyanobacteria growing on bark and wood in the lowland tropical forests of Singapore. Plant Ecology and Evolution, 143(1), 51–62.

35. Noga T., Kochman N., Peszek Ł., Stanek-Tarkowska J., Pajączek A. 2014. Diatoms (Bacillariophyceae) in rivers and streams and on cultivated soils of the Podkarpacie Region in the years 2007–2011. Journal of Ecological Engineering, 15(1), 6–25.

36. Noga T., Stanek-Tarkowska J., Rybak M., Kochman-Kędziora N., Peszek Ł., Pajączek A. 2016. Diversity of diatoms in the natural, mid-forest Terebowiec stream – Bieszczady National Park. Journal of Ecological Engineering, 17(4), 232–247.

37. Nováková J., Poulíčková A. 2004. Moss diatom (Bacillariophyceae) flora of the Nature Reserve Adršpašsko-Teplické Rocks (Czech Republic). Czech Phycology, Olomouc, 4, 75–86.

38. Ochyra, R., Żarnowiec, J., Bednarek-Ochyra H. 2003. Census Catalogue of Polish mosses. Polish Academy of Sciences, Institute of Botany, Kraków, 372 pp.

39. Plášek V. 2013. Bryophytes in forests. Field guide for foresters and valuators. Centrum Informacyjne Lasów Państwowych, Warszawa, 130 pp. (in Polish).

40. Podbielkowski Z. 1996. Algae – Glony. WSiP, Warszawa, 215 pp. (in Polish).

41. Poulíčková A., Hájková P., Krenková P., Hájek M. 2004. Distribution of diatoms and bryophytes on linear transects through spring fens. Nova Hedwigia, 78(3–4), 411–424.

42. Qin B., Zheng M., Chen X., Yang X. 2016. Diatom composition of epiphytic bryophytes on trees and its ecological distribution in Wuhan City. Chinese Journal of Ecology, 35(11), 2983–2990.

43. Round F.E. 1957. The diatom community of some Bryophyta growing on sandstone. Botanical Journal of the Linnean Society, 55(362), 657–661.

44. Rydin H. 2009. Population and community ecology of bryophytes In: B. Goffinet B., A.J. Shaw (Eds). Bryophyte Biology 2nd edition, Cambridge University Press, Cambridge, p. 393–444.

45. Salleh A., Milow P. 1999. Notes on *Trentepohlia dialepta* (Nylander) Hariot (Trentepohliaceae, Chlorophyta) and sporangia of some other species of Trentepohlia Mart. from Malaysia. Micronesica, 31, 675–692.

46. Siemińska J., Bąk M., Dziedzic J., Gąbka M., Gregorowicz P., et al. 2006. Red list of the algae in Poland – Czerwona lista glonów w Polsce. In: Z. Mirek et al. (Eds) Red list of plants and fungi in Poland – Czerwona lista roślin i grzybów Polski. Polish Academy of Sciences, Kraków, 35–52.

47. Schmidt J., Kricke R., Feige G.B. 2001. Measurements of bark pH with a modifiedflathead electrode. Lichenologist, 33, 456–460.

48. Smith A.J.E. 2004. The Moss Flora of Britain and Ireland 2nd edition. Cambridge University Press, Cambridge, New York, Melbourne, Madrid, Cape Town, Singapore, São Paulo, pp. 1012.

49. Stanek-Tarkowska J., Noga T. 2012a. The diatoms communities developing on dust soils under sweet corn cultivation in Podkarpackie region. Fragmenta Floristica et Geobotanica Polonjca, 19(2), 525–536.

50. Stanek-Tarkowska J., Noga T. 2012b. Diversity of diatoms (Bacillariophyceae) in the soil under traditional tillage and reduced tillage. Inżynieria Ekologiczna, 30, 287–296.

51. Stanek-Tarkowska J, Noga T, Pajączek A, Peszek Ł. 2013. The occurrence of *Sellaphora nana* (Hust.) Lange-Bert. Cavacini, Tagliaventi and Alfinito, *Stauroneis borrichii* (J.B. Petersen) J.W.G. Lund, *S. parathermicola* Lange-Bert. and *S. thermicola* (J.B. Petersen) J.W.G. Lund on agricultural

soils. Algological Studies, 142, 109–120.

52. Stanek-Tarkowska J., Noga T., Kochman-Kędziora N., Peszek Ł., Pajączek A., Kozak E. 2015. The diversity of diatom assemblages developed on fallow soil in Pogórska Wola (Southern Poland). Acta Agrobotanica, 68(1), 33–42.

53. Stanek-Tarkowska J., Noga T., Kochman-Kędziora N., Rybak M. 2016. Diatom assemblages growing on cropping soil in Pogórska Wola near Tarnów. Inżynieria Ekologiczna, 46, 128–134 (in Polish).

54. Štifterová A., Neustupa J. 2015. Community structure of corticolous microalgae within a single forest stand: evaluating the effects of bark surface pH and tree species. Fottea, 15(2), 113–122.

55. Ter Braak C.J.F., Šmilauer P. 2012. Canoco reference manual and user's guide. Software for ordination (version 5.0). Microcomputer Power, Ithaca, NY, USA. 496 pp.

56. Thompson R.H., Wujek D.H. 1997. Trentepohliales: *Cepaleuros*, *Phycopeltis* and *Stomatochroon*. Morphology, taxonomy and ecology. Enfield, Science Publishing.

57. Van de Vijver B., Beyens L. 1997. The epiphytic diatom flora of mosses from Strømnes Bay area, South Georgia. Polar Biology, 17(6), 492–501.

58. Veen A., Hof C.H.J., Kouwets F.A.C. & Berkhout T. 2015. Taxa Watermanagement the Netherlands (TWN) [Rijkswaterstaat Waterdienst, Informatiehuis Water] http://ipt.nlbif.nl/ipt/resource?r=checklist-twn.

59. Wirth V. 1995. Die Flechten Baden-Württembergs, Teil 1. Stuttgart, Verlag Eugen Ulmer, 527 pp.

60. Zimny H. 2006. Ecological evaluation of the state of the environment. Bioindication and biomonitoring. ARW A. Gregorczyk, Warszawa, 264 pp. (in Polish).

13

The Effect of Different Rates of Biochar and Biochar in Combination with N Fertilizer on the Parameters of Soil Organic Matter and Soil Structure

Martin Juriga[1], Vladimír Šimanský[1*], Ján Horák[2], Elena Kondrlová[2],
Dušan Igaz[2], Nora Polláková[1], Natalya Buchkina[3], Eugene Balashov[3]

[1] Department of Soil Science, Faculty of Agrobiology and Food Resources, Slovak University of Agriculture, Tr. A. Hlinku 2, 949 76 Nitra, Slovakia

[2] Department of Biometeorology and Hydrology, Horticulture and Landscape Engineering Faculty, Slovak University of Agriculture, Hospodárska 7, 949 01 Nitra, Slovakia

[3] Agrophysical Research Institute, 14 Grazhdansky prospekt, St. Petersburg, 195220, Russia

* Corresponding author's e-mail: vladimir.simansky@uniag.sk

ABSTRACT

Since biochar is considered to be a significant source of carbon, in this work we have evaluated the changes in soil organic matter (SOM) and soil structure due to application of biochar and biochar with N fertilization, and have considered the interrelationships between the SOM parameters and the soil structure. The soil samples were collected from Haplic Luvisol at the locality of Dolná Malanta (Slovakia) during 2017. The field experiment included three rates of biochar application (B0 – no biochar, B10 – biochar at the rate of 10 t ha^{-1}, B20 – biochar at the rate of 20 t ha^{-1}) and three levels of N fertilization (N0 – no nitrogen, N160 – nitrogen at the rate of 160 kg ha^{-1}, N240 – nitrogen at the rate of 240 kg ha^{-1}). The rate of biochar at 20 t ha^{-1} caused an increase in the organic carbon (C_{org}) content. The combination of both rates of biochar with 160 and 240 kg N ha^{-1} also caused an increase in C_{org}. In the case of B20 the extractability of humic substances carbon (C_{HS}) was 17.79% lower than at B0. A significant drop was also observed in the values of the extraction of humic acids carbon (C_{HA}) and fulvic acids carbon (C_{FA}) after the addition of biochar at a dose of 20 t ha^{-1} with 160 kg N ha^{-1}. However, both rates of biochar had a significant effect at 240 kg N ha^{-1}. After application of 20 t ha^{-1} of biochar the content of water-stable macro-aggregates (WSA_{ma}) significantly increased compared to control. This rate of biochar also increased the mean weight diameter (MWD_W) and the index of water-stable aggregates (Sw) and decreased the coefficient of vulnerability (Kv). The biochar at a rate of 20 t ha^{-1} with 240 kg N ha^{-1} the value of MWD_W increased and value of Kv decreased significantly. The contents of C_{org} and C_L correlated positively with WSA_{ma}, MWD_W and Sw and negatively with WSA_{mi} and Kv. The extraction of C_{HA} and C_{FA} was in negative relationship with MWD_W. We conclude that the application of biochar and biochar combined with N fertilizer had a positive influence on SOM and soil structure.

Keywords: biochar; soil organic matter; soil structure; nitrogen fertilizer

INTRODUCTION

Soil structure is one of the most important soil properties. It is defined as a spatial arrangement of soil particles with pores among them [Odes 1993]. According to Blanda et al. [2014] soil aggregates are basic soil structural units that control the dynamics of soil organic matter (SOM) and influence the soil's ability to sequestrate and stabilize organic carbon. Conceptually, aggregates are generally classified into macro-aggregates (>0.25 mm) and micro-aggregates (<0.25 mm) [Six et al. 2000]. The stability of soil aggregates is one of the most important elements of soil protection and conservation of its functions. Soil resistance to erosive agents and compaction increases with improvement of aggregate stability [Chaplot and Cooper 2015]. Aggregation is the result of reorganisation, flocculation and cementation of soil particles. As reported by Bronick and Lal [2005], aggregation is influenced by a number of factors. First example is when oxides of iron and aluminium can act as inorganic binders and second is when extracts of roots, fungal hyphae, bacteria and soil fauna are considered as cementing agents which connect soil particles into stable macro-aggregates [Odes 1993, Tisdall 1996].

SOM is the most important indicator of soil quality because of its effects on wide range of soil particles. It is considered to be a key element in the stabilization of soil aggregates. The dynamic of SOM is related to the formation and destruction of macro-aggregates. Soil aggregates control the dynamics of SOM [Six et al. 2000] which is the main source of soil organic carbon (SOC). SOC improves aggregation by bonding soil particles together and its effect depends on the rate of its decomposition. The intensification of agriculture has led to a significant decrease of organic matter content in agricultural soils in the last decades of the twentieth century. It has deteriorated the soil structure and the soil fertility [Bossuyt et al. 2004]. The ways how to enhance the stock of SOM or SOC in arable soils, are still being researched [Sainju et al. 2009]. A good option to increase carbon sequestration in the soil is production and application of biochar. The intentional and unintentional addition of biochar into soils, known as „Terra preta,“ has promoted soil fertility. These soils are among the most significant examples of biochar enriched soils by humans [Wang et al. 2016].

Biochar is a solid, C-rich product [Fisher and Glaser 2012] that arises during the thermal decomposition of different organic material in conditions with low or no oxygen. The properties of biochar mainly depend on the type of material used for its production and on the temperature of pyrolysis [Ahamd et al. 2014, Zlielinska et al. 2015]. Firstly, the biochar produced from manure usually has smaller surface area, than biochar produced from wood. Secondly, the higher temperature increases the content of carbon in biochar while the content of oxygen and hydrogen decreases. Biochar has the potential to enhance the chemical, physical and biological properties of soil [Hussian et al. 2016]. The addition of biochar can increase cation exchange capacity (CEC) and pH. Biochar can absorb nutrients but also heavy metals due to its high porosity and the presence of carboxyl and hydroxyl groups [Glaser et al. 2002, Joseph 2009]. In addition, biochar can increase soil porosity, reduced soil bulk density and improve soil retention capacity [Abel et al. 2013, Omondi et al. 2016]. The increase of soil aggregate stability and the content of macro-aggregates

have been shown in several studies [Zhang et al. 2015, Liu et al. 2014, Sun and Lu 2014].

Since biochar is material rich in C we expected that its application would increase the content of labile C in soil and the increase will correlate with the application rate of biochar which would also improve the condition of soil structure. The aim of this study was to evaluate: (i) the effects of different rates of biochar and biochar with N fertilizer on the parameters of soil organic matter, (ii) the parameters of the soil structure, (iii) the inter-relationships between measured parameters of the soil organic matter and the soil structure affected by biochar and biochar with N.

MATERIAL AND METHODS

The characteristics of the territory

The study was carried out at the experimental site of Slovak University of Agriculture in the Nitra region of Slovakia (Dolná Malanta, lat.48°10‘00“, lon.18°19‘00“). The area is located about 4 km from Nitra and it has flat terrain properties with a slight southwestern slope. From a geological point of view, the territory is located on the geological boundary of the mountain range of Tribeč and Danubian Lowland with an altitude 175–180 m a.s.l. The soil is classified as Haplic Luvisol and it has neutral pH (6.69). The territory is located in a warm agro-climatic zone with a mean annual air temperature of 10.2°C. The mean total annual rainfall is 539 mm.

The description of the experiment

The experiment was established in March 2014. The crop rotation consisted of spring barley (2014), corn (2015), spring wheat (2016) and corn (2017). In 2017 following treatments were used: 1. B0 – without biochar, 2. B10 –10 t ha^{-1} of biochar, 3. B20 – 20 t ha^{-1} of biochar, 4. B0N160 – without biochar but with 160 kg N ha^{-1}, 5. B10N160 – 10 t ha^{-1} of biochar and 160 kg N ha^{-1}, 6. B20N160 – 20 t ha^{-1} of biochar and 160 kg N ha^{-1}, 7. B0N240 – without biochar but with 240 kg N ha^{-1}, 8. B10N240 – 10 t ha^{-1} of biochar and 240 kg N ha^{-1}, 9. B20N240 – 20 t ha^{-1} of biochar and 240 kg N ha^{-1}. The LAD 27 was used as N fertilizer. The used biochar was made of grain husks and paper sludge at a ratio of 1:1 at the temperature of 500 °C. The composition of the biochar and its properties are shown in Table 1.

Table 1. Basic composition and properties of applied biochar

Ca	57 g kg^{-1}
Mg	3,9 g kg^{-1}
K	15 g kg^{-1}
N	0.7 g kg^{-1}
Total C	53.1%
Total N	1.4%
Ash	38.3%
pH	8.8
Size of biochar	1 – 5 mm
Surface area	21.7 m^{-2} g^{-1}

The collection of samples and analytical methods

The soil samples were collected in 2017 at monthly intervals from the beginning (April) to the end (September) of the corn growing season (from 38th to 43rd month since biochar application). The soil samples were collected from the depth 0–30 cm. Subsequently, roots and other large parts of plants were removed from the samples, they were transferred to the laboratory and air dried at room temperature. The fallowing parameters of soil organic matter and soil structure were evaluated:

- content of water-stable macro- (WSA$_{ma}$) and micro-aggregates (WSA$_{mi}$) (fractions: >5 mm, 5–3 mm, 3–2 mm, 2–1 mm, 1–0.5 mm, 0,5–0.25 mm a <0.25 mm) by Bachsayev´s method [Hraško et al. 1962],
- mean weight diameter of macro-aggregates (MWDd), which was calculated from the percentage representation of individual fractions of structural macro-aggregates obtained by sifting through a set of sieves:

$$MWD_d = \sum_{i=1}^{n} x_i w_i \qquad (1)$$

where:
  $i – 1, 2, 3 ....n$ – corresponds to each determined fraction
  xi – weighted average of the size fraction
  wi – percentage of sample on sieve,

- mean weight diameter of water-resistant macro-aggregates (MWD$_W$), which was calculated from the percentage of the individual fraction of water-resistant macro-aggregates obtained by sifting through a set of sieves in distilled water:

$$MWD_W = \sum_{i=1}^{n} x_i WSA \qquad (2)$$

where:
  $i – 1, 2, 3 ....n$ a– corresponds to each determined fraction
  xi – weighted average fraction size (mm)
  WSA – water-resistant aggregates,

- coefficient of vulnerability (Kv) [Valla et al. 2000] according to the equation:

$$K_v = \frac{MWD_d}{MWD_w} \qquad (3)$$

- index of water-resistant macro-aggregates (Sw) was calculated based on the grain composition and the percentage representation of water-resistant macro-aggregates:

$$Sw = \frac{WSA - 0.09 sand}{silt + clay} \qquad (4)$$

where:
  WSA – % content of water-resistant aggregates,

- content of total organic carbon (C$_{org}$) was determined oxidometrically [Dzadowiec and Gonet 1999a],
- content of labile carbon (C$_L$) [Loginow et al. 1987],
- group composition of humic substances [Dzadowiec and Gonet 1999b],
- colour quotient of humic substances (Q$_{HS}$) and colour quotient of humic acids (Q$_{HA}$).

Statistical analysis

The individual parameters of organic matter and soil structure were evaluated by statistical analysis through the Statgraphic Centurion XV Program I. (Statpoint Technologies, Inc., USA) using ANOVA single-factor analysis. LSD test with a significance level of $\alpha=0.05$ was used to compare the effect of biochar and N fertilization. The dependence between the parameters of soil structure and soil organic matter was evaluated by the correlation matrix.

RESULTS AND DISCUSSION

Total organic carbon (C$_{org}$) is an important indicator of SOM, which is widely used in the evaluation of the total amount of organic compounds in soils [Zhang et al. 2005, Visco et al. 2005]. Due to its predominantly aromatic structure, biochar is a relatively stable form of C [Pasakayastha et al. 2015]. Only a small part of biochar can be mineralized in the short time after application, especially if biochar is produced at lower pyrolysis temperatures [Mukome et al. 2015]. Our results indicate that the addition of biochar in overall increased the C_{org} content in the soil (Fig. 1) but the increase was statistically significant (28.69%) only with the higher biochar rate (B20) compared to the control (B0). When compared to B0N160 treatment, the content of C_{org} increased by 12.05% for B10N160 and by 23.59% for B20N160. Furthermore, the higher rate of biochar had stronger effect on C_{org} content in the soil than the lower rate of biochar (12.22% difference in the C_{org} content). Soil C_{org} content increased also when the combination of biochar (both rates) with 240 kg N ha^{-1} was used. The obtained value in B10N240 was 19.20% higher, while in B20N240 – 19.51% higher than in B0N240. The difference between the rates of biochar was not statistically significant. Similar findings have also been reported by Mavi et al. [2018] who recorded a significant C_{org} increase after the application of biochar and biochar with N, but only at higher doses of N (120 and 150 kg ha^{-1}). The lower rate of N (60 kg ha^{-1}) did not notably alter the content of C_{org}. Labile carbon (C_L) is an important component of SOM and it is considered to be a sensitive indicator of soil quality [Jiang and Xu 2006]. During pyrolysis, either labile or leached organic carbon is generated. These low-weight molecular organic compounds directly increase the content of C_L in soil. The addition of N into the soil reduces the content of carbon in organic matter due to a decrease of C:N ratio [Yan et al. 2007]. In conclusion, the addition of biochar alone or biochar with N did not have a significant effect on C_L content (Fig. 2).

Humic substances (HS) are considered to be stable fractions of soil organic matter. They represent heterogeneous components consisting of large macromolecules with functional groups formed by chemical and biochemical reactions. HS play an important role not only in soil fertility but also in the sequestration of C [Spaccini et al. 2002]. There is still lack of information about the effect of biochar on the chemical composition of HS. However, recent studies have shown that biochar can play an important role in the formation of HS [Jindo et al. 2016]. As a result of biochar and biochar with N application, the reduction in humic substances carbon (C_{HS}), humic acids carbon (C_{HA}) and fulvic acids carbon (C_{FA}) was observed (Table 2). Application of the higher rate of biochar resulted in the reduction of C_{HS} content by 17.79%, C_{HA} – by 15.98% and C_{FA} – by 19.69% compared to the control. After the addition of biochar with 160 kg N ha^{-1}, there was a significant decrease in C_{HS} content by 21.87%, C_{HA} – by 24.46% and C_{FA} – by 18.66% in B20N160 when compared to B0N160 treatment. But, when biochar with 240 kg N ha^{-1} was applied, the high effect was registered for both rates of biochar (B10N240 and B20N240): C_{HS} decreased by 23.32% and 24.43%, C_{HA} – by 15.61% and 23.75%, C_{FA} – by 31.77%

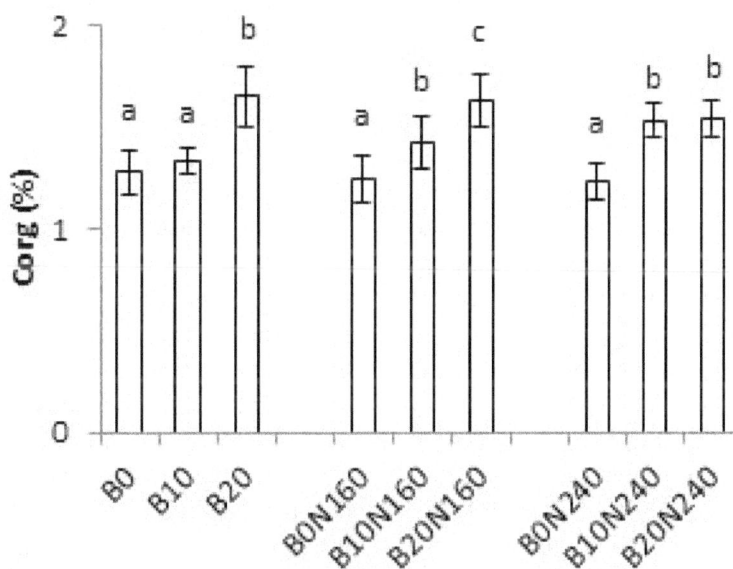

Figure 1. Contents of soil organic carbon. Treatments are stated in Material and methods. Different letters (a, b, c) between columna indicate that treatment means are significantly different at P<0.05 according to LSD test

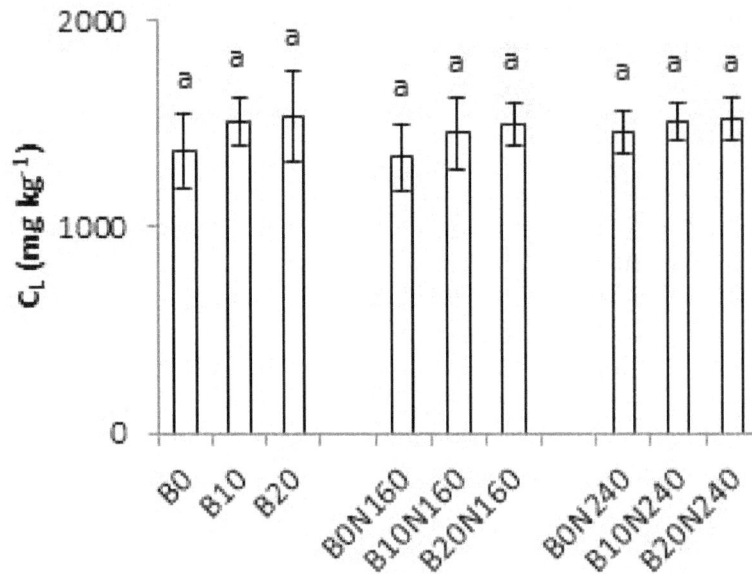

Figure 2. Contents of labile carbon. Treatments are stated in Material and methods. Different letters (a, b, c) between columns indicate that treatment means are significantly different at P<0.05 according to LSD test

Table 2. Statistic evaluation of the parameters of soil organic matter

Treatment	C_{HS}	C_{HA}	C_{FA}	$C_{HA}{:}C_{FA}$	Q_{HS}	Q_{HA}
	% from C_{org}					
B0	36.31±1.98[b]	19.59±2.01[b]	16.73±0.87[b]	1.18±0.14[a]	4.55±0.20[a]	3.72±0.17[a]
B10	36.81±3.82[b]	19.41±1.62[b]	17.40±2.86[b]	1.14±0.17[a]	4.37±0.20[a]	3.72±0.10[a]
B20	29.85+2.73[a]	16.46±2.49[a]	13.39±1.84[a]	1.25±0.25[a]	4.53±0.29[a]	3.78±0.10[a]
B0N160	39.65±3.45[c]	21.91±2.71[b]	17.74±2.67[b]	1.27±0.27[a]	4.60±0.27[a]	3.77±0.03[a]
B10N160	35.16±3.35[b]	19.78±2.00[ab]	15.98±3.22[ab]	1.25±0.31[a]	4.51±0.32[a]	3.80±0.13[a]
B20N160	30.98±2.11[a]	16.55±2.66[a]	14.43±1.91[a]	1.18±0.36[a]	4.45±0.28[a]	3.31±0.14[a]
B0N240	42.66±3.38[b]	22.36±1.70[b]	20.30±3.45[b]	1.13±0.24[a]	4.70±0.30[a]	3.87±0.15[a]
B10N240	32.71±1.47[a]	18.87±1.75[a]	13.85±1.57[a]	1.39±0.27[a]	4.52±0.27[a]	3.78±0.17[a]
B20N240	32.24±2.58[a]	17.05±2.12[a]	15.19±2.57[a]	1.16±0.30[a]	4.53±0.22[a]	3.79±0.15[a]

Treatments are stated in Material and methods. Different letters (a, b, c) between lines indicate that treatment means are significantly different at *P*<0.05 according to LSD test.

C_{HS} – content of humic substances carbon, C_{HA} – content of humic acids carbon, C_{FA} – content od fulvic acids carbon , $C_{HA}{:}C_{FA}$ – humic acids carbon to fulvic acids carbon ratio, Q_{HS} – colour qutient od humic substances, Q_{HA} – colour qoutient od humic acids

and 24.16%, respectively, compared to the control (B0N240). The significant differences were also identified between the rates of biochar after the application of biochar alone, where B20 treatment resulted in 18.91% (C_{HS}), 16.03% (C_{HA}) and 23.05% (C_{FA}) lower concentrations than B10. According to Zhao et al. [2017], the different effect of biochars on the content of C_{HA} and C_{FA} depended on the different pyrolysis temperatures during the biochar production (300°C, 400°C, 500°C, 600°C). All tested types of biochar initially had a beneficial effect on C_{HA} and C_{FA}, however after a short time a decrease of C_{HA} and C_{FA} was observed in the case of biochar produced at a lower temperature

(300°C, 400°C). An increase of soil microorganisms was stimulated by biochar application, which promoted the production of humic (HA) and fulvic acids (FA). Over time, parts of HA and FA are used by microorganisms as a result of the decline in the slightly mineralized sources of carbon. The humic acids carbon to fulvic acids carbon ratio ($C_{HA}{:}C_{FA}$) is the next of evaluated qualitative parameter of humus. The values greater that 1 characterise fertile soils [Rudkowska and Pikula 2013]. There was no any significant changes in $C_{HA}{:}C_{FA}$, colour quotient of humic substances (Q_{HS}) and colour quotient of humic acids (Q_{HA}) after the application of biochar or biochar with N (Table 2).

The average values of the soil structure parameters for the corn growing season [2017] are shown in Table 3. When evaluating the soil structure, the one of the most important parameters is content of water-stable macro-aggregates (WSA$_{ma}$). It represents the individual size groups of aggregates and their water resistance [Scott, 2000]. We concluded that the application of biochar and biochar in combination with N fertilizer had a positive effect on the WSA$_{ma}$ content. It increased with the application of biochar in the order B0<B10<B20. The application of biochar alone at a dose of 20 t ha^{-1} (B20) significantly increased the content of WSA$_{ma}$ by 10.41% compared to the control (B0). The content of macro-aggregates increased, while at the same time the content of water-stable micro-aggregates (WSA$_{mi}$) decreased with increasing rate of biochar or biochar with N. The significant effect was found only in the B20 treatment, where the value of WSA$_{mi}$ was 29% lower than in the control. According to Lu et al. [2014] the addition of a high rate of biochar increased the content of WSA$_{ma}$ by 31% compared to the control. On the other hand, Zhang et al. [2015] stated that the application of biochar had no significant effect on the value of WSA$_{ma}$. These results were obtained in the first year after the application of biochar. As Obia et al. [2016] wrote, a longer period was needed for the oxidation of applied biochar particles. The positive effect of fertilizer on soil aggregation has been proven in many studies [e.g. Chen et al. 2015; Wang et al. 2016; 2017]. Wang et al. [2014] found that N fertilization increased the content of macro-aggregates that were larger than 2 mm by 7% compared to control. However, the content of smaller macro-aggregates did not increase. In addition, the effect of fertilization was induced by an increase of root biomass and fungal hyphae in these size groups of aggregates. It corresponds to the statement by Bronick and Lal [2005] that plant roots and fungal hyphae are very important attributes for macro-aggregate formation. The mean weight diameter of aggregates gained by dry sieving (MWD) is a commonly used parameter to evaluate the stability of aggregates. It determines the representation of individual fractions of macro-aggregates and the extent of their stability [Amezketa 1999]. MWD$_d$ was not significantly altered by the application of biochar alone or biochar with N fertilizer in both rates of biochar and N fertilizer. Conversely, statistically significant changes were observed within the mean weight diameter of water-stable macro-aggregates gained by wet sieving (MWD$_W$). Compared to the controls, the value of this parameter increased after B20 treatment by 37.11% and after B20N240 treatment by 41.23%. No significant effect of the B20N160 or B20N240 treatments on MWD$_{WSA}$ was observed. Next, the coefficient of vulnerability (Kv) and the index of water-stable aggregates (Sw) were evaluated. The calculated values of Kv and Sw also confirmed that application of biochar and biochar with N had a beneficial effect on soil structure. The results showed that the fertilized treatments were characterized by a lower value of Kv and a higher value of Sw than the control. The higher rate of biochar decreased the value of Kv and increased value of Sw. However in the case of Kv, the decrease was significant only with B20N240 treatment (by 43.09%) compared to the control B0N240. In the case of Sw, a sig-

Table 3. Statistical evaluation of the soil structure parameters

Treatments	WSA$_{ma}$	WSA$_{mi}$	MWD$_d$	MWD$_W$	Kv	Sw
	(%)		(mm)			
B0	71.9±4.38[a]	28.1±4.38[b]	3.04±0.88[a]	0.61±0.09[a]	4.94±1.24[a]	0.83±0.05[a]
B10	78.3±7.33[ab]	21.8±7.33[ab]	3.02±0.76[a]	0.86±0.20[ab]	3.71±1.85[a]	0.91±0.09[ab]
B20	80.3±6.34[b]	19.7±6.34[a]	2.77±0.97[a]	0.97±0.34[b]	3.24±1.64[a]	0.93±0.07[b]
B0N160	71.3±9.11[a]	28.7±9.11[a]	2.67±0.46[a]	0.69±0.25[a]	4.31±1.72[a]	0.83±0.11[a]
B10N160	73.6±5.34[a]	26.3±5.34[a]	2.45±0.30[a]	0.76±0.22[a]	3.56±1.57[a]	0.85±0.06[a]
B20N160	79.3±10.94[a]	20.7±10.94[a]	2.74±0.47[a]	0.93±0.35[a]	3.47±1.75[a]	0.92±0.13[a]
B0N240	73.5±9.28[a]	26.5±9.28[a]	3.09±0.68[a]	0.67±0.20[a]	5.06±2.20[b]	0.85±0.11[a]
B10N240	75.3±9.43[a]	24.7±9.43[a]	2.77±0.38[a]	0.80±0.31[ab]	3.96±1.61[ab]	0.87±0.11[a]
B20N240	83.1±6.12[a]	16.9±6.12[a]	2.93±0.47[a]	1.14±0.35[b]	2.88±1.30[a]	0.96±0.07[a]

Treatments are stated in Material and methods. Different letters (a, b, c) between lines indicate that treatment means are significantly different at P<0.05 according to LSD test.
WSA$_{ma}$ – water-stable macro-aggregates, WSA$_{mi}$ – water-stable micro-aggregates, MWD$_d$ – mean weight diameter, MWD$_W$ – mean weight dimeter of water-stable aggregates, Kv – coefficient of vulnerability, Sw – index of water-stable aggregates.

nificant increase was observed in B20 (by 10.75%) compared to B0. The positive effect of biochar on soil aggregation has been demonstrated by several studies [Lu et al. 2014; Germida 2015; Obia et al. 2016; Sing and Cowie 2014]. Biochar can affect aggregation via many mechanisms. First, Glaser et al. [2002] pointed at the bonding of applied particles of biochar with soil particles by the carboxyl and hydroxyl groups that are present on the surface of biochar. Next, the biochar also increases the hydrophobicity of soil particles that results in the increase of aggregate stability [Lu et al. 2014]. Biochar promotes the development of soil microorganisms via various mechanisms, thereby; it also contributes to the increased formation and the stability of aggregates [Lehmann et al. 2011].

The correlations between the evaluated parameters of soil structure and soil organic matter are shown in Table 4. The organic matter is a key element in stabilizing soil aggregates. The dynamics of SOM is related to the formation and disintigration of macro-aggregates [Six et al. 2000]. The beneficial effect of SOM on the formation and stabilization of aggregates has been demonstrated in studies by several authors [Six et al. 2002; Chaney and Swift, 1984; Spaccini et al., 2002]. In our study it was shown that C_{org} and C_L had a strong relationship with the majority of soil structure parameters (WSA_{ma}, WSA_{mi}, MWD_W, Kv and Sw). Both of these quantitative parameters of SOM correlated positively with WSA_{ma} and negatively with WSA_{mi}. For example, Burreto et al. [2009] found a positive relationship between the content of organic carbon and the shares of macro-aggregates. Next, the relationship between C_L and WSA_{ma} has confirmed the results of other studies [Polláková et al. 2017; Šimanský et al. 2016]. An increase of macro-aggregate stability and a higher content C_{org} and C_L was also confirmed by their positive correlation with MWD_W and Sw but it had a negative correlation with Kv. At the same time, we con-

cluded that the humic substances did not participate in the stabilization of macro-aggregates. The negative correlation of C_{HS} with WSA_{ma} and positive correlation of WSA_{mi} were found. In addition, significant negative correlations of MWD_W with C_{HA} and C_{FA} were demonstrated. When it comes to the $C_{HA}:C_{FA}$, no significant relationship with any parameters of soil structure was shown. Q_{HA} was in a negative correlation with Kv.

CONCLUSION

Biochar improved the quantitative parameters of SOM. The content of C_{org} increased with the rate of biochar, after the application of biochar alone or in combination with a lower rate of N. The values of C_L, $C_{HA}:C_{FA}$, Q_{HS} and Q_{HA} were not significantly altered by the addition of either biochar or N fertilizer. The values of C_{HS}, C_{HA} and C_{FA} were simultaneously reduced with biochar in rate of 20 t·ha^{-1} with 160 kg N·ha^{-1} and biochar in both rates with 240 kg N·ha^{-1}.

The results of our study confirmed the positive effect of biochar application on all evaluated parameters of soil structure, except MWD_d. The higher dose of biochar, the better the soil structure was. It resulted in the higher values of WSA_{ma}, MWD_W and Sw and in the lower values of WSA_{mi} and Kv. None of the rates of biochar with 160 kg N·ha^{-1} had a significant effect on the parameters of soil structure. Biochar application with higher rate of N resulted in higher values of WSA_{ma} and MWD_W and lower value of vulnerability of soil structure.

The results showed that C_{org} and CL positively correlated with WSA_{ma}, MWD_W, Sw and negatively with WSA_{mi} and Kv. C_{HS} had a significantly positive relationship with WSA_{mi} and a negative relationship with WSA_{ma}, MWD_W and Sw. C_{HA} and C_{FA} negatively correlated with MWD_W and at the same time the Q_{HA} negatively correlated with MWD_d and Kv.

Table 4. The values of correlation coefficients between the parameters of soil organic matter and soil structure

Parameters	WSA_{ma}	WSA_{mi}	MWD	MWD_W	Kv	Sw
C_{org}	0.295*	-0.295*	n.s.	0.353**	-0.377**	0.298*
C_L	0.396**	-0.396**	n.s.	0.432***	-0.433***	0.395**
C_{HS}	-0.276*	0.276*	n.s.	-0.374**	n.s.	-0.276*
C_{HA}	n.s.	n.s.	n.s.	-0.288*	n.s.	n.s.
C_{FA}	n.s.	n.s.	n.s.	-0.314*	n.s.	n.s.
$C_{HA}:C_{FA}$	n.s.	n.s.	n.s.	n.s.	n.s.	n.s.
Q_{HS}	n.s.	n.s.	n.s.	n.s.	n.s.	n.s.
Q_{HA}	n.s.	n.s.	-0.463***	n.s.	-0.359**	n.s.

n = 54; *** = P≤0.001; ** = P≤0.01; * = P≤0.05; n.s. = P>0.05.

Acknowledgments

This study was supported by the Slovak Grant Agency VEGA, No. 1/0604/16 and No. 1/0136/17, KEGA, No. 026SPU-4/2017 and Slovak Research and Development Agency under the contract No. APVV-15-0160.

REFERENCES

1. Abel S., Peters A., Trinks S., Schonsky H., Facklam M., Wessoiek G. 2013. Impact of biochar and hydrochar addition on water retention and water repellency of sandy soil. Geoderma, 202–203, 183–191.

2. Ahmad M., Rajapaksha A.U., Lim J.E., Zhang M., Bolan N., Mohav D., Vithanage M., Lee S.S., Ok Y.S. 2014. Biochar as a sorbent for contaminant management in soil and water. A review. Chemosphere, 99, 19–33.

3. Amezketa E. 1999. Soil aggregate stability: a review. Journal of Sustainable Agriculture, 14, 83–151.

4. Blaud A., Chevallier T., Virto I., Pablo A.L., Chenu C., Brauman A. 2014. Bacterial community structure in soil microaggregates and on particulate organic matter fractions located outside on inside soil macroaggregates. Pedopshere, 57, 191–194.

5. Bossuyt H., Six J., Heudrix P.F. 2004 Protection of soil carbon by microaggregates within earthworm casts. Soil Biology and Biochemistry, 37, 251–258.

6. Bronick C.J., Lal R. 2005. The soil structure and land management: a review. Geoderma, 124, 3–22.

7. Burreto R.C., Madari B.E., Maddock J.E.L., Machado L.O.A., Torres E., Frauchini J., Costa A.R. 2009. The impact of soil management on aggregation, carbon stabilization and carbon loss as CO_2 in the surface layer of a Rhodic Ferralsol in Southern Brazil. Agricutural Ecosystem and Environment, 132, 243–251.

8. Dziadowiec H., Gonet S.S. 1999a. Estimation of soil organic carbon by Tyurin's method. Methodical guide-book for soil organic matter studies, 120, 7–8 (in Polish).

9. Dziadowiec H., Gonet S.S. 1999b. Estimation of fractional composition of soil humus by Kononova-Bielcikova's method. Methodical guide-book for soil organic matter studies, 120, 31–34, (in Polish).

10. Chaney K., Suift R.S. 1984. The influence of organic matter on aggregate stability in some British soils. European Journal of Soil Science, 35, 223–230.

11. Chaplot V., Cooper M. 2015. Soil aggregate stability to predict organic carbon outputs from soils. Geoderma, 243–244, 205–213.

12. Chen X., Li Z., Liu M., Jiang Ch., Che Y. 2015. Microbial community and functional diversity associated with different aggregate fractions of a paddy soil fertilized with organic manure and/or NPK fertilizer for 20 years. Journal of Soil and Sedimens, 15, 292–301.

13. Germida J., Gupta V.S.R. 2015. Soil aggregation: Influence on microbial biomass and application of biological process. Soil Biological and Biochemistry, 80, A3–A9.

14. Glaser B., Lehmann J., Zech W. 2002. Ameliorating physical and chemical properties of highly weathered soils in the tropic with charcoal- a review. Biology and Fertility of Soils, 35, 219–230.

15. Fischer D., Glaser B. 2012. Synergisms between compost and biochar for sustainable soil amelioration. In: Kumar S. (Ed), Management of Organic Waste, Tech Europe, Rijeka, pp. 167–198.

16. Hraško J., Červenka L., Facek Z., Komár J., Němeček J., Pospíšil J., Sirový V. 1962. Soil analyses. SVPL, Bratislava, (in Slovak).

17. Hussian M., Farooq M., Nawaaz A., Sadi A.M., Salaiman Z.M., Algmandi S.S., Amyara U., Ok Y.S., Siddique K.H.M. 2016. Biochar for crop production: potential benefits and risks. Journal of Soils and Sediments, 17, 685–716.

18. Jiang K., Xu Q.F. 2006. Abundance and dynamics of soil labile carbon pools under different types of forest vegetation. Pedosphere, 16, 505–511.

19. Jindo K., Sonoki T., Matsumoto K., Canellas L., Roig A., Sanchez-Monedero M.A. 2016. Influence of biochar addition on the humic substances of composting manures. Waste Management, 49, 545–552.

20. Joseph S.D., Arbestain M.C., Lin Y., Munroe P., Chia C.H., Hook J., Zwieten L., Kimber S., Cowie A., Singht B.P., Lehmenn J., Foidl N., Smernik R.J., Amonette, J.E. 2009. An investigation into the reaction of biochar in soil. Australian Journal of Soil Research, 48, 501–505.

21. Lehmann J., Rilling M., Thies J., Masiello C.A., Hockadaj W.C., Crowley D. 2011. Biochar effect on soil biota – A review. Biology and Biochemistry, 43, 1812–1836.

22. Liu X., Chen X., Jing Y., Li Q., Zhang J., Huang Q. 2014. Effect of biochar amendment on rapessd and swelt polutor yields and water stable aggregates in upland red soil. Catena, 123, 45–51.

23. Loginow W., Wisniewski W., Gonet S.S., Ciescinska B. 1987. Fractionation of organic carbon based on susceptibility to oxidation. Polish Journal of Soil Science, 20, 47–52.

24. Lu S.G., Sun F.F., Zono Y.T. 2014. Effect of rice hust biochar and coal fly ash on some physical properties of expansive clayey soil (Vertisol). Catena, 114, 37–44.

25. Mavi M.S., Singh G., Singh B.P., Serhon B.S., Choudrary O.P., Sagi S., Berry R. 2018. Interactive effects of rice-residue biochar and N-fertilizer on

soil structure functions and crop biomass in contrasting soils. Journal of Soil Science and Plant Nutrition, 107, 718–729.

26. Mukome F.N.D., Parikh S.J. 2015. Chemical, physical and surface characterization of biochar. In: Ok Y.S., Uchimiya M.S., Chang S.X., Blan N. (Eds), Biochar: production, characterization and application. CRC Press/Taylor and Francis Group, pp 400–407.

27. Oades J.M. 1993. The role of biology in the formation, stabilization and degradation of soil structure. Geoderma, 56, 377–400.

28. Obia A., Mulder J., Martines V., Conrelissen G., Borresen T. 2016. In situ effects of biochar on aggregation, water retention and porosity in light-textured tropical soils. Soil and Tillage Research, 155, 35–44.

29. Omondi M.O., Xia X., Nahayo A., Liu X., Kora P.K., Pan G. 2016. Quantification of biochar effects on soil hydrological properties using meta-analysis of literature data. Geoderma, 274, 28–34.

30. Parakayastha T.J., Kumari S., Pathak H. 2015. Characterisation, stability, and microbial effect of four biochars produced crop residues. Geoderma, 239–240, 293–303.

31. Polláková N., Šimanský V., Kravka M. 2017. The influence of soil organic matter fractions in aggregates stabilization in agricultural and forest soils of selected Slovak and Czech hilly lands. Journal of Soils and Sediment, 18(8), 2790–2800.

32. Rutkowska A., Pikula D. 2013. Effect of crop rotation and nitrogen fertilization on the quality and quantity of soil organic matter. Quality Assessment, 25, 249–257.

33. Sainju U.M., Tonthan T.C., Lenssen A.W., Evans R.K. 2009. Tillage and cropping sequence impacts on nitrogen cycling in dryland farming in eastern Montana, USA. Soil and Tillage Research, 103, 332–341.

34. Scott H.D. 2000. Soil physics: agriculture and environmental application. Wiley-Blackwell, London.

35. Šimanský V., Horák J., Igaz D., Jonczak J., Markiewicz M., Felber R., Rizmiya E.Y., Lukac M. 2016. How dose of biochar and biochar with nitrogen can improve the parameters of soil organic matter and soil structure? Biologia, 71, 989–995.

36. Singh B.P., Cowie A.L. 2014. Long-term influence of biochar on native organic carbon mineralisation in carbon clayey soil. Scientific Reports, 4, 1–9.

37. Six J., Caunt R.T., Paustian K., Paulie A. 2002. Stabilization mechanisms of soil organic matter: Implications for C-saturation of soils. Plant and Soil, 241, 155–176.

38. Six J., Elliot E.T., Paustian K. 2000. Soil macroaggregates turnover and microaggregates formation: a mechanism for C sequestration under no-tillage

agriculture. Soil Biology and Biochemistry, 32, 2099–2103.

39. Spaccini R., Piccolo A., Conte P., Haberhauder G., Gerzabek M.H. 2002. Increased soil organic carbon sequestration through hydrophobic production by humic substances. Soil Biology and Biochemistry, 34, 1839–1851.

40. Sun F., Lu S. 2014. Biochars improve aggregate stability, water retention and pore–space properties of clayey soil. Journal of Plant Nutrition and Soil Science, 177, 26–33.

41. Tisdall J.M. 1996. Formation of soil aggregates and accumulation of soil organic matter. In: Carter M.R., Steward B.A. (Eds), Structure and organic matter storage in agricultural soils, Boca Raton, CRC/Leuis Publisher, 481–487.

42. Valla M., Kozák J., Ondáček V. 2000. Vulnerability of aggregates separated from selected anthrosols developed on reclaimed dumpsites. Rostlinna Vyroba, 46, 563–568.

43. Wang D., Fonte S.J., Parikh S.J., Six J., Scow K.M. 2017. Biochar additions can enhance soil structure and the physical stabilization of C in aggregates. Geoderma, 303, 110–117.

44. Wang J., Xiong Z., Kozyakov Y. 2016. Biochar stability in soil ultra-analysis of composition and primming effect. Bioenergy, 8, 512–513.

45. Wang Y., Wang Z.L., Zhang Q., Hu N., Li Z., Lou Y., Li Y., Xue D., Chen Y., Wu Ch., Zou Ch.B., Kuzyakov Y. 2018. Long-term effects of nitrogen fertilization on aggregation and localization of carbon, nitrogen and microbial activities in soil. Science of the Total Environment, 624, 1113–1139.

46. Yan D., Wang D., Yang L. 2007. Long-term effect of chemical fertilizer, straw, and manure on labile organic matter fractions in a paddy soil. Biology and Fertility of Soils, 44, 93–101.

47. Zhang Q., Du Z.L., Lou Y., Me X. 2015. A one-year short-term biochar application improved carbon accumulation in large macroaggregate fractions. Catena, 127, 26–31.

48. Zhang Q., Worsnop D.R., Canagaratha M.R., Jimenez J.L. 2005. Hydrocarbon-like and oxygenated organic aerosoils in Pittsburgh: insights into sources and processes of organic aerosols. Atmospheric Chemistry and Physics, 5, 3289–3311.

49. Zhao S., Ta N., Li Z., Yang Y., Zhang X., Liu D., Zhang A., Wang X. 2017. Varying pyrolysis temperature impacts application effects of biochar on soil labile organic carbon and humic substances. Applied Soil Ecology, 116, 399–409.

50. Zlielinska A., Olesczuk P., Charmas B., Zieba J.S., Pasieczua-Patkowska S. 2015. Effect of sewage sludge properties on the biochar characteristic. Journal of Analyticall and Applies Pyrolysis, 112, 201–213.

14

The Effects of Water Extracts from Lemon Balm on Pea Leaf Weevil and Black Bean Aphid Behaviour

Milena Rusin[1*], Janina Gospodarek[1]

[1] Department of Agricultural Environment Protection, University of Agriculture, al. Mickiewicza 21, 31-120 Krakow, Poland

* Corresponding author's e-mail: milena_rusin@wp.pl

ABSTRACT

The objective of this study was to determine the effects of various concentrations of water extracts prepared from the fresh or dry matter of lemon balm on *Sitona lineatus L.* and *Aphis fabae* Scop. behaviour. The assessment pertaining to the feeding intensity of beetles was carried out by measuring the surface of feeds caused by *S. lineatus*. While examining the effect of extracts on *A. fabae*, the mortality of wingless female and aphid larvae was determined. In the studies on the olfactory reaction glass olfactometer "Y-tube" and 4-armed arena olfactometer were used. The results of the experiment showed that the water extract prepared from dry matter of lemon balm with 2% concentration limited the feeding of both female and male of *S. lineatus*. The increase in the mortality of the black bean aphid females and larvae was obtained only after applying the extracts from fresh and dry matter at highest concentrations. The evident deterrent reaction of the odour substances obtained from the lemon balm plants towards the beetles of *S. lineatus*, could find application in ecological farms via introducing the plant as an accompanying crop to the main crops. The winged individuals of *A. fabae* did not react to the abovementioned factor.

Keywords: water extracts, *Melissa officinalis* L., olfactometer, biological control

INTRODUCTION

Application of chemical means of pest control is increasingly contributing to the emergence of pest resistance to the active substances contained in these compounds, resulting in their reduced effectiveness [Hansen 2008, Pimentel et al. 2009, Wawrzyniak et al. 2015]. Furthermore, these compounds pose the threat to all elements of the natural environment [Tscharntke et al. 2005]. In line with the principles of integrated plant protection, priority should be given to non-chemical methods where the use of insecticides is limited to the necessary minimum. One of the non-chemical methods of reducing the numbers of pests of cultivated plants is the use of water extracts prepared from fresh or dry parts of herb plants. The olfactory stimuli are very important in the life of insects. The odour of a host plant could be the principal factor determining their behaviour and the possibility to locate their host plant. For this reason, the substances which prevent the identification of the host plant by the pest, or have a deterrent effect on the latter are increasingly often applied in plant protection [Korczyński and Koźmiński 2007].

Lemon balm (*Melissa officinalis* L.) is a perennial herb of the mint family (Lamiaceae), used in many applications. It is used in medicine because of its antimicrobial and anti-inflammatory properties [Rostami et al. 2012], and it also functions as a natural anti-oxidant [Koksal et al. 2011, Saeb et al. 2011]. Owing to their allelopathic properties, the water extracts from the plant can be used in protection against weeds [Kato-Noguchi 2001]. It is thus justified to test the lemon balm as a natural means for the protection against pests.

The objective of the presented study was to determine the effects of various concentrations of the water extracts prepared from the fresh matter (FM) and dry matter (DM) of the lemon balm upon the feeding by pea leaf weevil (*Sitona lineatus* L.) and on the mortality of the black bean aphid (*Aphis fabae* Scop.). Furthermore, their reactions to the odour of the lemon balm were studied by the use of olfactometer.

MATERIAL AND METHODS

The experiment was conducted in the laboratory, in six replicates. The extracts from dry matter of *Melissa officinalis* L. were prepared at concentration assumed conventionally as 2%, 5% and 10% (dried plants + cold redistilled water in the ratios of 2 : 100, 5 : 100 and 10 : 100) and at the concentrations of 10%, 20% and 30% for fresh matter (fresh above-ground parts of plants + cold redistilled water in the ratios of 10 : 100, 20 : 100 and 30 : 100). The extracts were stored in the dark for the period of 24 hours, and then filtered through filter papers and immediately used to conduct the experiment. The test was performed on Petri dishes, and the substrate consisted of moist filter paper. Plant leaves (pea in case of *Sitona lineatus* L., broad bean in the case of *Aphis fabae* Scop.) were soaked for 3 seconds in adequate plant extracts and in distilled water used as control, and then dried at room temperature. In each dish, a single leaf of a plant, suitable for a specific object was placed and then pests were introduced – one adult of *S. lineatus* (male and female separately), six wingless females of *A. fabae* and ten larvae of aphid.

The assessment of the feeding intensity of beetles was carried out by measuring the surface of feeds caused by *S. lineatus* at 12 hour intervals. In addition, the values of palatability index as the ratio of the percentage area of leaves consumed in individual objects to the percentage area of leaves consumed in the control was calculated. Furthermore, an absolute deterrence index, which takes into account the relationship between the area of leaves consumed in the individual objects and the area of leaves consumed in the control, was established:

$$Bwd = [(K–T) : (K+T)] \cdot 100 \qquad (1)$$

where: *Bwd* – absolute deterrence index.

K – area of leaves consumed in control [mm^2].

T – area of leaves consumed in individual objects [mm^2] [Kiełczewski 1979].

While examining the effect of *M. officinalis* extracts on *A. fabae*, the mortality of wingless female and aphid larvae was determined at 12 hour intervals.

In the studies on the olfactory reaction of the abovementioned insects, for *S. lineatus* glass olfactometer "Y-tube" was used and for *A. fabae* – 4-armed arena olfactometer, applied in multiple choice tests. There are commonly used for the evaluation of odour preferences in insects [Vet et al. 1983, Schaller and Nentwig 2000, Ukeh and Umoetok 200, Ranjith 2007]. "Y- tube" had one incoming arm and two test arms. The area of the olfactometer has the central field and four test arms. The air, cleaned in a carbon filter was forced in by a pump, and directed to each test arms. Then, the air stream was flowing through the source of odour i.e. glass container with either 30 g of fresh matter of lemon balm together with either a wet circle of filter paper (to ensure appropriate air humidity) or only wet filter paper (control) in the case of "Y- tube". In the case of four-arm arena, air was pumped into two arms through separate containers, containing lemon balm and wet filter paper, and through two more, containing only wet filter paper (control). An imago of pests was placed in the outlet of incoming arm in the case of "Y- tube" or in the central part of central field of four-arm arena and its behaviour was observed for 10 minutes, recording the number of incursions into particular test arms of the olfactometer (with or without the odor derived from lemon balm). The experiment with pest insects was performed in 12 repetitions.

The obtained results were then subjected to analysis with STATISTICA 10.0 software. The significance of differences between the means were tested by univariate analysis of variance, and the means were differentiated by Fisher's LSD test at $\alpha = 0.05$. The Student's t-test for independent groups (the grouping variable was either the presence or the lack of test plant odour influx) was used to determine the statistical significances of differences in the results obtained with the use of an olfactometer.

RESULTS AND DISCUSSION

The extracts prepared from dry matter of the lemon balm at the two highest concentrations (5% and 10%) at all dates of observations, led to a significant reduction in the feeding by males of pea leaf weevil (Table 1). Similar regularities were observed in the case of the extract from dry matter, with the concentration of 2% but only up to 96 hours following the beginning of experiment. After the experiment was concluded, the dry matter extract with the concentration of 10% resulted in the decrease in the surface area of eaten-up places in pea leaves, resulting from male feeding by nearly 30% compared with the control

Table 1. The effect of extracts from *Melissa officinalis* L. on the surface area of places eaten-up in pea leaves by females and males of *Sitona lineatus* L. [mm^2]

Object	12 h	24 h	36 h	48 h	60 h	72 h	84 h	96 h	108 h	120 h
					Males					
C	11.6 c*	26.1 b	45.5 b	72.9 b	105.2 bc	184.3 bc	214.2 bc	240.8 bc	267.5 bc	294.8 b
DM 2%	1.3 ab	2.8 a	5.3 a	26.4 a	46.2 a	116.2 a	129.7 a	158.9 a	206.0 ab	241.3 ab
DM 5%	0.3 a	2.5 a	2.8 a	21.4 a	45.2 a	104.9 a	125.9 a	146.6 a	189.0 a	218.1 a
DM 10 %	1.3 ab	2.8 a	7.2 a	22.0 a	45.2 a	112.4 a	131.6 a	158.3 a	183.4 a	206.6 a
FM 10%	4.7 abc	15.7 ab	36.7 ab	55.9 ab	79.8 ab	141.6 ab	148.5 ab	181.5 ab	202.8 ab	222.2 ab
ŚW 20%	9.7 bc	31.4 b	43.0 ab	61.5 b	93.9 abc	181.8 bc	211.6 b	236.8 bc	244.2 bc	248.7 ab
FM 30%	5.3 abc	22.9 ab	38.3 ab	55.3 ab	132.2 c	193.1 c	243.0 c	266.9 c	284.2 c	298.3 b
					Females					
C	14.4 c	32.3 c	53.1 c	96.7 b	155.7 b	243.0 c	282.9 c	356.7 c	400.7 c	426.5 c
DM 2%	6.0 ab	13.8 ab	24.2 ab	38.9 a	68.8 a	95.1 ab	111.5 ab	121.8 ab	140.4 ab	185.0 ab
DM 5%	1.6 a	7.9 a	15.1 a	43.3 a	79.1 a	162.0 abc	183.7 bc	195.9 ab	222.6 ab	230.3 ab
DM 10 %	1.9 a	7.9 a	13.5 a	22.9 a	31.7 a	48.7 a	52.8 a	63.1 a	71.0 a	82.2 a
FM 10%	8.2 abc	24.8 bc	44.3 bc	69.1 ab	98.9 ab	177.4 bc	197.5 bc	239.3 bc	258.7 bc	269.6 bc
FM 20%	9.4 bc	22.0 abc	35.8 abc	54.6 ab	82.3 ab	121.2 abc	153.9 abc	178.7 ab	206.0 ab	217.8 ab
FM 30%	10.4 bc	19.5 abc	32.3 abc	40.5 a	49.6 a	65.3 ab	79.1 ab	120.3 ab	146.0 ab	182.3 ab

C – control, DM – dry matter, FM – fresh matter.
* Values for individual terms of observations marked by different letters are statistically different (α = 0.05).

object. No significant effect of any of the extracts prepared from the fresh matter upon the analysed feature was noted.

The females of pea leaf weevil showed more intensive feeding than the males of the pest. After 120 hours, the area of places eaten-up by females was greater by more than 130 mm^2 than that in the case of males. Similarly as in the case of males, the dry matter extracts contributed to a major reduction in feeding by females (the differences were statistically significant at all dates of observations, except for the extract with 5% concentration after 72 and 84 hours after the inception of the experiment). The extracts from fresh matter at the highest concentrations (20 and 30%) after 96 and 48 hours, respectively, also led to significant reduction of female feeding.

In all objects, the absolute deterrence index for the females of pea leaf weevil reached positive values, which testifies to the inhibiting effect of the applied extracts towards the feeding by the studied pest (Fig. 1). In the case of males, after applying the fresh matter extract at 30% concentration, a negative value of the index was noted indicating the stimulating effect of food. The strongest deterrent effect for both males and females appeared in the dry matter extract with 10% concentration. In the case of the extracts prepared from the fresh matter, the increases in the absolute deterrence index in females were noted along the increases in concentrations, whereas a reverse

relationship was found in males. In all objects, the values of the absolute deterrence index were definitely higher for females than that for males.

In males, the palatability index was the highest when the extract from the fresh matter with 30% concentration was applied whereas in females – in the extract from fresh matter but at 10% concentration (Fig. 2). In all objects, the value of the analysed index was higher in males than in females when the same kind and concentration of extract was applied. The lowest values of the palatability index for both sexes were noted after the extract with 10% concentration, prepared from the dry matter of lemon balm plants was applied.

The extract prepared from the fresh matter of the highest concentration (10%) resulted in a significant increase in the mortality of the wingless females of the black bean aphid after 48 hours from the inception of the experiment, and this status was maintained up to the end of the experiment (Table 2). After 72 hours, the extract from the fresh matter at 30% concentration also resulted in the increased mortality; however, the effect was nearly fourfold weaker that the action of the extract from dry matter at 10% concentration. The remaining extracts prepared from either dry or fresh matter had no significant effect on the analysed feature.

Similar regularities were observed in the case of the larvae of the black bean aphid. The extracts from dry and fresh matter applied in the highest

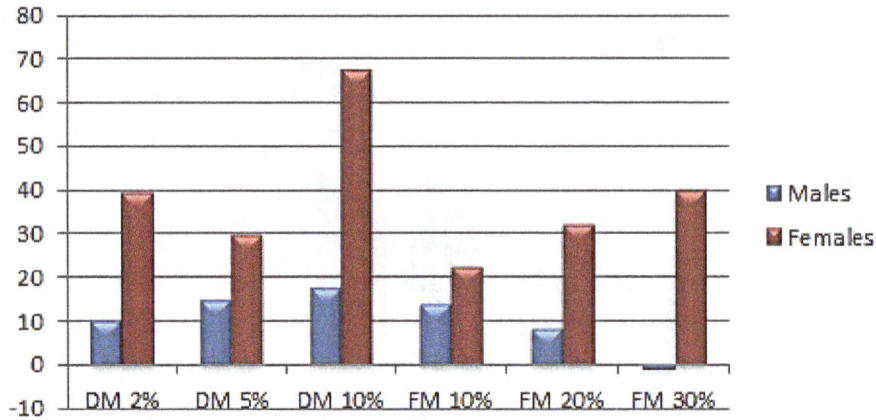

Fig. 1. Absolute deterrence index. Symbols as in Table 1

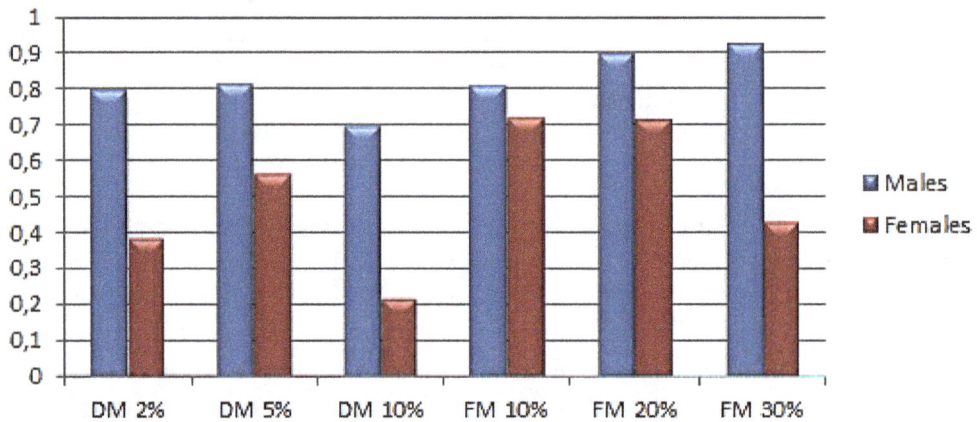

Fig. 2. Palatability index. Symbols as in Table 1

Table 2. The effect of extracts from *Melissa officinalis* L. on mortality of wingless females and larvae of *Aphis fabae* Scop. [%]

Object	12 h	24 h	36 h	48 h	60 h	72 h	84 h	96 h	108 h	120 h
Wingless females										
C	0.0 a*	0.0 a	0.0 a	0.0 a	0.0 a	0.0 a	2.8 a	2.8 a	5.6 a	5.6 a
DM 2%	0.0 a	0.0 a	0.0 a	0.0 a	0.0 a	2.8 a	8.3 ab	11.1 ab	13.9 ab	16.7 ab
DM 5%	0.0 a	0.0 a	0.0 a	2.8 a	2.8 a	2.8 a	11.1 ab	13.9 ab	19.4 ab	19.4 ab
DM 10 %	0.0 a	0.0 a	0.0 a	19.4 b	33.3 b	41.7 c	50.0 c	66.7 c	75.0 b	100.0 c
FM 10%	0.0 a	0.0 a	0.0 a	0.0 a	0.0 a	0.0 a	2.8 a	5.6 ab	5.6 a	11.1 ab
ŚW 20%	0.0 a	0.0 a	0.0 a	0.0 a	2.8 a	2.8 a	2.8 a	2.8 a	5.6 a	13.9 ab
FM 30%	0.0 a	0.0 a	0.0 a	2.8 a	5.6 a	13.9 b	16.7 b	19.4 b	22.2 b	25.0 b
Larvae										
C	0.0 a	0.0 a	0.0 a	0.0 a	0.0 a	1.7 a	1.7 a	6.7 a	11.7 a	27.8 a
DM 2%	0.0 a	0.0 a	1.7 ab	1.7 ab	3.3 ab	3.3 ab	6.7 ab	10.0 a	20.0 ab	28.3 ab
DM 5%	0.0 a	0.0 a	1.7 ab	3.3 ab	3.3 ab	6.7 ab	11.7 abc	18.3 ab	36.7 bc	48.3 bc
DM 10 %	0.0 a	0.0 a	5.0 b	6.7 b	8.3 b	13.3 b	26.7 c	51.7 c	73.3 d	83.3 d
FM 10%	0.0 a	0.0 a	0.0 a	0.0 a	0.0 a	1.7 a	6.7 ab	16.7 ab	21.7 ab	30.0 ab
ŚW 20%	0.0 a	0.0 a	0.0 a	0.0 a	3.3 ab	5.0 ab	11.7 abc	20.0 ab	31.7 abc	38.3 abc
FM 30%	0.0 a	0.0 a	1.7 ab	5.0 ab	5.0 ab	6.7 ab	18.3 bc	31.7 b	43.3 c	56.7 c

Symbols as in. **Table 1**. *Values for individual terms of observations marked by different letters are statistically different ($\alpha = 0.05$).

concentrations resulted in a significant increase in the mortality of the studied pest (after 36 and 84 hours after the inception of experiment, respectively). Furthermore, the extract from dry matter with 5% concentration, in the two last dates of observations, also contributed to a significant increase among the larvae of the black bean aphid.

Among the available scientific publications, there are few studies on the effects of the water extracts prepared from the lemon balm upon the feeding by the pests of cultivated crops. Only Hiiesaar et al. [2000] demonstrated in their research that the water extract from that plant resulted in an evident drop in the number of eggs laid by the females of *Trialeurodes vaporariorum* (West.). There were also some studies conducted on the effect of volatile oils obtained from the lemon balm plants, and the effect of alcohol extracts from that plant. Pavela [2004] found that the alcohol extract from the lemon balm plants showed weak repellent and deterrent effects towards the adult individuals of the Colorado potato beetle. Rafiei-Karahroodi et al. [2011] demonstrated that the volatile oil obtained from the lemon balm had a toxic effect on the larvae of *Plodia interpunctella* Hübner and contributed to the increased mortality among eggs of the pest. The aforementioned authors had also proven that with the increased concentration of volatile oil, its adverse effect towards the pest is also increased. Other authors confirmed the effectiveness of using volatile oils from the plants of Lamiaceae family in limiting the feeding by the pests of stored products [Ebadollahi 2011, Maede et al. 2011, Popović et al. 2013], *Aphis fabae* Scop. and *Brevicoryne brassicae* L. [Nottingam 1991] as well as *Frankliniella occidentalis* Pergande [Picard et al. 2012].

The studies on the effects of water extracts obtained from other herb plants upon the feeding by pea leaf weevils demonstrate that in the case of the wormwood (*Artemisia absinthium* L.), the reduction in feeding by the females of the pea leaf weevil was obtained after the application of the extract from dry matter of the plant in at least 5% concentration whereas towards the pea leaf weevil males, only the extract from dry matter with concentration of no less than 10% was effective [Rusin et al. 2016 b]. Biniaś et al. [2016 a] demonstrated that the extracts from the fennel (*Foeniculum vulgare* Mill.) seeds did not affect the feeding by the females of pea leaf weevil although they could significantly limit the feeding by the males of that pest. Against the background, the water extracts from the dry matter of lemon balm

showed a stronger effect on the studied pests, as they limited the feeding of pea leaf weevils even in the lowest concentration (2%). In turn, the extracts from the fresh matter were effectively limiting the feeding of the females of the studied pest only in the highest concentration (30%).

The dry matter extracts obtained from wormwood, show an inhibiting effect on the feeding by the females of pea leaf weevils, as indicated by the positive values of the absolute deterrence index obtained in the study by Rusin et al. [2016 b]. Similar relationships were also noted in the presented experiment after applying the extract from the lemon balm, and the values of the studied index were decisively higher than those obtained after applying the extracts from the wormwood. The aforementioned authors had also demonstrated that the extracts from fresh matter can have a stimulating effect on feeding by males, which also corresponds to the results of the presented experiment regarding the fresh matter extract at 30% concentration. In the presented experiment, the highest values of the absolute deterrence index were obtained after applying the dry matter extract from lemon balm plants at 10% concentration (67 for females, and 17 for males). In the studies by Biniaś et al. [2016 a], the value of that index for pea leaf weevils, after applying the extract from the fennel seeds at the same concentration, amounted to 19 for females, and 27 for males, respectively. For males, the value of palatability index in the presented experiment was the highest after applying the extract obtained from the fresh matter with 30% concentration (more than 0.9), and in the case of females – after applying the fresh matter extract at 10% concentration (0.7). In the experiment mentioned earlier [Biniaś et al. 2016 a], the values of the index in question after applying the extract obtained from the seeds of the fennel at the same concentrations, fell into the range of 0.5–0.6, for both males and females.

The high concentrations of extracts from the dry and fresh matter of the wormwood (10% and 30%, respectively) [Rusin et al. 2016 b] and from the tarragon (*Artemisia dracunculus* L.) (10% for dry matter, and 20% and 30% for fresh matter) [Rusin et al. 2016 a], result in the increased mortality among the wingless females and larvae of black bean aphid (at 100% level for females, and nearly 50% for larvae in the case of the extracts prepared from the wormwood, and at 100% level for both females and larvae in the case of the extracts from the tarragon, after 96 hours of experiment). On the other hand, the high concentrations of extracts prepared from the mountain savory (*Satureja montana* L.) (10% for dry matter, and 20% and 30% for fresh matter) increase the mortality of the females and larvae of the black bean aphid, up to respective 100% and 78.8% in the case of dry matter, and to more than 80% and to around 70% for fresh matter extracts, in 96 hours after the application of extract [Rusin et al. 2016 c]. Generally, also in the presented experiment, only the highest concentrations of the extracts from the lemon balm resulted in the increased mortality in the black bean aphids, but the values were lower. For the objects DM 10% and FM 30%, after 96 hours of experiment, they amounted to the respective 66.7% and 19.4% for wingless females, and 51.7% and 31.7% for larvae.

A great number of authors emphasize the fact that the olfactory stimuli can play a significant role in limiting the feeding by pests [Koschier et al. 2002; Koschier and Sedy 2003, Katerinopoulos et al. 2005]. In the presented experiment, an evident negative reaction to odour substances obtained from the fresh matter of lemon balm plants was noted in females (t=2.14, P=0.048) and males (t=3.62, P=0.002) of the pea leaf weevil (Table 3). The pests selected the arm of the olfactometer where the odour of lemon balm was pumped nearly two- to fourfold less frequent than the control arm. No significant reaction of the winged females of black bean aphids towards the odour substances obtained from *M. officinalis* (t=1.53, P=0.149) was found. In their studies, Rusin et al. [2016] also demonstrated the strong deterrent reaction of the odour substances obtained from the wormwood plants, towards the adult beetles of pea leaf weevils (males and females) but did not find such a relationship in the case of winged females of *Acyrthosiphon pisum* Harris. Similarly as in the results obtained in the presented experiment, Biniaś et al. [2016 b] did not find an evident repellent effect of odour substances derived from the common sage (*Salvia officinalis* L.) towards the black bean aphid.

CONCLUSIONS

1. The water extract prepared from dry matter of lemon balm with 2% concentration contributed to limiting the feeding of both female and male of pea leaf weevil. In the case of females, a similar relationship was observed using the fresh matter extracts with 20% and 30% concentration, however, at later dates of observations.
2. The increase in the mortality of the black bean aphid females and larvae was obtained only after applying the extracts from fresh and dry matter at highest concentrations (30% and 10%, respectively).
3. The evident deterrent effect of the odour substances obtained from the lemon balm plants towards the beetles of pea leaf weevil (both females and males), could find the application in ecological farms via introducing the plant as an accompanying crop to the main crops. The winged individuals of the black bean aphid did not react to the abovementioned factor.

Acknowledgments

This scientific publication was financed by the Ministry of Science and Higher Education of the Republic of Poland

REFERENCES

1. Biniaś B., Gospodarek J., Rusin M. 2016 a. Effect of fennel water extracts on reduction of feeding of pea leaf weevil. J. Ecol. Eng., 17(5), 192–197.
2. Biniaś B., Gospodarek J., Rusin M. 2016 b. The effect of extracts from sage (Salvia officinalis L.) on black bean aphid (Aphis fabae Scop.) and Colorado potato beetle (Leptinotarsa decemlineata Say.) behaviour. Zesz. Probl. Post. Nauk Rol., 586, 135–145 (in Polish).
3. Ebadollahi A. 2011. Susceptibility of Two Sitophilus species (Coleoptera: Curculionidae) to Essential Oils from Foeniculum vulgare and Satureja

Table 3. Responses of pests to odors derived from *Melissa officinalis* L. fresh matter expressed as a number of incursions per one insect into selected areas of Y-tube olfactometer (*Sitona lineatus* L.) or four-armed arena (*Aphis fabae* Scop.).

Pest	Control	*Melissa officinalis* L.
Aphis fabae Scop. – winged females	0.75	0.38
Sitona lineatus L. – females *	0.89	0.44
Sitona lineatus L. – males *	0.88	0.22

* differences significant at α = 0.05, in other cases differences not proven statistically.

hortensis. Ecologia Balkanica, 3 (2), 1–8.

4. Hansen L. M. 2008. Occurrence of insecticide resistant pollen beetles (Meligethes aeneus F.) in Danish oilseed rape (Brassica napus L.) crops. Bull. OEPP/EPPO Bull., 38 (1), 95–98.

5. Hiiesaar K., Metspalu L., Kuusik A. 2000. Insect-plant chemical interaction: the behavioural effects evoked by plant substances on greenhouse pests. Transactions of the Estonian Agricultural University, 209, 46–49.

6. Kato-Noguchi H. 2001. Effects of lemon balm (Melissa officinalis L.) extract on germination and seedling growth of six plants. Acta Physiol. Plant., 23 (1), 49–53.

7. Katerinopoulos H.E., Pagona G., Afratis A., Stratigakis N., Roditakis N. 2005. Composition and insect attracting activity of the essential oil of Rosmarinus officinalis. J. Chem. Ecol., 31, 111–122.

8. Korczyński I., Koźmiński R. 2007. Response of large pine weevil Hylobius abietis (L.) (Coleoptera, Curculionidae) beetles to the smell of alcoholic extract from plants of selected species. Acta Sci. Pol., 6 (1), 27–31.

9. Koschier E.H., Sendy K.A. 2003. Labiate essential oils affecting host selection and acceptance of Thrips tabaci Lindeman. Crop. Prot., 22 (7), 929–934.

10. Koschier E.H., Sendy K.A., Novak J. 2002. Influence of plant volatiles on feeding damage caused by the onion thrips Thrips tabaci Lindeman. Crop Prot., 22(7), 929–934.

11. Koksal E., Bursal E., Dikici E., Tozoglu F., Gulcin I. 2011. Antioxidant activity of Melissa officinalis leaves. J. Med. Plant. Res., 5 (2), 217–222.

12. Maede M., Hamzeh I., Hossein D., Majid A., Reza R.K. 2011. Bioactivity of essential oil from Satureja hortensis (Laminaceae) against three stored-product insect species. AJB, 10 (34), 6620–6627.

13. Nottingham S.F., Hardie J., Dawson G.W., Hick A.J., Pickett J.A., Wadhams L.J., Woodcock C.M. 1991. Behavioral and electrophysiological responses of Aphids to host and nonhost plant volatiles. J. Chem. Ecol., 17 (6), 1231–1242.

14. Pavela R. 2004. Repellent effect of ethanol extracts from plants of the family Lamiaceae on Colorado Potato Beetle adults (Leptinotarsa decemlineata Say). Natl Acad. Sci. Lett., 27 (5–6), 195–203.

15. Picard I., Hollingsworth R.G., Salmieri S., Lacroix M. 2012. Repellency of essential oils to Frankliniella occidentalis (Thysanoptera: Thripidae) as affected by type of oil and polymer release. J. Econ. Entomol. 105 (4), 1238–1247.

16. Pimentel M.A.G., Faroni L.R.D'A., Guedes R.N.C., Sousa A.H., Tótola M.R. 2009. Phosphine resistance in Brazilian populations of Sitophilus zeamays Motschulsky (Coleoptera: Curculioni-

dae). J. Stored Prod. Res. 45 (l), 71–74.

17. Popović A., Šućur J., Orčić D., Štrbac P. 2013. Effects of essential oil formulations on the adult insect Tribolium castaneum (Herbst) (Col., Tenebrionidae). JCEA, 14 (2), 659–671.

18. Rafiei-Karahroodi Z., Moharramipour S., Farazmand H., Karimzadeh-Esfahani J. 2011. Insecticidal effect of six native medicinal plants essential oil on Indian meal moth, Plodia Interpunctella Hübner (Lep.: Pyralidae). Mun. Ent. Zool., 6 (1), 339–345.

19. Ranjith A.M. 2007. An inexpensive olfactometer and wind tunnel for Trichogramma chilonis Ishii (Irichogrammatidae:Hymenoptera). J. Trop. Agri., 45 (1–2), 63–65.

20. Rostami H., Kazemi M., Shafiei S. 2012. Antibacterial activity of Lavandula officinalis and Melissa officinalis against some human pathogenic bacteria. Asian J. Biochem., 7 (3), 133–142.

21. Rusin M., Gospodarek J., Biniaś B. 2016 a. Effect of aqueous extracts from tarragon (Artemisia dracunculus L.) on feeding of selected crop pests. J. Res. Appl. Agric. Eng., 61(4), 143–146.

22. Rusin M., Gospodarek J., Biniaś B. 2016 b. Effect of water extracts from Artemisia absinthium L. on feeding of selected pests and their response to the odor of this plant. JCEA, 17 (1), 188–206.

23. Rusin M., Gospodarek J., Biniaś B. 2016 c. The effect of water extracts from winter savory on black bean aphid mortality. J. Ecol. Eng., 17(1), 101–105.

24. Saeb K., Gholamrezaee S., Asadi M. 2011. Variation of antioxidant activity of Melissa officinalis leaves extracts during the different stages of plant growth. Biomed. Pharmacol. J., 4 (2), 237–243.

25. Schaller M., Nentwig W. 2000. Olfactory orientation of the seven-spot ladybird beetle, Coccinella septempunctata (Coleoptera: Coccinellidae): Attraction of adults to plants and conspecific females. Eur. J. Entomol. 97, 155–159.

26. Tscharntke T., Klein A. M., Kruess A., Dewenter I.S., Thies C. 2005. Landscape perspectives on agricultural intensification and biodiversity – ecosystem service management. Ecol. Letters, 8 (8), 857–874.

27. Ukeh D.A., Umoetok S.B.A., 2007. Effects of host and non-host plant volatiles on the behavior of the Lesser Grain Borer, Rhyzopertha dominica (Fab.). J. Entomol., 4 (6), 435–443.

28. Vet L.E.M., Van Lenteren J.C., Meelis E. 1983. An airflow olfactometer for measuring olfactory responses of hymenopterous parasitoids and other small insects. Physiol. Entomol., 8, 97–106.

29. Wawrzyniak M., Wrzesińska D., Lamparski R., Piesik D. 2015. Effect of dried material from peppermint (Mentha piperita L.) on the development and the fecundity of grain weevil (Sitophilus granarius L.). Zesz. Probl. Post. Nauk Roln., 580, 141–148 (in Polish).

15

The use of Coconut Shells for the Removal of Dyes from Aqueous Solutions

Tomasz Jóźwiak[1*], Urszula Filipkowska[1], Paula Bugajska[1], Tomasz Kalkowski[1]

[1] Department of Environmental Engineering, Faculty of Environmental Sciences, University of Warmia and Mazury in Olsztyn, ul. Warszawska 117, 10–720 Olsztyn, Poland

[*] Corresponding author's e-mail: tomasz.jozwiak@uwm.edu.pl

ABSTRACT

The main purpose of the work was to check the possibility of using coconut shells for the removal of the dyes popular in the textile industry from aqueous solutions. The sorption abilities of an unconventional sorbent were tested against four anionic dyes: Reactive Black 5, Reactive Yellow 84, Acid Yellow 23, Acid Red 18 as well as two cationic dyes: Basic Violet 10 and Basic Red 46. The scope of research included investigation pertaining to the effect of pH on the effectiveness of sorption of dyes, conducted in order to determine the time of equilibrium of sorption and determine the maximum sorption capacity of coconut shells with respect to pigments. The most favorable pH of sorption for the anionic dyes and Basic Violet 10 was pH 3 and for Basic Red 46 – pH 6. The equilibrium time of sorption was the shortest in the case of acidic dyes (Acid Yellow 23/ Acid Red 18 – 45 min), while the longest in the case of alkaline dyes (Basic Red 46 – 90 min, Basic Violet 10 – 180 min). The sorption capacity of coconut shells in relation to anionic dyes was for Reactive Black 5 – 0.82 mg/g, Reactive Yellow 84 – 0.96 mg/g, Acid Yellow 23 – 0.53 mg/g and for Acid Red 18 – 0.66 mg/g. The tested sorbent showed much higher sorption capacity with respect to the cationic dyes, i.e. Basic Violet 10 (28.54 mg/g) and Basic Red 46 (68.52 mg/g).

Keywords: sorption, coconut shells, anionic dyes, cationic dyes

INTRODUCTION

The post-production sewage generated in dyeing, textile or paper factories often contains high concentrations of dyes. Due to the low susceptibility of dyes for biodegradation, the decolorization of industrial wastewater with traditional, biological methods of wastewater treatment is usually ineffective [Robinson et al. 2001]. There is a high risk that the colored substances that have not been removed from sewage may enter the natural environment, causing its degradation. The perspective of the environment contaminated by dyes should encourage entrepreneurs to use effective technologies for decolorizing the industrial wastewater.

Currently, the scientific community holds the opinion that sorption is the most economical and the most environmentally friendly method of wastewater decolorization. During the sorption, no toxic intermediates are formed, in contrast to the oxidation methods (ozonation, oxidation with NaOCl) [Wijannarong et al. 2013]. Sorption also does not cause sediment formation or sewage salinity, like the precipitation methods do (coagulation and electrocoagulation) [Liang et al. 2014, Nandi and Patel 2013].

Decolorization of sewage by sorption does not require the use of complicated installations, as is the case with membrane methods. The sorption costs and its effectiveness mainly depend on the type of sorbent. The most frequently used sorbents for the decolorization of wastewater are various types of activated carbons. Due to their high price, cheaper substitutes are currently being sought. High hopes are associated with the use of waste materials from the agro-food industry as commonly available alternatives to the commercial sorbents.

The coconut shells are easily available waste material from the food industry. Their popularity, which translates into a low price of raw material is the result of wide range of applications of coconut. This paper explores the possibility of using coconut shells for the removal of dyes popular in the textile industry from aqueous solutions.

MATERIALS

Coconut shells

Coconuts from coconut palm (Cocos nucifera), originating from the Philippines, were purchased at a local hypermarket in Olsztyn (Poland). Due to the monotypic nature of the Cocos genus, the coconuts available on the market always have a similar composition, regardless of the country of origin. The shells of coconuts usually contain 46% lignin, 14% cellulose and 32% hemicellulose [Cagnon et al. 2009].

Dyes

Six dyes popular in the textile industry, including 4 anionic and 2 cationic dyes were used in the research. Among the anionic dyes there were 2 reactive dyes (Reactive Black 5, Reactive Yellow 84) and 2 acid dyes (Acid Red 18, Acid Yellow 23). The tested cationic dyes were Basic Violet 10 and Basic Red 46. The characteristics of the dyes used in this work are summarized in Table 1.

METHODOLOGY

The coconut shells were ground in a laboratory grinder and sifted through the sieves with a mesh diameter of 5 mm and then 3 mm. A 3–5 mm diameter fraction was placed in 2M H_2SO_4. After 24 h, the shells were rinsed with distilled water, then placed in 2 M NaOH for 24 h. Next, the shells were drained and washed with distilled water (until pH ~ 7 was obtained in the leachate). The coconut shells (CS) were ready for study after drying at 105°C.

The coconut shells (CS) were weighed in an amount of 1 g each and added to the series of conical flasks (250 mL). Then, dye solutions (10 mg/L – 100 mL) at pH 2–11 were added to the flasks. Then, the flasks were placed on a laboratory shaker (150 r.p.m.) with vibration amplitude of 25 mm. After 60 min, the samples were taken from the flasks (10 mL) to determine the concentration of dye in the solution.

CS in an amount of 10 g each were weighed and added to the beakers (1000 mL). Afterwards, the dye solutions (10 mg/L – 500 mL) with the

Table 1. Characteristics of dyes used in this work

Reactive Black 5 – (RB5)		Acid Yellow 23 – (AY23)		Basic Violet 10 – (BV10)	
$C_{26}H_{21}N_5Na_4O_{19}S_6$		$C_{16}H_9N_4Na_3O_9S_2$		$C_{28}H_{31}ClN_2O_3$	
molar mass	991 [g/mol]	molar mass	534 [g/mol]	molar mass	479 [g/mol]
character	anionic	character	anionic	character	cationic
Reactive Yellow 84 – (RY84)		Acid Red 18 – (AR18)		Basic Red 46 – (BR46)	
$C_{56}H_{38}Cl_2N_{14}Na_6O_{20}S_6$		$C_{20}H_{11}N_2Na_3O_{10}S_3$		$C_{18}H_{21}BrN_6$	
molar mass	1701 [g/mol]	molar mass	605 [g/mol]	molar mass	401 [g/mol]
character	anionic	character	anionic	character	cationic

optimum pH, determined in 4.1 were added to the beakers.

Then, the beakers were placed on a magnetic stirrer (150 rpm). During the sorption, samples (5 mL) were taken from the solutions at specified intervals to determine the dye concentration.

CS (1 g) were weighed and added to the conical flasks (250 mL). Next, dye solutions (100 mL) with optimal pH, determined in 4.1., with conc. 5–50 mg/L (anionic dyes) or 10–500 mg/L (cationic dyes) were added to the flasks and then the flasks were placed on a laboratory shaker (150 r.p.m.). After the sorption time determined in 4.2, the samples (10 mL) were taken from the flasks to determine the concentration of the dye remaining in the solution.

The amount of dye adsorbed on the sorbent was calculated from the equation 1:

$$Qs = \frac{(Co - Cs) \cdot V}{m} \qquad (1)$$

Three sorption isotherms were used to determine the sorption capacity of CS: Langmuir (2), Langmuir 2 (Langmuir double isotherm) (3) and Freundlich (4).

$$Qs = \frac{q_{max} \cdot K_c \cdot C}{1 + K_c \cdot C} \qquad (2)$$

$$Qs = \frac{b_1 \cdot k_1 \cdot C}{1 + k_1 \cdot C} + \frac{b_2 \cdot k_2 \cdot C}{1 + k_2 \cdot C} \qquad (3)$$

$$Qs = K \cdot C^n \qquad (4)$$

where: Q_s – mass of the sorbed dye [mg/g]
C_o – initial dye concentration [mg/L]
C_s – concentration of dye after sorption [mg/L]
V – volume of the solution [L]
m – mass of the sorbent [g.d.m.]
q_{max} – maximum sorption capacity in the Langmuir equation [mg/g]
K_c – constant in the Langmuir equation [L/mg]
C – concentration of dye remaining in solution [mg/L]
$b_1; b_2$ – maximum capacity of active sites (type I and II) [mg/g]
$k_1; k_2$ – constants in the Langmuir 2 equation [L/mg]
K – constant of sorption equilibrium in Freundlich model
n – parameter of heterogeneity [-]

RESULTS AND DISCUSSION

The effect of pH on the effectiveness of dye sorption

The sorption effectiveness of reactive dyes RB5 and RY84 on the tested sorbent was the highest at pH 2 and decreased with the increasing pH of the solution, obtaining the lowest value at pH 11. The maximum decrease in the sorption efficiency of RB5 and RY84 was found in the pH range 2–4 (Fig. 1a,d).

The decrease of the sorption efficiency of reactive dyes, together with the increase in pH, has also been confirmed in the sorption studies of RB5 and RY84 on compost [Jóźwiak et al. 2013], chitosan sorbents [Filipkowska and Jóźwiak 201] wheat bran [Annadurai et al. 2002] and modified starches [Wang et al. 2010].

As in the case of reactive dyes, the sorption effectiveness of acid dyes on CS was decreased with the increasing pH of the solution. The effect of pH increase on the reduction of binding efficiency of AY23 and AR18 was most evident in the narrow pH range 2–3. (Fig. 1b,e). At pH 11, the sorption of AY23 and AR18 on CS practically did not occur. A very similar tendency was also observed in the studies on AY23 and AR18 sorption on sawdust [Shokoohi et al. 2010].

The increased sorption efficiency of anionic dyes at low pH results from the change in surface charge of the sorbent. Probably, in the low pH, functional groups present in the CS structure (-OH) are protonated, owing to which the sorbent gains a positive charge. The sorbent, which has a positive charge, attracts the anionic electrostatic dyes, supporting their sorption. The interaction between the dye and the nut shell is the stronger the lower the pH of the solution (Fig. 1a,b,d,e).

The sorption efficiency of BV10 on the CS was the highest at pH 3 and decreased with the increasing pH. The biggest decrease in the sorption efficiency of BV10 was found in the range of pH 3–5, and in the range of pH 8–11 (Fig. 1c). The effectiveness of BR46 sorption at the initial pH range of 2–6 was increased along with pH, obtaining the highest value at pH 6. A further increase in pH in the pH range 6–8 decreased the binding efficiency of RB46 on the tested sorbent. A characteristic feature of BR46 solutions is decolorization at pH> 8. For this reason, the effectiveness of BR46 sorption at pH 9–11 has not been included in the graph (Fig. 1f).

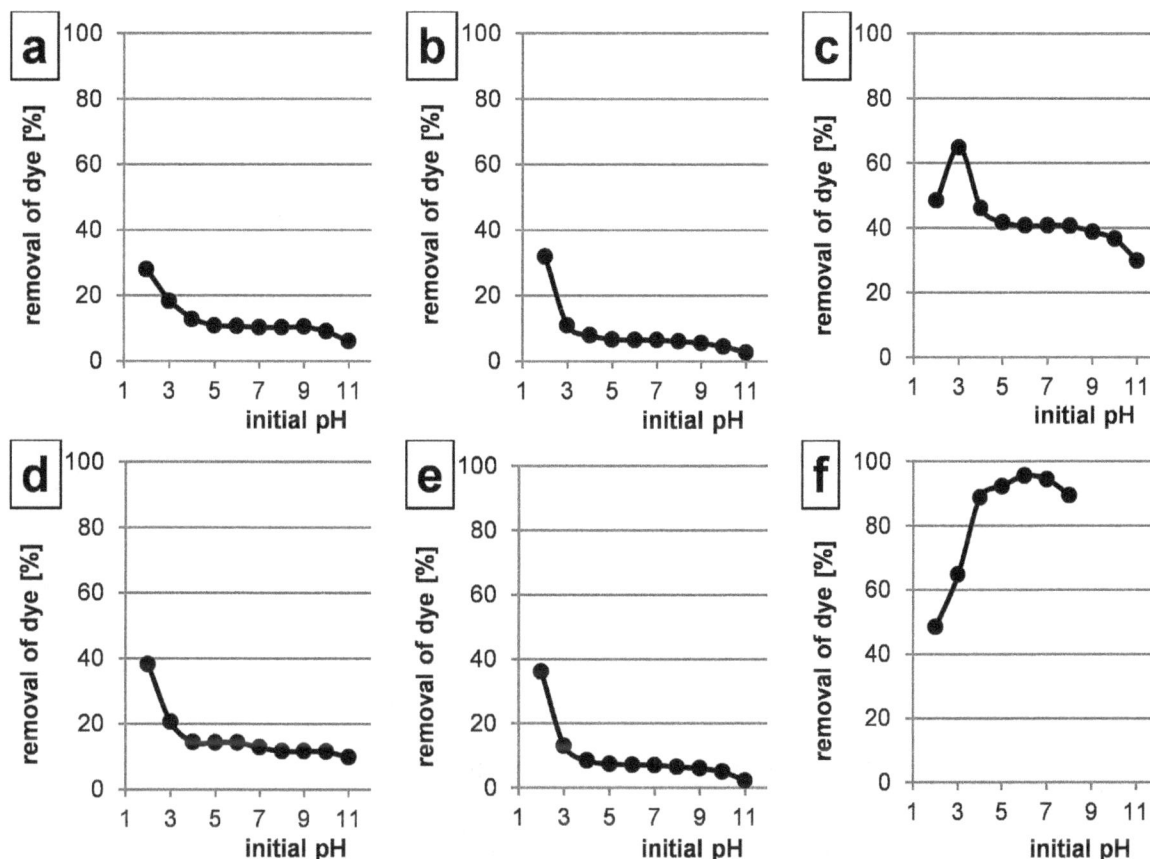

Fig. 1. The effect of pH on the effectiveness of sorption of dyes on CS: a) RB5, b) AY23,
c) BV10, d) RY84, e) AR18, f) BR46

The differences between the optimal pH of sorption of BV10 and BR46 may result from the presence of a carboxylic functional group in the BV10 structure. Generating a local negative charge, the -COOH group (-COO⁻ + H⁺) caused that under certain conditions, at low pH, BV10 behaved like anionic dyes. Because BR46 does not have anionic functional groups, its sorption at low pH was ineffective.

Determination of the equilibrium time of sorption

For reactive dyes RB5 and RY84, irrespective of the initial concentration of the dye, the equilibrium time of sorption was 60 min (Fig. 2 a, d). The highest intensity of sorption of dyes was noted in the first phase of the process.

Similar time of sorption of reactive dyes on biosorbents was obtained in the studies on sorbtion of RB5 on cotton seed shells (60 min) [Uçar et al. 2012] and rapeseed stems (60 min) [Hamzeh et al. 2012].

The equilibrium time of sorption of AR18 and AY23 acid dyes on the tested sorbent was 45 min (Fig. 2b, e). Similarly to the sorption of reactive dyes, the intensity of acid dye binding was highest in the first phase of sorption. The shorter time to obtain the sorption equilibrium of AR18 and AY23 with respect to RB5 and RY84 could result from the lower molar mass of acid dyes, compared to reactive dyes. Presumably, smaller sizes of acid dye molecules allowed them to penetrate into more difficultly available sorption centers located in the deeper layers of the sorbent in a shorter time.

The equilibrium time of sorption of cationic dyes on CS was longer than in the case of anionic dyes. For BR46, the sorption equilibrium was obtained after 90 min of sorption (Figure 2f). The same equilibrium time was also established in the research on Basic Red 46 sorption on pine needles [Deniz and Karaman 2011], as well as fir sawdust [Laasri et al. 2007].

Out of the tested dyes, the longest time of CS sorption, equaling 180 min, was obtained by BV10 (Figure 2c). The BV10 sorption equilibrium time, in comparison to BR46, may have resulted from its higher molar mass and the presence of the -COO⁻ group in the BV10 structure.

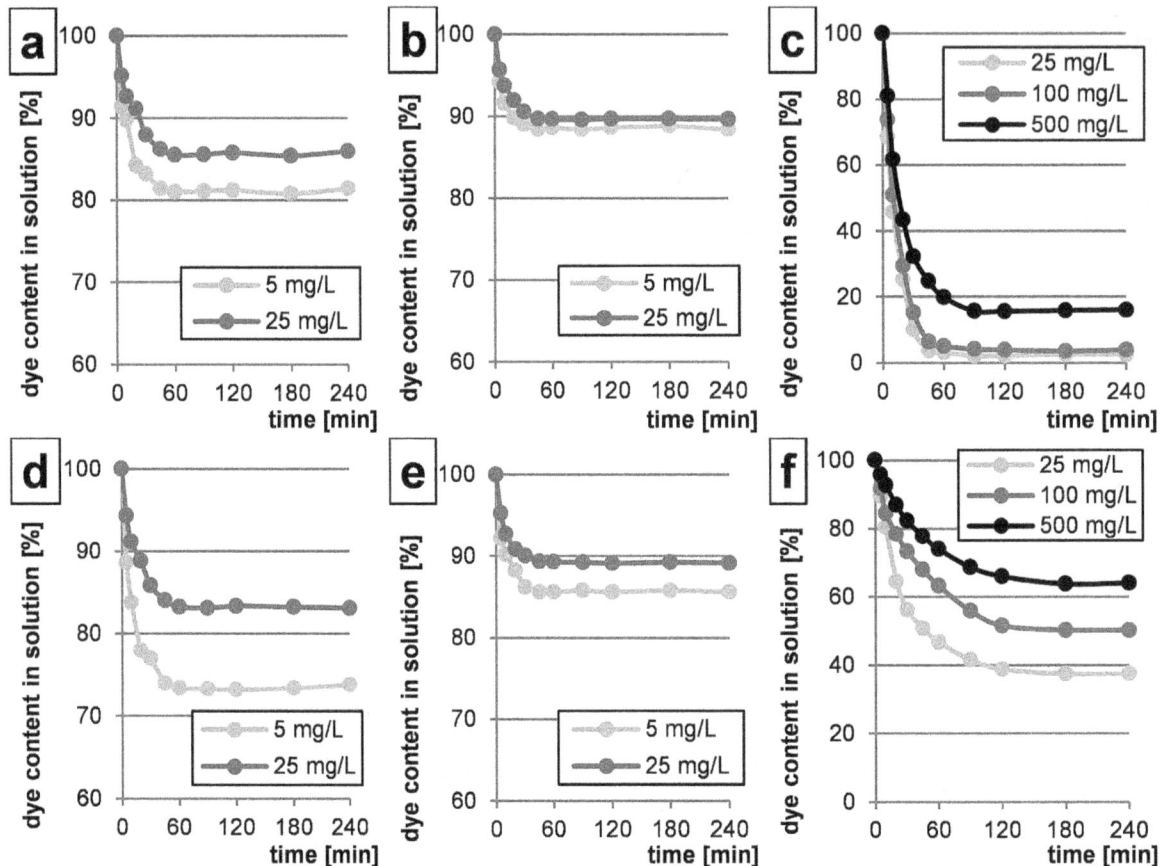

Fig. 2. Percent changes in the concentration of dyes: a) RB5, b) AY23, c) BV10, d) RY84, e) AR18, f) BR46 during sorption using CS as a sorbent

The sorption equilibrium time level of 180 min was also obtained in the studies on Basic Violet 10 sorption on coffee grounds [Shen and Gondal 2013], as well as the leaves of *A. nilotica* [Santhi et al. 2014].

Sorption capacity

On the basis of the coefficient of determination (R^2), it was established that the Langmuir isotherm and the Langmuir double isotherm describe the sorption of CS dyes better than the Freundlich isotherm (Table 2).

In the case of sorption of anionic dyes on CS, the constants determined from the Langmuir and Langmuir 2 isotherm (Q_{max}, K/K_c, R^2) obtain identical values. This may suggest the presence of only one type of sorption center for anionic dyes on CS. The discussed type of sorption center may be a small amount of the displaced hydroxyl groups of polysaccharides and lignins contained in the sorbent.

In turn, the sorption of cationic dyes on CS is better described by the Langmuir 2 isotherm. This

probably indicates the possibility of occurrence on the sorbent of at least two types of active sites relative to cationic dyes. Presumably, the sorption centers may be the primary and secondary hydroxyl groups of cellulose, hemicellulose and lignin, contained in the CS structure.

The sorption capacity of CS in relation to reactive dyes RB5 and RY84 was 0.82 mg/g and 0.96 mg/g. In the case of acid dyes, the obtained sorption capacity was lower and for AY23 and AR18 it was 0.53 mg/g and 0.66 mg/g, respectively (Table 2). The values of K_c constants denoting the degree of dye affinity for the sorbent for all anionic dyes are similar.

CS sorbs the reactive dyes more efficiently than the acidic dyes, because RB5 and RY84 have reactive functional groups. The vinyl sulfone group RB5 or the chlorotriazine group RY84 could support the sorption of dyes on CS by chemisorption.

The repeatedly tested sorbent showed a higher sorption capacity with respect to cationic than anionic dyes. CS in the amount of 1 g was able to bind 28.54 g BV10 or 68.52 g BR46 (Table 2).

Table 2. Constants designated from the isotherms: (Langmuir, Langmuir 2 and Freundlich)

Models (izotherms)	Constants in sorption model	DYE					
		RB5	RY84	AY23	AR18	BV10	BR46
Langmuir	Q_{max} [mg/g s.m.]	0.819	0.962	0.526	0.661	25.825	49.426
	K_c [l/g s.m.]	0.035	0.038	0.043	0.030	0.007	0.059
	R^2	0.992	0.975	0.938	0.981	0.994	0.976
Langmuir 2	Q_{max} [mg/g s.m.]	0.819	0.962	0.526	0.661	28.544	68.515
	b_1 [mg/g s.m.]	0.413	0.481	0.263	0.330	1.994	54.031
	k_1 [L/mg]	0.035	0.038	0.043	0.030	0.086	0.015
	b_2 [mg/g s.m.]	0.406	0.481	0.263	0.331	26.550	14.484
	k_2 [L/mg]	0.035	0.038	0.043	0.030	0.005	0.378
	R^2	0.992	0.975	0.938	0.981	0.996	0.993
Freundlich	K [-]	0.057	0.072	0.045	0.039	0.816	6.102
	n [L/g s.m.]	0.581	0.569	0.544	0.606	0.547	0.457
	R^2	0.979	0.957	0.894	0.965	0.986	0.981

The sorption of cationic dyes on the tested sorbent was promoted by its acidic nature, which was explained in section 4.1.

The k1 and k2 constants determined from the Langmuir 2 model indicate a higher affinity of BR46 for the sorbent compared to BV10. Lower sorption capacities of the tested sorbent with respect to BV10 may result from the greater weight of the dye as well as the presence of a carboxyl group that generates a strong local negative charge.

CONCLUSIONS

The sorption effectiveness of dyes on CS depends on the type of dye. The effectiveness of sorption on the tested sorbent characterizing the cationic dyes is many times higher than that of anionic dyes.

The condition for obtaining satisfactory results of dye sorption is the correction of the pH of the solution in which the sorption takes place. The sorption efficiency of anionic dyes is the highest in the pH range 2–3. Optimal pH of the sorption of cationic dyes should be determined individually for each dye.

The Equilibrium time of dye sorption on CS depends on the type of dyes. The sorption equilibrium time for CS increases in the following series: acidic dyes <reactive dyes <cationic dyes (BR46 <BV10). The equilibrium time may be influenced by: the nature of the dye, its molar mass and functional groups.

Probably, CS have only one type of active site relative to anionic dyes. Relative to cationic dyes, CS presumably have at least two types of sorption centers.

Acknowledgements

This study was financed under Project No. 18.610.008–300 of the University of Warmia and Mazury in Olsztyn, Poland.

REFERENCES

1. Annadurai G., Juang R.S., Lee D.J. 2002. Use of cellulose-based wastes for adsorption of dyes from aqueous solutions. Journal of Hazardous Materials, 92, 263–274.

2. Cagnon B., Py X. Guillot A., Stoeckli F., Chambat G. 2009. Contributions of hemicellulose, cellulose and lignin to the mass and the porous properties of chars and steam activated carbons from various lignocellulosic precursors. Bioresource Technology, 100, 292–298.

3. Deniz F., Karaman S. 2011. Removal of Basic Red 46 dye from aqueous solution by pine tree leaves. Chemical Engineering Journal, 170, 67–74 .

4. Filipkowska U., Jóźwiak T. 2013. Application of chemically-cross-linked chitosan for the removal of Reactive Black 5 and Reactive Yellow 84 dyes from aqueous solutions. Journal of Polymer Engineering, 33, 735–747.

5. Hamzeh Y., Ashori A., Azadeh E., Abdulkhani A. 2012. Removal of Acid Orange 7 and Remazol Black 5 reactive dyes from aqueous solutions using a novel biosorbent. Materials Science and Engineering: C, 32, 1394–1400.

6. Jóźwiak T., Filipkowska U., Rodziewicz J., Mielcarek A., Owczarkowska, D. 2013). Zastosowanie

kompostu jako taniego sorbentu do usuwania bar-wników z roztworów wodnych. Annual Set The Environment Protection, 15, 2398–2411.

7. Laasri L., Elamrani M.K., Cherkaoui O. 2007. Removal of two cationic dyes from a textile efflu-ent by filtration-adsorption on wood sawdust. En-vironmental Science and Pollution Research, 14, 237–240.

8. Liang C.Z., Sun S.P., Li F.Y., Ong Y.K., Chung T.S. 2014. Treatment of highly concentrated wastewater containing multiple synthetic dyes by a combined process of coagulation/flocculation and nanofiltra-tion. Journal of Membrane Science, 469, 306–315.

9. Nandi B.K., Patel S. 2013. Effects of operational parameters on the removal of brilliant green dye from aqueous solutions by electrocoagulation. Ara-bian Journal of Chemistry, 10 (S2), 2961–2968.

10. Robinson T., McMullan G., Marchant R., Nigam P. 2001. Remediation of dyes in textile effluent: a critical review on current treatment technologies with a proposed alternative. Bioresource Technol-ogy, 77, 247–255.

11. Santhi T., Prasad A.L., Manonmani S. 2014. A comparative study of microwave and chemically

treated Acacia nilotica leaf as an eco friendly ad-sorbent for the removal of rhodamine B dye from aqueous solution. Arabian Journal of Chemistry, 7, 494–503.

12. Shen K., Gondal M.A. 2013. Removal of hazard-ous Rhodamine dye from water by adsorption onto exhausted coffee ground. Journal of Saudi Chemi-cal Society, 21, 120–127

13. Shokoohi R., Vatanpoor V., Zarrabi M., Vatani A. 2010. Adsorption of Acid Red 18 (AR18) by Ac-tivated Carbon from Poplar Wood – A Kinetic and Equilibrium Study. E-Journal of Chemistry, 7, 65–72.

14. Uçar D., Armağan B. 2012. The removal of reac-tive black 5 from aqueous solutions by cotton seed shell. Water Environment Research, 84, 323–327.

15. Wang Z., Xiang B., Cheng R., Li Y. 2010. Behav-iors and mechanism of acid dyes sorption onto diethylenetriamine-modified native and enzymatic hydrolysis starch. Journal of Hazardous Materials, 183, 224–232.

16. Wijannarong S. Aroonsrimorakot S., Thavipoke P., Kumsopa C., Sangjan S. 2013. Removal of Reac-tive Dyes from Textile Dyeing Industrial Effluent by Ozonation Process. APCBEE Procedia, 5, 279–282.

16

The use of Macroelements from Municipal Sewage Sludge by the Multiflora Rose and the Virginia fanpetals

Jacek Antonkiewicz[1*], Barbara Kołodziej[2], Elżbieta Jolanta Bielińska[3], Katarzyna Gleń-Karolczyk[4]

[1] Department of Agricultural and Environmental Chemistry, University of Agriculture in Krakow, Poland

[2] Department of Industrial and Medicinal Plants, University of Life Sciences in Lublin, Poland

[3] Institute of Soil Science, Environment Engineering and Management, University of Life Sciences in Lublin, Lublin, Poland

[4] Department of Agricultural Environment Protection, University of Agriculture in Krakow, Poland

* Corresponding author's e-mail: rrantonk@cyf-kr.edu.pl

ABSTRACT

Municipal sewage sludge contains many valuable nutrients which can be used in the cultivation of energy crops. Application of large doses of sewage sludge can be a cause of environmental pollution, especially with nutrients. The multiflora rose and the Virginia fanpetals are plants with high nutritional requirements. The use of municipal sewage sludge in the cultivation of energy crops will allow recycling the nutrients from this organic waste. The aim of the study was to evaluate the use of macroelements from municipal sewage sludge by the multiflora rose var. "Jatar" (*Rosa multiflora* Thunb. ex Murray) and the Virginia fanpetals (*Sida hermaphrodita* Rusby). Four levels of sewage sludge fertilization were applied in the 6-year field experiment: 0, 10, 20, 40, 60 Mg DM sludge · ha[-1]. Sewage sludge was applied once before planting energy crops. Due to the low potassium content in sewage sludge, a single supplementary fertilization with 100 kg K · ha[-1] in the form of 40% potassium salt (KCl) was applied on each plot. The study involved the evaluation of the yield, uptake and use by energy plants of N, P, K, Ca, Mg, and Na from sewage sludge. It was found that the increasing doses of sewage sludge significantly raised the multiflora rose and the Virginia fanpetals biomass yields. The yield of the Virginia fanpetals was one and a half times higher than that of the multiflora rose. The increasing doses of sewage sludge significantly raised the contents and uptake of N, P, K, Ca, Mg, and Na by these plants. The highest uptake of macronutrients by the multiflora rose and the Virginia fanpetals crops was determined for 60 Mg DM · ha[-1] fertilization dose. The results show that the Virginia fanpetals used N, P, K, Ca, Mg, and Na from the sewage sludge to a greater extent than the multiflora rose. The analyses indicate that due to the greater yields, bioaccumulation and uptake of macronutrients, Virginia fanpetals is more effective in the 'purification' of the substrate from excess nutrients that may pose a threat to the environment.

Keywords: multiflora rose, Virginia fanpetals, macroelements, phytoremediation, municipal sewage sludge

INTRODUCTION

The application of municipal sewage sludge in the cultivation of energy crops is one of the alternative methods of its disposal [Nahm, Morhart 2018; Kołodziej et al. 2015]. The high content of organic matter (more than 50% on average) is the property of sewage sludge that favors its natural management [Helios et al. 2014]. What is more, municipal sewage sludge contains almost all nutrients necessary for plants. Cultivation of energy crops usually requires high contents of ni-

trogen and phosphorus, which can be greater than in natural fertilizers [Zapałowska et al. 2017]. In turn, the amount of potassium in sewage sludge is lower than in organic fertilizers, composts, and natural fertilizer [Awasthi et al. 2017]. Additionally, after its application to the soil, municipal sewage sludge improves its physical, chemical and biological properties [Abdul Khaliq et al. 2017]. Unfortunately, sewage sludge may also contain significant amounts of heavy metals that can make it inadequate for natural management [Regulation 2015]. The excessive application of sewage sludge may result in soil contamination with nutrients that may contribute to water eutrophication [Lee et al. 2018]. Previous studies show that the management of municipal sewage sludge in the cultivation of energy and industrial plants may be an alternative method of nutrient recycling from this waste [Antonkiewicz 2010; Nahm, Morhart 2018].

With appropriate fertilization, energy crops are characterized by high yields and effectiveness in using nutrients [Kołodziej et al. 2016]. Previous studies showed that plants such as the Virginia fanpetals and the multiflora rose are very effective in the extraction of heavy metals from sewage sludge [Antonkiewicz et al. 2017; Korzeniowska, Stanisławska-Glubiak 2015]. Therefore, studies were carried out in order to determine the suitability of these species for phytoextraction of nutrients from municipal sewage sludge.

The aim of the study was to evaluate, under field conditions, which species are most effective in the use (phytoextraction) of N, P, K, Ca, Mg, and Na from municipal sewage sludge. The phytoextraction potential of the tested energy crops was assessed taking into account the yield, content, uptake, and balance of macroelements, stating the increase, decrease and use of these components.

The selection of the energy plant species with the best efficiency and suitability for biomass production for energy purposes and with high phytoextraction capabilities was important for the implementation of the above-mentioned study objective [Antonkiewicz et al. 2017; Korzeniowska, Stanisławska-Glubiak 2015]. This constitute a proposal of a solution for the development of wastelands, marginal soils, and post-industrial areas, where there is usually no agricultural production or cultivation of consumer crops, and such development can bring a positive ecological and economic effect [Nahm, Morhart 2018; Schröder et al. 2018].

MATERIAL AND METHODS

The studies on the effect of increasing doses of municipal sewage sludge on the use of macroelements by the multiflora rose and the Virginia fanpetals were carried out in 2008–2013 in a municipal wastewater treatment plant in Janów Lubelski [50°43'17.7''N 22°22'08,0"E] located in southeastern Poland. This study is a continuation of the research on the extraction of heavy metals from

municipal sewage sludge by the afore-mentioned plant species [Antonkiewicz et al. 2017].

Soil and municipal sewage sludge

The soil on which the experiment was set up is classified as clay loam (CL) (Table 1) [Polish Soil Classification 2011; Soil Survey Staff 2014]. The soil had slightly acidic pH, low available phosphorus and potassium contents, and very low magnesium content. The content of Cr, Ni, Cu, Zn, Cd, and Pb in the soil did not exceed the limit values for the reclamation of municipal sewage sludge [Regulation 2016].

The municipal sewage sludge came from the municipal wastewater treatment plant in Janów Lubelski and this organic waste, catalog number 19 08 05 [Waste Catalogue 2014], was stabilized and sanitized before use. The municipal sewage sludge was applied once in late autumn 2007 – it was mixed with a 20 cm surface layer of soil. Due to the low potassium content in sewage sludge, a single supplementary fertilization with 100 kg K $\cdot$ ha^{-1} in the form of 40% potassium salt (KCl) was applied on each plot. No phosphorus fertilization was used in the field experiment, because the element content in the municipal sewage sludge satisfied the energy plants' demand for this component. The determined contents of Cr, Ni, Cu, Zn, Cd, Pb, and Hg in the municipal sewage sludge did not exceed the limit values for the reclamation of this waste [Regulation 2015]. No microbiological contamination was discovered in the sewage sludge used in the experiment.

Design and conditions of the experiment

The two-factor field experiment was set up using the randomized block method on 14.4 m^2 plots, in three replicates. The first experimental factor was the dose of municipal sewage sludge. The experimental design consisted of 5 treatments: 1) control; 2) 10 Mg DM; 3) 20 Mg DM; 4) 40 Mg DM; and 5) 60 Mg DM of municipal sewage sludge/1 ha. The second experimental factor comprised two species of energy plants: the multiflora rose var. "Jatar" (*Rosa multiflora* Thunb. ex Murr.) and the Virginia fanpetals (*Sida hermaphrodita* Rusby).

Woody cuttings (25 cm long) of the multiflora rose and roots sections (8–12 cm long with several buds) of Virginia fanpetals were planted on April 22, 2008 at a spacing of 0.75×0.8 m and 0.75×0.4 m, respectively.

The determination of dry matter yield, macroelements content and soil enzymatic activity

Each year (2008–2013), the tested crops were harvested in autumn, at the turn of October and November. After the annual harvest, the plant material from each plot was dried at 70°C in a forced air circulation dryer, and, subsequently, the air-dry mass was determined. The samples of the analyzed plants were subjected to dry mineralization in a muffle furnace at 450°C [Ostrowska et al. 1991].

After microwave digestion in a mixture of concentrated HCl and HNO_3 (3:1, v/v), the contents of elements (P, K, Na, Mg, and Ca) determined in the air-dried samples of soil and municipal sewage sludge were similar to the total contents [Ostrowska et al. 1991]. After mineralization of plant and soil material, the contents of these elements were determined using an ICP-OES (Inductively Coupled Plasma – Optical Emission Spectroscopy) emission spectrometer [Jones and Case 1990]. Total nitrogen content in the tested plants (plant material) and municipal sewage sludge was determined by the Kjeldahl distillation method [Ostrowska et al. 1991].

The soil pH in 1 mol · dm^{-3} KCl was determined with potentiometric method, the available P and K content was determined using the Egner-Riehm method, while the available Mg content was determined according to the Schachtschabel method (Ostrowska et al. 1991). Each year, during the vegetation season, in May, soil samples were collected from each plot (each repetition) from 0.20 cm depth, using the Egner's soil probe sampler, in order to assess the enzymatic activity of the soil. The analyses of the soil also involved determinations pertaining to the activities of enzymes which play a key role in the stable mineralization of organic matter and in supplying nutrients to the roots of energy crops. The activity of the studied enzymes was determined using the following methods: the activity of dehydrogenases with a TTC (triphenyl tetrazolium chloride) substrate using the Thalmann method (Thalmann 1968); the activity of acid phosphatase and alkaline phosphatase using the Tabatabai and Bremner method (Tabatabai and Bremner 1969); the activity of urease using the Zantua and Bremner method (Zantua and Bremner 1975); the activity of protease using the Ladd and Butler method (Ladd and Butler 1972). The activity of dehydrogenases was given in cm^3 H$_2$, necessary for reducing TTC to TFP (triphenyl phormo-

san); of phosphatases – in mmols of p–nitrophenol (PNP) produced from sodium 4-nitrophenylphosphate; urease – in mg N-NH$_4^+$ generated from hydrolyzed urea; protease – in mg tyrosine developed from sodium caseinate. The results of the analyses pertaining to the enzymatic activity of the soil were presented in the paper as means for the six years of studies, i.e. for the 2008–2013 period.

Analytical quality control

The ICP-OES Optima 7300 DV, atomic emission spectrometer from Perkin Elmer Company was used for the determination of macroelements in plant and soil materials. Determinations in each of the analyzed samples were carried out in three replications. The quantitative analysis mode was used for the data acquisition of the samples. The scanning of each single sample was repeated three times to gather reasonably good results. During measurements, care was taken to avoid the memory effect and therefore a wash-out time of 0.5 min was used. The accuracy of the analytical methods was verified based on certified reference materials: CRM IAEA/V – 10 Hay (International Atomic Energy Agency), CRM – CD281 – Rey Grass (Institute for Reference Materials and Measurements), CRM023–050 – Trace Metals – Sandy Loam 7 (RT Corporation).

Computations and statistical analysis

Due to the cultivation of various plant species and the variability of conditions in individual years, the contents of N, P, K, Na, Mg, and Ca in the total plant yield is presented as the weighted average from 2008–2013. The macroelement uptake (U) was computed by multiplying the dry matter yield (Y) and macroelement content (C) according to the formula: $U = Y \cdot C$. The macroelement balance (B) was computed as a difference between the amount of elements introduced (I) with the sewage sludge dose and the amount of macronutrients uptaken (U) with the plant yield, according to the formula: $B = I - U$. The simplified balance did not include the supply of macroelements from atmospheric precipitation, mineralization of organic matter, and leaching of macroelements into the soil profile. The phytoremediation of macroelements presented in the balance constitutes the percentage of macronutrient uptake by plants in relation to the amounts introduced into the soil with municipal sewage sludge.

The statistical analysis of the results was carried out using the Microsoft Office Excel 2003 spreadsheet and the Statistica v. 10 PL package. The statistical evaluation of the variability of the results was carried out using the two-way analysis of variance. The significance of differences between mean values was verified based on the Tukey's t-test at the significance level of $\alpha \leq 0.05$. The value of Pearson's linear correlation coefficient (r) was calculated for some relations (parameters) at $p \leq 0.05$. A maximum 5% dispersion of measurements in a chemical analysis was assumed in the study.

RESULTS

Plant yield

The dry mass yields of the multiflora rose and the Virginia fanpetals obtained in individual years of investigations were presented in an earlier publication [Antonkiewicz et al. 2017].

Plant yielding was an important indicator of the plant's response to the sewage sludge applied. The average yield of energy crops obtained from several years (2008–2013) ranged from 3.42 to 14.78 Mg DM · ha^{-1} and depended on the sewage sludge dose and plant species (Figure 1). Close correlations between the dose of sewage sludge and the average yield of energy crops (r=0.771375) were demonstrated in the field experiment.

During the multi-year research cycle, it was shown that a single application of 10–60 Mg DM · ha^{-1} of sewage sludge significantly increased plant yields compared to the control. The application of the largest dose of sewage sludge (60 Mg DM · ha^{-1}) resulted in an increase in the average yield of the multiflora rose and Virginia fanpetals by over 264% and 90%, respectively, in relation to the control (Figure 1). The study showed that the multiflora rose responded to the application of sewage sludge with a higher yield increase compared to the Virginia fanpetals.

On the other hand, when comparing the yield of energy crops, one could notice that the yield potential of the Virginia fanpetals was greater than that of the multiflora rose. The greatest difference in plant yielding was noted in the control treatment where neither mineral nor organic fertilizations were applied. The difference between the species in the control treatment was over 128%. Subsequent doses of sewage sludge decreased the differences in yields between the tested species (for the highest dose of sewage sludge, the difference between these species was over 18%). The study showed that the largest yield-forming effect for the multiflora rose and the Virginia fanpetals was achieved in the treatment where sewage sludge at a dose of 60 Mg·ha^{-1} DM was applied (Figure 1).

Macroelement content in plants

The chemical analysis showed that the content of N, P, and Na in the municipal sewage sludge was over 72.7; 34.7; and 8.3-fold higher, respectively, compared to the topsoil (Table 1). In

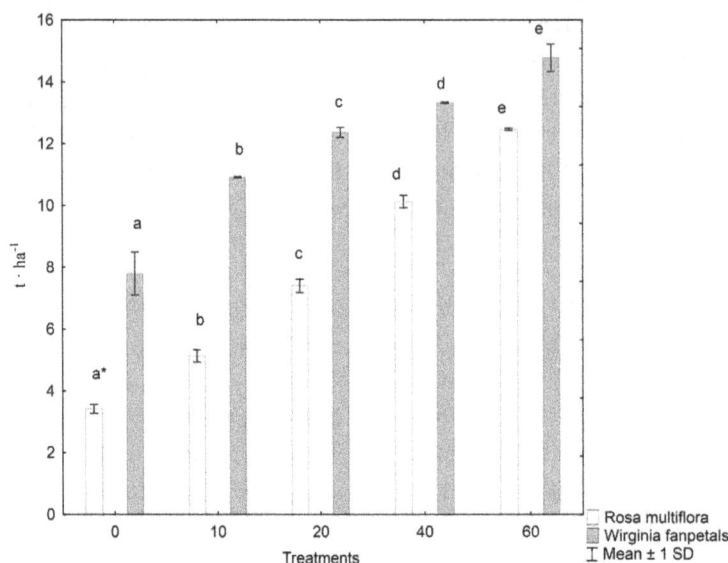

Figure 1. The average yield of the energy crops

contrast, the total contents of K, Mg, and Ca in soil were over 1.0; 0.3; 2.1-fold higher that these contents determined in sewage sludge. The investigations revealed that the applied municipal sewage sludge was a potential source of N, P, and Na for energy crops. The total contents of K, Mg, and Ca determined in the soil can be taken into account for the assessment of their bioavailability for plants, due to their presence in hardly accessible forms (e.g. soil minerals).

The macroelement content in the energy plants obtained in the control treatment was the lowest compared to the treatments with sewage sludge applied (Table 2). The use of increasing doses of sewage sludge (10–60 Mg DM · ha^{-1}) significantly raised the content of macroelements in energy plants.

The largest increases in the content of macroelements in energy crops were recorded after the use of sewage sludge at a dose of 60 Mg DM · ha^{-1}. In the case of the multiflora rose, the increase in macroelement content was: 74% for N, 95% for P, 58% for K, 37% for Na, 33% for Mg and Ca, respectively, relative to the control. For the Virginia fanpetals biomass, the highest increase in macronutrients was: 82% for N, 81% for P, 49% for K, 40% for Na, 77% for Mg, and 53% for Ca,

respectively, relative to the control. The investigations showed that regardless of the plant species, the application of the highest dose of sewage sludge resulted in the increase in the contents of the tested elements: the P and N content was raised the most, then Mg and K, and the lowest content increase was recorded for Ca and Na.

At the highest dose of sewage sludge, the Virginia fanpetals accumulated more N, Mg, Ca, and Na than the multiflora rose. The multiflora rose exhibited higher contents of P and K than the Virginia fanpetals. The research proved that increasing doses of sewage sludge differentiated the content of macroelements in the tested energy plant species.

The analysis of the Pearson's linear correlation coefficient (r) revealed close relationships between the sewage sludge dose and the content of N, P, K, Na, Mg, Ca in energy plants (r = 0.859526 – 0.966288). Additionally, a strong link between the amount of plant biomass and the content of the macroelements was discovered in these plants (r = 0.761727 – 0.954657). The correlations mentioned above indicate a significant effect of sewage sludge on the quality of biomass, evaluated in terms of energy as well as phytoremediation.

Table 1 Selected physical and chemical properties of the soil before experiment establishment and chemical composition of municipal sewage sludge used

Parameter	Unit	Content in the soil layer		Content in sewage sludge
		0–20 cm	20–40 cm	
Fraction 2–0.05 mm		32	23	-
Fraction 0.05–0.002 mm	%	39	45	-
Fraction <0.002 mm		29	32	-
pH$_{KCl}$		6.29	6.44	6.04
Organic matter	g · kg^{-1} DM	14.5	14.1	594.0
Available phosphorus (P)		30.9	29.6	2.25
Available potassium (K)	mg · kg^{-1} DM	91.3	60.6	Bdl*
Available magnesium (Mg)		27.6	24.6	0.28
Total nitrogen (N)		1.01	-	74.5
Total phosphorus (P)		0.35	-	12.5
Total potassium (K)	g · kg^{-1} DM	3.83	-	1.90
Total sodium (Na)		0.16	-	1.50
Total magnesium (Mg)		3.71	-	2.80
Total calcium (Ca)		8.95	-	2.90
Total chromium (Cr)		9.66	9.89	25.4
Total nickel (Ni)		6.39	6.31	14.8
Total copper (Cu)		3.20	3.60	111
Total zinc (Zn)	mg · kg^{-1} DM	31.97	31.00	1005
Total cadmium (Cd)		<0.27	<0.27	2.35
Total lead (Pb)		13.67	13.63	42.9

*Bdl – Below detection level

Table 2 Weighted average of macroelement content in energy plants (g · kg^{-1} DM)

Treatments	N	P	K	Na	Mg	Ca
Sludge dose (Mg DM · ha^{-1})	Rosa multiflora					
0	9.4	0.6	3.0	0.2	0.7	7.4
10	12.6	0.6	3.2	0.2	0.8	7.9
20	14.4	0.7	3.7	0.2	0.8	8.5
40	15.8	0.7	4.1	0.3	0.9	9.4
60	16.3	1.1	4.8	0.3	1.0	9.8
Mean	13.7	0.7	3.7	0.2	0.8	8.6
CV (%)*	19.4	28.9	18.9	12.8	11.9	12.5
Sludge dose (Mg DM · ha^{-1})	Sida hermaphrodita					
0	8.8	0.4	2.5	0.2	0.4	3.4
10	10.5	0.5	2.7	0.2	0.4	3.8
20	13.7	0.5	2.9	0.3	0.6	4.2
40	14.4	0.6	3.2	0.3	0.7	4.7
60	16.1	0.7	3.7	0.3	0.8	5.2
Mean	12.7	0.6	3.0	0.3	0.6	4.3
CV (%)*	21.8	21.4	14.5	17.4	26.2	16.1
Sludge dose (Mg DM · ha^{-1})	Mean for dose of the sewage sludge					
0	9.1	0.5	2.7	0.2	0.6	5.4
10	11.5	0.6	3.0	0.2	0.6	5.8
20	14.0	0.6	3.3	0.3	0.7	6.4
40	15.1	0.6	3.6	0.3	0.8	7.0
60	16.2	0.9	4.2	0.3	0.9	7.5
LSD for dose**	0.40	0.05	0.18	0.01	0.03	0.34
LSD for species	0.63	0.08	0.28	0.01	0.05	0.54
LSD for interaction	0.89	0.11	0.40	0.02	0.07	0.77

* CV – Variability Coefficient
**LSD – Least Significant Differences

Macroelement uptake by plants

Table 3 shows the macroelement uptake by energy plants as the sum of data from the entire experiment period, i.e. 2008–2013.

The amount of macroelements taken up by the tested species depended on the yield and the content of elements in individual species (Figure 1, Table 2). Increasing doses of municipal sewage sludge significantly raised the uptake of N, P, K, Na, Mg, and Ca with the yield of energy crops (Table 3). The lowest macroelement uptake with the yield of multiflora rose and Virginia fanpetals was determined in the control treatment where no sewage sludge was applied. On the other hand, the largest macroelement uptake with the yield of the tested species was found in the treatment with the maximum dose of sewage sludge. The increase in the N, P, K, Na, Mg, and Ca uptake by the multiflora rose, after the application of 60 Mg DM · ha^{-1} of sewage sludge was over

534%; 611%; 477%; 401%; 394%, and 387%, respectively, higher than for the control treatment. In the case of Virginia fanpetals, the increase in the N, P, K, Na, Mg, and Ca uptake at the highest dose of sewage sludge was over 245%; 244%; 183%; 166%; 235%, and 190%, respectively, compared to the control treatment. The investigations showed that the increasing doses of sewage sludge made the multiflora rose respond with more intensive macroelement uptake compared to the Virginia fanpetals (Table 3).

A comparison of the macroelement uptake at the highest dose of sewage sludge revealed a greater uptake of N and Na with the biomass yield by the Virginia fanpetals than by the multiflora rose. On the other hand, the multiflora rose took up more P, K, Mg, and Ca than the Virginia fanpetals. The study showed that the greatest difference in the amount of macroelements taken up by the tested plant species was recorded in the control treatment where no sewage sludge was applied.

Table 3. Total macroelement uptake by energy plants (kg · ha^{-1})

Treatments	N	P	K	Na	Mg	Ca
Sludge dose (Mg DM · ha^{-1})	Rosa multiflora					
0	192.0	11.6	61.7	4.4	14.8	150.9
10	388.3	18.9	98.9	6.8	24.1	244.5
20	636.5	29.6	162.0	10.6	36.8	379.0
40	959.2	43.6	247.6	15.9	55.9	569.4
60	1218.4	82.5	356.2	22.1	73.1	735.5
Mean	678.9	37.3	185.3	12.0	40.9	415.8
CV (%)*	57.0	70.8	60.1	55.4	54.0	53.5
Sludge dose (Mg DM · ha^{-1})	Sida hermaphrodita					
0	412.9	19.3	115.8	11.4	20.3	159.9
10	684.7	32.0	179.8	14.0	26.0	246.6
20	1013.0	39.3	218.6	21.0	47.0	312.4
40	1148.2	45.5	257.6	24.4	56.3	375.3
60	1426.9	66.5	327.8	30.3	68.1	465.1
Mean	937.2	40.5	219.9	20.2	43.5	311.8
CV (%)*	39.3	40.3	33.8	35.5	43.1	35.0
Sludge dose (Mg DM · ha^{-1})	Mean for dose of the sewage sludge					
0	302.4	15.5	88.7	7.9	17.5	155.4
10	536.5	25.5	139.3	10.4	25.0	245.5
20	824.8	34.5	190.3	15.8	41.9	345.7
40	1053.7	44.6	252.6	20.1	56.1	472.3
60	1322.6	74.5	342.0	26.2	70.6	600.3
LSD for dose**	25.77	3.29	12.17	0.62	2.22	20.53
LSD for species	40.75	5.20	19.25	0.98	3.51	32.45
LSD for interaction	57.62	7.36	27.22	1.39	4.97	45.90

* CV – Variability Coefficient
**LSD – Least Significant Differences

Significant correlations between the dose of sewage sludge and the macroelement uptake by the tested species was demonstrated (r = 0.836856 – 0.971657). The uptake of macroelements by plants was also strongly correlated with the mean yield of the tested species (r = 0.582655 – 0.963922). The study also showed significant relationships between the content of macroelements in plants and their uptake (r = 0.813716 – 0.90496).

Simplified balance and macroelement phytoremediation

The use of municipal sewage sludge in the post-industrial areas, poor in nutrients will increase the amount of macroelements available there.

Understanding the macroelement cycle in the soil-plant system will allow a better assessment of the use of these components from organic waste. Such an assessment can be made on the basis of a simplified balance of macroelements and phytoremediation.

The macroelement balance was determined by the introduction of components with sewage sludge and the total uptake of macroelements with the plant yield (Table 4). In control treatments where no sewage sludge was applied, the total balance of macroelements was always negative. The negative balance resulted from the simplified balance which did not include the supply of macroelements from external sources (amounts available in soil, mineralization of organic matter, atmospheric precipitation).

In the treatments fertilized with sewage sludge, a negative balance for K and Ca was also determined. This balance indicates that K and Ca were taken up by the tested species in larger amounts, compared to the amounts introduces with a sewage sludge dose.

Table 4. Simplified balance of macroelements after six years of research

Treatments	Introduced	Uptake	Balance	Recovery	Uptake	Balance	Recovery
	kg · ha⁻¹			%	kg · ha⁻¹		%
		Rosa multiflora			Sida hermaphrodita		
Sludge dose (Mg DM · ha⁻¹)				N			
0	0	192	-192	0	413	-413	0
10	745	388	357	52	685	60	92
20	1490	637	853	43	1013	477	68
40	2980	959	2021	32	1148	1832	39
60	4470	1218	3252	27	1427	3043	32
Sludge dose (Mg DM · ha⁻¹)				P			
0	0	12	-12	0	19	-19	0
10	125	19	106	15	32	93	26
20	250	30	220	12	39	211	16
40	500	44	456	9	45	455	9
60	750	83	667	11	66	684	9
Sludge dose (Mg DM · ha⁻¹)				K			
0	0	62	-62	0	116	-116	0
10	119	99	20	83	180	-61	151
20	138	162	-24	117	219	-81	158
40	176	248	-72	141	258	-82	146
60	214	356	-142	166	328	-114	153
Sludge dose (Mg DM · ha⁻¹)				Na			
0	0	4	-4	0	11	-11	0
10	15	7	8	45	14	1	93
20	30	11	19	35	21	9	70
40	60	16	44	27	24	36	41
60	90	22	68	25	30	60	34
Sludge dose (Mg DM · ha⁻¹)				Mg			
0	0	15	-15	0	20	-20	0
10	28	24	4	86	26	2	93
20	56	37	19	66	47	9	84
40	112	56	56	50	56	56	50
60	168	73	95	44	68	100	41
Sludge dose (Mg DM · ha⁻¹)				Ca			
0	0	151	-151	0	160	-160	0
10	29	244	-215	843	247	-218	850
20	58	379	-321	653	312	-254	539
40	116	569	-453	491	375	-259	323
60	174	735	-561	423	465	-291	267

The increasing doses of sewage sludge caused the positive balance for N, P, Na, and Mg. This positive balance of the above-mentioned macroelements resulted from the greater supply of these components with sewage sludge compared to the amounts taken up by the tested plant species. The greatest balance difference for macroelements was determined in the treatments in which the highest dose of sewage sludge was applied. The balance difference resulted from the amount of macronutrients introduced into the soil.

The largest N, P, Na, and Mg phytoremediation was recorded in treatments fertilized with the smallest doses of sewage sludge (10 Mg DM · ha⁻¹). The increasing doses of sewage sludge systematically decreased phytoremediation in comparison with the lowest dose of sewage sludge. The lowest macroelement phytoremediation probably resulted from the size of the sewage sludge dose, yielding, and uptake of these components by the tested species.

Among the tested plant species, the Virginia fanpetals used N, P, K, Na, and Mg (phytoremediation) to a greater extent than the multiflora rose, which was related to the greater yielding and uptake of these elements. In the case of Ca, greater phytoremediation of this element was determined for the multiflora rose than for the Virginia fanpetals.

It was found that among the evaluated macroelements, the percentage of Ca and K phytoremediation was the highest (more taken up than introduced). The second highest phytoremediation was exhibited by Mg, Na, and N, and the last P.

Soil enzymatic activity

The experiment also involved the examination of the soil enzymatic activity after applying the increasing doses of sewage sludge (Table 5). A significant effect of increasing the doses of sewage sludge on soil microorganisms and their enzymatic activity was revealed. The lowest enzymatic activity was determined in the control treatment.

In the case of the multiflora rose, at the highest sewage sludge dose (60 Mg DM · ha^{-1}), we discovered an over 2.6, 2.0, 2.1, 2.0, 1.5-fold increase in the soil enzymatic activity for dehydrogenase, urease, protease, acid and alkaline phosphatases, respectively, compared to the control. In turn, for the Virginia fanpetals, the soil enzymatic activity increased over 2.9, 1.9, 2.1, 2.6, 2.4-fold, respectively, in comparison with the control treatment.

Table 5. Soil Enzymatic activity (average from 2008–2013)

Treatments	Dehydrogenases activity (cm^3 H$_2$ · kg^{-1} · d^{-1})	Urease activity (mg N-NH$_4^+$ · kg^{-1} · h^{-1})	Protease activity (mg of tyrosine · kg^{-1} · h^{-1})	Acid phosphatase activity (mmol PNP · kg^{-1} · h^{-1})	Alkaline phosphatase activity (mmol PNP · kg^{-1} · h^{-1})
Sludge dose (Mg DM · ha^{-1})	Rosa multiflora				
0	4.7	7.9	10.2	23.1	17.3
10	4.8	9.7	10.8	24.2	17.6
20	5.6	11.7	16.4	35.1	22.2
40	8.4	16.7	24.4	72.2	51.7
60	17.0	24.1	32.5	69.4	43.3
Mean	8.1	14.0	18.9	44.8	30.4
CV (%)*	59.8	43.4	47.1	50.1	49.3
Sludge dose (Mg DM · ha^{-1})	Sida hermaphrodita				
0	4.3	7.5	10.5	26.3	18.1
10	5.4	9.3	13.1	31.8	20.6
20	7.6	12.1	17.5	43.4	28.4
40	11.7	16.3	27.0	98.2	66.6
60	16.9	22.2	32.6	95.7	63.0
Mean	9.2	13.5	20.2	59.1	39.3
CV (%)*	52.6	41.1	44.0	58.7	59.6
Sludge dose (Mg DM · ha^{-1})	Mean for dose of the sewage sludge				
0	4.5	7.7	10.4	24.7	17.7
10	5.1	9.5	12.0	28.0	19.1
20	6.6	11.9	17.0	39.2	25.3
40	10.0	16.5	25.7	85.2	59.1
60	16.9	23.2	32.6	82.5	53.1
LSD for dose**	0.52	0.80	1.40	7.74	5.64
LSD for species	0.82	1.26	2.22	12.23	8.92
LSD for interaction	1.16	1.79	3.14	17.30	12.62

* CV – Variability Coefficient

**LSD – Least Significant Differences

The investigations showed that regardless of the plant species, the application of the highest dose of sewage sludge resulted in the highest increase in the activity of dehydrogenase, and the lowest in the case of alkaline phosphatase. The field experiment revealed that the Virginia fanpetals stimulated greater activity of dehydrogenase, protease, as well as acid and alkaline phosphatases than the multiflora rose. On the other hand, the multiflora rose stimulated greater activity of urease than the Virginia fanpetals.

Large amounts of nutrients, organic colloids, and soil microorganisms were introduced with the municipal sewage sludge, which had a stimulating effect on the soil enzymes. This phenomenon was confirmed by significant correlations between the sewage sludge dose and soil enzymatic activity (r = 0.816675 – 0.979479). The field experiment also showed a close link between the activity of soil enzymes and the yield of the tested plants (r = 0.721015 – 0.793485). The activity of soil enzymes was correlated with the content of macroelements in plants (r = 0.5497994 – 0.982658) as well. Significant relationships were also demonstrated between the activity of soil enzymes and the uptake of macroelements by the tested plant species (r = 0.626945 – 0.948615).

DISCUSSION

The use of the municipal sewage sludge in degraded, marginal, and post-industrial areas improves the physicochemical properties of these lands [Kicińska et al. 2018; Korzeniowska, Stanisławska-Glubiak 2015]. Sewage sludge, apart from heavy metals, constitutes a potential source of macroelements which can be used in the cultivation of energy and industrial plants [Kołodziej et al. 2015; 2016; Zapałowska et al. 2017].

Yield

Our study proved that sewage sludge significantly increases yielding of the tested plants, which is confirmed by the earlier publications [Kołodziej et al. 2015; 2016]. The studies of Abdul Khaliq et al. [2017] and Helios et al. [2014] also confirm that the application of the municipal sewage sludge results in the energy plant yielding increase. Other studies showed that compared to other energy plants, the Virginia fanpetals had the highest biomass yields [Nahm, Morhart 2018]. The size of the Virginia fanpetals and multiflora

rose biomass yields was comparable to the one obtained by the above-mentioned authors. According to Borkowska and Molas [2012, 2013], the Virginia fanpetals planted on sewage sludge has high yield potential.

Macroelement content

A significant increase in the macroelement content in the Virginia fanpetals and multiflora rose was found after the application of sewage sludge. According to Borkowska and Molas [2012, 2013] as well as Krzywy-Gawrońska [2012], the increase in the content of macroelements in energy plants is determined by the dose of sewage sludge and mineral fertilization, which was also confirmed by our study. Pogrzeba et al. [2018] indicated that mineral fertilization and the addition of microorganisms increased the macroelement content in the selected energy crops, which was confirmed by us. As stated by Wierzbowska et al. [2016] and Krzywy-Gawrońska [2012], increasing the doses of sewage sludge raised the contents of N, P, K, Na, Ca, and Mg in the Virginia fanpetals, which confirms that this species is very effective in the use of macronutrients from these wastes. On the other hand, the study of Helios et al. [2014] reveled that sewage sludge used in the cultivation of prairie cordgrass (*Spartina pectinata* Link.) did not increase the macroelement content in this species. The lack of effect of sewage sludge on the increase of macronutrients in prairie cordgrass most probably resulted from a small dose of sewage sludge (1.4–4.2 Mg DM · ha⁻¹) compared to the doses used in our study (10–60 Mg DM · ha⁻¹), [Helios et al. 2014].

Uptake

Our study confirmed that the increasing doses of sewage sludge significantly raised the macroelement uptake by the tested plants compared to the control. Similar relationships were demonstrated by Wierzbowska et al. [2012], who proved that sewage sludge stimulates a greater uptake of macroelements by the Virginia fanpetals. The studies of Helios et al. [2014] confirmed that increasing doses of sewage sludge raises the macroelement uptake by the prairie cordgrass. Higher uptake of macroelements from sewage sludge by perennial and energy crops results, inter alia, from a long growing season (systematic uptake) and high yield potential [Korzeniowska, Stanisławska-Glubiak 2015; Kocoń, Jurga 2017; Sienkiewicz et al. 2018]. According to Arduini et al. [2018], the converted sewage sludge (biosolid) affects the systematic uptake of macronutrients by crops. This author demonstrated that macronutrients derived from sewage sludge were taken up by plants in larger amounts than from mineral fertilizers. The greater uptake of macronutrients by plants results from the improvement of physicochemical properties of soil after the application of sewage sludge [Arduni et al. 2018; Schröder et al. 2018]. The analysis of world literature carried out by Nahm and Morhart [2018] indicated that the Virginia fanpetals is suitable for the effective uptake of nitrogen, and thus the removal of excess nutrients from the soil.

Balance

The presented balance shows that sewage sludge is a potential source of macroelements, especially N and P [Kacprzak et al. 2017; Zapałowska et al. 2018]. Sewage sludge can contain significant amounts of Na, Mg, and Ca – the elements originating from hygienisation and sewage treatment technology [Kacprzak et al. 2017]. The study showed that the balance of macroelements in treatments fertilized with sewage sludge was positive, which resulted from the amount of these components supplied with this waste. Other studies confirmed that sewage sludge used in large amounts affects the positive balance of these components in the soil [Wierzbowska et al. 2016]. A negative balance was determined for K. This is mainly due to the fact that sewage sludge was poor in this element, which is confirmed by other studies [Sienkiewicz et al. 2018; Tontti et al. 2017]. As indicated by Lee et al. [2018], sewage sludge constitutes a large source of macroelements that can be recovered for the environment. In general, sewage sludge can be used as an alternative source of macroelements to minimize the use of mineral fertilizers [Lee et al. 2018]. The high phytoremediation of macronutrients by the Virginia fanpetals was also recorded in the study of Wierzbowska [2016]. The effective use of nutrients from municipal sewage sludge is crucial for sustainable development and, at the same time, minimizes the effect of excessive nutrient applications on the environment [Sassenrath et al. 2013].

Enzymatic activity

Microorganisms support the use of soil macronutrients by plants [Pogrzeba et al. 2018]. Our ex-periment yielded high values of the activity of the tested enzymes, which confirms the effectiveness of the use of sewage sludge as an organic fertilizer in the cultivation of energy crops [Kołodziej et al. 2015; 2016; Wolna-Murawka et al. 2018]. The research carried out by Symanowicz et al. [2018] confirmed that mineral and organic fertilizations increase the soil enzymatic activity, and thus raise the macroelement uptake by plants.

In our study, the activity of soil enzymes increased significantly (progressively) along with the dose of sewage sludge, which was related to the amount of carbon substrates available for microorganisms and enzymes [Bielińska et al. 2015; Wolna-Murawka et al. 2018]. Our investigations and studies conducted by other authors confirmed that the Virginia fanpetals secretes root slime which stimulates the enzymatic activity of soil [Wielgosz 1999]. What is more, the addition of microorganisms to the soil had beneficial effect on the growth and physiological activity of the Virginia fanpetals, increasing, inter alia, the biomass yield [Piotrowski et al. 2016]. The correlations between the activity of soil enzymes and the content and uptake of macroelements demonstrated in the experiment confirm the importance of the intensity of removing nutrients from the soil environment [Pogrzeba et al. 2018; Wierzbowska et al. 2016].

CONCLUSIONS

1. The Virginia fanpetals had a higher yield potential in comparison with the multiflora rose. On the other hand, the multiflora rose responded to the application of sewage sludge with a higher yield increase compared to the Virginia fanpetals.

2. The sewage sludge applied to the soil significantly increased the macroelement content in the tested plants. Higher content of N, Mg, Ca, and Na was found in the Virginia fanpetals than in the multiflora rose. In turn, the multiflora rose was characterized with higher contents of P and K compared to the Virginia fanpetals.

3. The increasing doses of sewage sludge significantly raised the macroelement uptake by plants. The highest dose of sewage sludge reveled greater uptake of N and Na with the biomass yield by the Virginia fanpetals than by the multiflora rose. On the other hand, the multiflora rose took up more P, K, Mg, and Ca than the Virginia fanpetals.

4. The evaluation of the percentage phytoremediation revealed that the Virginia fanpetals used N, P, K, Na, and Mg to a greater extent than the multiflora rose. In contrast, the multiflora rose accumulated more Ca compared to the Virginia fanpetals.

5. On the basis of the yield, uptake, and phytoremediation, it can be concluded that the Virginia fanpetals is a plant with a great potential for the recycling of macroelements from the municipal sewage sludge.

REFERENCES

1. Abdul Khaliq, S.J., Al-Busaidi A., Ahmed M., Al-Wardy M., Agrama H., Choudri B.S. 2017. The effect of municipal sewage sludge on the quality of soil and crops. International Journal of Recycling of Organic Waste in Agriculture, 6, 4, 289–299. https://doi.org/10.1007/s40093-017-0176-4

2. Arduini I., Cardelli R., Pampana S. 2018. Biosolids affect the growth, nitrogen accumulationand nitrogen leaching of barley. Plant, Soil and Environment, 64, 3, 95–101. https://doi.org/10.17221/745/2017-PSE

3. Antonkiewicz J. 2010. Effect of sewage sludge and furnace waste on the content of selected elements in the sward of legume-grass mixture. Journal of Elementology, 15, 3, 435–443. DOI: 10.5601/jelem.2010.15.3.435–443

4. Antonkiewicz J., Kołodziej B., Bielińska E. 2017. Phytoextraction of heavy metals from municipal sewage sludge by Rosa multiflora and Sida hermaphrodita. International Journal of Phytoremediation, 19, 4, 309–318. http://dx.doi.org/10.1080/15226514.2016.1225283

5. Awasthi M.K., Wang M.J., Pandey A., Chen H.Y., Awasthi S.K., Wang Q., Ren X., Lahore A.H., Li D.S., Li R.H., Hang Z.Q. 2017. Heterogeneity of zeolite combined with biochar properties as a function of sewagesludge composting and production of nutrient-rich compost. Waste Management, 68, 760–773. https://doi.org/10.1016/j.wasman.2017.06.008

6. Bielińska E.J., Futa B., Baran S., Żukowska G., Olenderek H. 2015. Soils enzymes as bio-indicators of forest soils health and quality within the range of impact of Zakłady Azotowe 'Pulawy' SA. Sylwan, 159, 11, 921–930.

7. Borkowska H., Molas R. 2012. Two extremely different crops, Salix and Sida, as sources of renewable bioenergy. Biomass and Bioenergy, 36, 234–240. http://dx.doi.org/10.1016/j.biombioe.2011.10.025

8. Borkowska H., Molas R. 2013. Yield comparison of four lignocellulosic perennial energy crop species. Biomas and Bioenergy, 51, 145–153. http://dx.doi.org/10.1016/j.biombioe.2013.01.017

9. Helios W., Kozak M., Malarz W., Kotecki A. 2014. Effect of sewage sludge application on the growth, yield and chemical composition of prairie cordgrass (Spartina pectinata Link.). Journal of Elementology, 19, 4, 1021–1036, DOI: 10.5601/jelem.2014.19.3.725

10. Jones J.B., Case V.W. 1990. Soil testing and plant analysis. 3rd ed. Soil Science Society of America SSSA, Chapter 15.

11. Kacprzak M., Neczaj E., Fijałkowski K., Grobelak A., Grossem A., Worwag M., Rorat A., Brattebo H., Almås Å., Singh B.R. 2017. Sewage sludge disposal strategies for sustainable development. Environmental Research, 156, 39–46. DOI: https://doi.org/10.1016/j.envres.2017.03.010

12. Kocoń A, Jurga B. 2017. The evaluation of growth and phytoextraction potential of Miscanthus x giganteus and Sida hermaphrodita on soil contaminated simultaneously with Cd, Cu, Ni, Pb, and Zn. Environmental Science and Pollution Research, 24, 5, 4990–5000. https://doi.org/10.1007/s11356-016-8241-5

13. Kołodziej B., Antonkiewicz J., Stachyra M., Bielińska E.J., Wiśniewski J., Luchowska K., Kwiatkowski C. 2015. Use of sewage sludge in bioenergy production – A case study on the effects on sorghum biomass production. European Journal of Agronomy, 69, 63–74. http://dx.doi.org/10.1016/j.eja.2015.06.004

14. Kołodziej B., Stachyra M., Antonkiewicz J., Bielińska E., Wiśniewski J. 2016. The effect of harvest frequency on yielding and quality of energy raw material of reed canary grass grown on municipal sewage sludge. Biomass and Bioenergy, 85, 363–370. http://dx.doi.org/10.1016/j.biombioe.2015.12.025

15. Korzeniowska J., Stanisławska-Glubiak E. 2015. Phytoremediation potential of Miscanthus x giganteus and Spartina pectinata in soil contaminated with heavy metals. Environmental Science and Pollution Research, 22, 15, 11648–11657. https://doi.org/10.1007/s11356-015-4439-1

16. Kicińska A., Kosa-Burda B., Kozub P. 2018. Utilization of a sewage sludge for rehabilitating the soils degraded by the metallurgical industry and a possible environmental risk involved. Human and Ecological Risk Assessment. https://doi.org/10.1080/10807039.2018.1435256

17. Krzywy-Gawrońska E. 2012. The effect of industrial wastes and municipal sewage sludge compost on the quality of Virginia fanpetals (Sida hermaphrodita Rusby) biomass Part 1. Macroelements content and their uptake dynamics. Polish Journal of Chemical Technology, 14, 2, 9–15. DOI: 10.2478/v10026-012-0064-7

18. Ladd N., Butler J.H.A. 1972. Short-term assays of soil proteolytic enzyme activities using proteins

and dipeptide derivatives as substrates. Soil Biol. Biochem., 4, 19–30.

19. Lee C.G., Alvarez P.J.J., Kim H.G., Jeong S., Lee S., Lee K.B., Lee S.H., Choi J.W. 2018. Phosphorous recovery fromsewagesludge using calcium silicate hydrates. Chemosphere, 193, 1087–1093. https://doi.org/10.1016/j.chemosphere.2017.11.129

20. Nahm M., Morhart C. 2018. Virginia mallow (Sida hermaphrodita (L.) Rusby) as perennial multipurpose crop: biomass yields, energetic valorization, utilization potentials, and management perspectives. Global Change Biology Bioenergy, 10, 6, 393–404. DOI: https://doi.org/10.1111/gcbb.12501

21. Ostrowska A., Gawliński S., Szczubiałka Z. 1991. Methods of analysis and assessment of soil and plant properties. A Catalgoue. Publisher: Institute of Environmental Protection – National Research Institute, Warsaw, pp 334.

22. Piotrowski K., Romanowska-Duda Z., Grzesik M. 2016. Cyanobacteria, Asahi SL and Biojodis as stimulants improving growth and development of the Sidahermaphrodita L. Rusby plant under changing climate conditions. Przemysl Chemiczny, 95, 8, 1569–1573. DOI: 10.15199/62.2016.8.31.

23. Pogrzeba M., Rusinowski S., Krzyżak J. 2018. Macroelements and heavy metals content in energy crops cultivated on contaminated soil under different fertilization–case studies on autumn harvest. Environmental Science and Pollution Research, 25, 12, 12096–12106. https://doi.org/10.1007/s11356–018–1490–8

24. Polish Soil Classification. 2011. Soil Science Annual, 62, 3, 1–193. http://www.ptg.sggw.pl

25. Regulation. 2015. Regulation of the Minister of the Natural Environment on municipal sewage sludge dated 6 February 2015. Journal of Laws of Poland, Item 257. http://isap.sejm.gov.pl/DetailsServlet?id=WDU20150000257

26. Regulation. 2016. Regulation of the Minister of the Natural Environment on how to conduct land surface pollution assessment dated 1 September 2016. Journal of Laws of Poland, Item 1395. http://isap.sejm.gov.pl/DetailsServlet?id=WDU20160001395

27. Sassenrath G.F., Schneider J.M., Gaj R., Grzebisz W., Halloran J.M. 2013. Nitrogen balance as an indicator of environmental impact: Toward sustainable agricultural production. Renewable Agriculture and Food Systems, 28, 3, 276–289. https://doi.org/10.1017/S1742170512000166

28. Schröder P., Beckers B., Daniels S., Gnädinger F., Maestri E., Marmiroli N., Mench M., Millan R., Obermeier M.M., Oustriere N., Persson T., Poschenrieder C., Rineau F., Rutkowska B., Schmid T., Szulc W., Witters N., Sæbø A. 2018. Intensify production, transform biomass to energy and novel goods and protect soils in Europe – A vision how to mobilize marginal lands. Science of The Total Environment, 616–617, 1101–1123.

https://doi.org/10.1016/j.scitotenv.2017.10.209

29. Siekiewicz S., Wierzbowska J., Kovacik P., Krzebietke S., Zarczyński P. 2018. Digestate as a substitute of fertilizers in the cultivation of Virginia fanpetals. Fresenius Environmental Bulletin, 27, 6, 3970–3976.

30. Soil Survey Staff. 2014. Keys to Soil Taxonomy, 12th ed. USDA-Natural Resources Conservation Service, Washington, DC.

31. Symanowicz B., Kalembasa S., NiedbałaM., Toczko M., Skwarek K. 2018. Fertilisation of pea (Pisum sativumL.) with nitrogen and potassium and its effect on soil enzymatic activity. Journal of Elementology, 23, 1, 57–67. DOI. 10.5601/jelem.2017.22.1.1395

32. Tabatabai M.A., Bremner J.M. 1969. Use of p-nitrophenyl phosphate for assay of soil phosphatase activity. Soil Biol. Biochem. 1, 301–307.

33. Thalmann A. 1968. Zur methodik der bestimmung der Dehydrogenaseaktivit~tt im Boden mittels Triphenyltetrazoliumchlorid (TTC). Landwirtsch Forsch, 21, 249–258.

34. Tontti T., Poutiainen H., Heinonen-Tanski H. 2017. Efficiently treated sewagesludge supplemented with nitrogen and potassium is a good fertilizer for cereals. Land Degradation & Development, 28, 2, 742–751. https://doi.org/10.1002/ldr.2528

35. Waste Catalogue. 2014. Regulation of the Minister of the Natural Environment on catalog of wastes dated 9 December 2014. Journal of Laws of Poland, Item 1923. http://isap.sejm.gov.pl/DetailsServlet?id=WDU20140001923

36. Wielgosz E. 1999. Aktywność mikrobiologiczna i enzymatyczna w glebie brunatnej pod uprawą ślazowca pensylwańskiego (Sida hermaphrodita Rusby) i topinambura (Helianthus tuberosus). Ann. UMCS., sect. E, 54, 21, 173–185.

37. Wierzbowska J., Sienkiewicz S., Krzebietke S., Sternik P. 2016. Sewage sludge as a source of nitrogen and phosphorus for Virginia fanpetals. Bulgarian Journal of Agricultural Science, 22, 5, 722–727.

38. Wolna-Maruwka A., Sulewska H., Niewiadomska A., Panasiewicz K., Borowiak K., Ratajczak K. 2018. The influenceof sewage sludge and a consortium of aerobic microorganisms added to the soil under a Willow plantation onthe biological indicators of transformation of organic nitrogen compounds. Polish Journal of Environmental Studies, 27, 1, 403–412. https://doi.org/10.15244/pjoes/74184

39. Zantua M.I., Bremner J.M. 1975. Comparison of methods of assaying urease activity in soils. Soil Biol. Biochem. 7, 291–295.

40. Zapałowska A., Puchalski C., Hury G., Makarewicz A. 2017. Influence of fertilization with the use of biomass ash and sewage sludge on the chemical composition of Jerusalem artichoke used for energy-related purposes. Journal of Ecological Engineering, 18, 5, 235–245. https://doi.org/10.12911/22998993/76214

17

Formulation of Biochar-Compost and Phosphate Solubilizing Fungi from Oil Palm Empty Fruit Bunch to Improve Growth of Maize in an Ultisol of Central Kalimantan

Gusti Irya Ichriani[1,2], Syehfani[3], Yulia Nuraini[3], Eko Handayanto[4*]

[1] Postgraduate Programme, Faculty of Agriculture, Brawijaya University, Jl. Veteran, Malang 65145, Indonesia

[2] Faculty of Agriculture, University of Palangka Raya, Jl. Yos Sudarso, Kota Palangka Raya 74874, Central Kalimantan, Indonesia

[3] Department of Soil Science, Faculty of Agriculture, Brawijaya University, Jl. Veteran, Malang 65145, Indonesia

[4] Research Centre for Management of Degraded and Mining Lands, Brawijaya University, Jl. Veteran, Malang 65145, Indonesia

[*] Corresponding author's e-mail: handayanto@ub.ac.id

ABSTRACT

The efficiency of phosphorus uptake by plants in an Ultisol soil is very low because most of the soil phosphorus is precipitated by Al and Fe. Oil palm empty fruit bunches can be used as basic materials of biochar and compost, and as sources of isolates of phosphate solubilizing fungi. This study was aimed at elucidating the effect of application of phosphate-solubilizing fungi with biochar and the compost produced from oil palm empty fruit bunches on the growth and yield of maize an Ultisol of Central Kalimantan. This study consisted of two experiments. The first experiment was inoculation of four isolates of phosphate solubilizing fungi isolated from of oil palm empty fruit bunches, i.e. *Acremonium* (TB1), *Aspergillus* (TM7), *Hymenella* (TM1) and *Neosartorya* (TM8) to 'biocom' media (mixture of biochar and compost generated from oil palm empty fruit bunches) to obtain phosphate-solubilizing fungi that can adapt to the media. In the second experiment, the best results in the first experiment were applied to an Ultisol soil planted with maize. The results showed that the isolates that were best adapted to biocom media were *Aspergillus*-TB7 with 60:40 proportion (60% biochar + 40% compost) and *Neosartorya*-TM8 with 70:30 proportions (60% biochar + 40% compost). The use of the first experiment results in the second experiment showed that the application of biocom plus *Neosartorya*-TM8 (BTM) on an Ultisol soil significantly improved the growth and yield of maize, as well as its the phosphorus uptake and uptake efficiency .

Keywords: biochar, phosphate-solubilizing fungi, oil palm empty fruit bunch, Ultisol

INTRODUCTION

The area of Ultisols in Indonesia reaches 45.8 million ha or 25% of the land area of Indonesia (Subagyo et al., 2004). In general, the available P content in Ultisols is low due to P fixation by Al and Fe (Osorio, 2014). The availability of P in Ultisol can be easily increased by the application of inorganic P fertilizer. However, only a small proportion of the added P is eventually taken up by the plant and the remainder (almost 75–90%) is precipitated by Fe, Al, and Ca complexes in the soil (Gyaneshwar et al., 2002). The low efficiency of P-use of cultivated crops on farms does not only result in larger P fertilizer applications, but also causes environmental problems, such as water eutrophication (Chang and Yang, 2009; Kang et al., 2011).The low content of soil organic matter in Ultisols (Prasetyo and Suriadikarta, 2006) also causes low soil buffering capacity that results in low fertilizer efficiency. Isgitani et al. (2005) reported that only 5–20% of the applied inorganic P fertilizer of 150–200 kg P ha^{-1} was taken up by the plants.

One source of organic material commonly used by local farmers to increase the availability of P in Ultisols in Central Kalimantan includes bunches of empty oil palm fruit (Ariani, 2009). This is because the area of oil palm plantation in Central Kalimantan is about 1.09 million ha (Central Bureau of Statistics, 2015). In addition to being a source of organic material, oil palm empty fruit bunches are also habitat for microorganisms, including the P-solubilizing fungi, bacteria, and actinomycetes (Sundara et al., 2002) that are able to convert the unavailable P to $H_2PO_4^-$ and HPO_4^{2-} that are available for crops (Coutinho et al., 2012). The phosphate-solubilizing microorganisms that are capable of converting the unavailable P into available forms for plants can also serve as biological fertilizers to increase the available P content (Narsian and Patel, 2000; Delvasto, 2006; Khan et al., 2007; Zhu et al., 2012). Many studies have shown that the growth and uptake of P by plants can be enhanced by the inoculation of phosphate-solubilizing fungi, either through pot experiments (Mittal et al., 2008), or field experiments (Duponnois et al., 2005).

In an earlier study, Ichriani et al. (2017) obtained 4 isolates of phosphate-solubilizing fungi from the compost of oil palm empty fruit bunches that were identified as *Acremonium kiliens*, *Aspergillus oryza*, *Hymenella Fr.*, and *Neosartorya fischeri*. The application of the four phosphate-solubilizing fungi isolates on liquid Pikovskaya medium containing tricalcium phosphate ($Ca_2(PO_4)_2$ could increase the available P by 451%, 400%, 216% and 114%, respectively, on day 5 (Ichriani et al., 2017) The phosphate-solubilizing fungi produce organic acid compounds which can dissolve the P compound fixed by the metal compound by forming a complex metal compound (Sharma et al., 2013; Fitriatin et al., 2014).

The application of phosphate-solubilizing fungi on the soil requires a carrier medium as a substrate for the life of the fungi. Suitable biological fertilizer carriers must meet the following criteria: (1) they must be available in powder or granule form; (2) should be able to support the growth and survival of microorganisms, and easily release functional microorganisms into the soil; (3) must have strong moisture absorption capability, good aeration characteristics, and excellent pH buffering capacity; (4) must be non-toxic and environmentally friendly; (5) should be easily sterilized, manufactured and handled in the field, and have good storage quality; and (6) should be cheap (Stephens and Rask, 2000; Rebah et al., 2002; Rivera-Cruz et al., 2008). Since fungi are commonly found in the areas containing organic substrates, oil palm empty fruit bunches may be used as carrier medium for phosphate-solubilizing fungi. However, the addition of fresh organic matter to the soil can increase carbon emissions into the atmosphere due to the decomposition of organic matter (Widowati et al., 2011). One alternative to reducing the carbon emissions due to decomposition of organic materials and at the same time maintaining soil fertility is with biochar (Crombie et al., 2015). Biochar is the result of biomass heating in pyrolysis installations at temperature of > 700°C under low oxygen conditions (Cheng et al., 2007). The physical and chemical properties of biochar depend on raw materials and pyrolysis conditions (Han et al. 2013, Mukome et al. 2013). The aromatic structure of biochar contributes to the long existence of biochar in the soil (Baldock and Smernik, 2002), which makes biochar a material for carbon sequestration, maintain nutrient elements (Mao et al., 2012), and suitable for use as soil amendments (Novak et al., 2009; Spokas et al., 2011). According to Šimanský and Klimaj (2017), biochar is a soil amendment that can increase the soil pH, and the greatest effect on soil pH is after application of 10 t ha^{-1} with a combination of 40 kg of nitrogen ha^{-1} (Šimanský and Klimaj, 2017). In addition to the potential physical and chemical impacts of biochar on soil productivity, biochar also benefits soil biology, such as selection for plant growth promoting bacteria or fungi (Graber et al., 2010). One of the mechanisms that cause biochar to have a positive impact on soil function is because biochar serves as a protection for fungal hyphae due to its porous nature (Warnock et al., 2007).

Several research results indicate that biochar is used as a solid carrier medium for inoculums of *Azospirilum sp.* (Kuppusamy et al., 2011), *Entorobacter cloacae* (Hale et al., 2014), and arbuscular mycorrhizae (Nurbaity et al., 2009). The use of biochar from oil palm empty fruit bunches for microbial carrier media for *Bulkhorideria nodosa* G.52 and *Trichoderma sp.* has been performed on several soil types (Ichriani et al., 2016). The mixture of oil palm empty fruit bunch biochar and peat was also used as solid carrier media for *Trichoderma harzianum* (Kresnawaty et al., 2012). Douds et al. (2014) stated that biochar could be used as a fungus medium in the inoculums production system. Thus, biochar can be used as a carrier medium for phosphate-solubilizing fungi. Collaboration of microbes and biochar produced from oil palm empty bunches improves the soil properties (Ichriani et al., 2016). Senoo et al. (2002) reported that compost and biochar carrying agents could maintain the vitality of microorganisms. Therefore, biochar and compost are good carriers for microbial inoculums (Somarathne et al., 2013). However, the information on the use of biochar of oil palm empty fruit bunches as a carrier medium for phosphate-solubilizing fungi is still very limited. The collaboration of oil palm empty fruit bunch biochar and phosphate-solubilizing fungi is expected to aid in the storage and supply of nutrients. The presence of phosphate-solubilizing fungi also helps to increase the availability of P in Ultisols to improve the crop production. The purpose of this study was to study the effect of phosphate-solubilizing fungi application with biochar and compost carrier media from oil palm empty fruit bunches on the growth and yield of maize grown on an Ultisol from Central Kalimantan.

MATERIALS AND METHODS

The materials used in this study were an Ultisol, biochar of oil palm empty fruit bunches, compost of oil palm empty fruit bunches, and isolates of phosphate-solubilizing fungi. The compost of oil palm empty fruit bunches was obtained from oil palm plantations PT. Surya Inti Sawit Kahuripan (Makin Group), Parenggean Subdistrict, Kotawaringin Timur District, Central Kalimantan. The soil was collected from farmers' land in Gunung Emas District, Central Kalimantan. The biochar of oil palm empty fruit bunches was made from oil palm empty fruit bunches which was heated pyrolysis at 400°C for 6–7 hours. The characteristics of soil (top soil, 0–30 cm) are as follows: clay loam (24% sand, 47% silt, 29% clay), 13.87% available water content, pH (H_2O) 4.3, 0.52% organic-C, 156.47 ppm total P, 2.13 ppm available P, 137.65 ppm total K, 15.25 ppm available K, exchangeable cations of Ca, Mg, K, Na, Al and H, respectively, 1.80, 0.45, 0.06, 0.21, 3.34, and 6.31 me $100g^{-1}$, CEC 5.64 me $100g^{-1}$, and 45% base saturation. The characteristics of oil palm empty fruit bunch compost are as follows: pH (H_2O) 6.7, 17.30%, organic C, 1.56% total N, 3,700 ppm total P, 1,100 ppm total K, exchangeable cations of Ca, Mg, K, and Na, respectively 3.19, 1.21, 0.52, and 0.83 me $100g^{-1}$, cation exchange capacity 31.95 me $100g^{-1}$, 170 ppm Zn,

168 ppm Cu, 3.1 ppm Fe, 44 ppm Co, 1190 ppm Mn, and 2.18 ppm Cd. The characteristics of biochar are as follows: pH 9.9, 61% C, 23.86% O, 1.78% Si, 1.18% Cl, 10.48% K, and 1.5% Ca.

Experiment 1: Adaptation of phosphate-solubilizing fungi on media of biochar and compost from oil palm empty fruit bunches

Four isolates of phosphate-solubilizing fungi from Ichriani et al. (2017) i.e. *Acremonium* (TB1), *Aspergillus* (TB7), *Hymenella* (TM1) and *Neosartorya* (TM8) were inoculated on mixtures of oil palm empty fruit bunch biochar and oil palm empty fruit bunch compost with proportion: 100% biochar – 0% compost (B0), 90% biochar – 10% compost (B1), 80% biochar – 20% compost (B2); 70% biochar – 30% compost (B3), and 60% biochar – 40% compost (B4). Henceforth, the carrier medium will be called 'biocom' (B). The total weight for each biocom carrier medium was 10 g. Prior to the inoculation of the phosphate solubilizing fungi, the 'biocom' was sterilized for 2 hours in boiling water, then after 24 hours, the biocom was reheated for 2 hours. The density of each inoculated phosphate-solubilizing fungi per biocom was 10^8 conidia $mL^{-1}10 g^{-1}$ of biocom media. The biocom that has been inoculated with phosphate solubilizing fungi was placed in a closed container and placed in a sterile room during the trial period. Twenty treatments (four isolates and five biocoms) were prepared in a completely randomized design with four replicates. The ability of the phosphate-solubilizing fungi to survive on the biocom medium was observed by observing the population of the fungi by the method of pouring and growing in PDA (Potato Dextrose Agar) medium at 2, 4, 8, 12 and 16 weeks after inoculation. The biocom media which had the highest population of phosphate-solubilizing fungi at the end of the observation period (week 4) was observed for microstructure using Scanning Electron Microscophy-Energy Dispersive X-ray (SEM-EDX) method. The measurements of pH and organic C of the carrier media were carried out at 4 and 8 weeks after inoculation.

Experiment 2: Effect of biocom + phosphate-solubilizing fungi application on maize growth

The treatments tested for maize plant growth were control (T), biochar (B), biocom 60:40 plus *Aspergillus*-TB7 (BTB), and biocom 70:30 plus

Neosartorya-TM8 (BTM). The treatments were arranged in a completely randomized design with five replicates. The dosage used in each polybag was the 12 kg of soil, basic fertilizer (200 kg N ha^{-1}, 150 kg P$_2$O$_5$ ha^{-1}, 150 kg K$_2$O ha^{-1}, 10 ppm Mg, 10 ppm Ca, 10 ppm Zn, 6 ppm Cu, 0.9 ppm B, and 0.9 ppm Mo) (Santi and Goenardi, 2012), 15 t biochar (B) ha^{-1}, or 15 t biocom + FPF (BTB or BTM) ha^{-1} (Sukartono et al. 2011, Uzoma et al. 2011). The content of P in the P fertilizer used is 17.87%. The observations consisted of vegetative and generative growth indicators, biomass, and P uptake by maize. The P uptake (g plant^{-1}) was calculated based on total P-content in maize plant biomass. Total P in plant was determined by HNO$_3$-HClO$_4$ wet extraction and continued measurements with spectrophotometer at 693 nm wavelength. The data obtained were subjected to analysis of variance followed by Duncan's Multiple Range Test (DMRT) at 0.05 level of significance. The efficiency of P fertilization (EhP) was calculated with the formula of Dobermann (2007): Ehp = [(SPp-SKp) / Hpp] × 100%, where Ehp is the nutrient uptake efficiency of P, SPp is the uptake of P nutrient in the plant fertilized with P, SKp is the uptake of P nutrient in plant not fertilized with P, and HPp is the nutrient content of P in the P fertilizer given to the plant.

RESULTS AND DISCUSSION

Adaptation of phosphate solubilizing fungi in oil palm empty fruit bunch biocom media

Isolates of phosphate solubilizing fungi applied to carrier media in the form of oil empty palm empty bunch biocom had different life and growing ability. The results of observation of the population involving each carrier medium 2, 4, 8, 12, and 16 weeks after inoculation are presented in Figure 1. The highest population of *Acremonium*-TB1 and *Hymenella*-TM1 was achieved at 8 weeks after inoculation, while the highest population of *Aspergillus*-TB7 and *Neosartorya*-TM8 was achieved 4 weeks after inoculation. However, the *Aspergillus*-TB7 and *Neosartorya*-TM8 populations within 4 weeks of inoculation were higher than those of *Acremonium*-TB1 and *Hymenella*-TM1. This shows that *Aspergillus*-TB7 and *Neosartorya*-TM8 are more capable of living and adapting in solid carrier media of oil palm empty fruit bunch biocom. One of the mechanisms that cause biochar to have a positive

impact on function is because biochar serves as a hiding place for fungal hyphae due to its porous nature (Warnock et al., 2007). Hadi et al. (2014) reported that biochar applications into the soil increased the fungi population by more than 22.2%. The phosphate-solubilizing *Aspergillus* species have been studied as the most highly adapted fungi to the growing environment and thus have a high ability of increasing the availability of soil P for improving the plant growth (El-Azouni, 2008; Mittal et al., 2008; Ogbo, 2010; Jain et al., 2012). After experiencing the highest population, there was a trend of isolate population decline. At 16 weeks after inoculation, *Aspergillus*-TB7 and *Neosartorya*-TM8 populations decreased as in the period of 2 weeks after inoculation.

The carrier medium containing only biochar without compost (B0) did not provide a good growth medium for all isolates, since it had the smallest population of phosphate-solubilizing fungi. The low nutrients contained in biochar media alone caused improper growth of the fungi. The changes of cellulose and lignin structure of oil palm empty fruit bunch into complex structures in the form of crystalline structures and C = C aromatic rings might have occurred during the process of slow pyrolysis of fresh oil palm empty fruit bunch material into biochar. According to Joseph et al. (2010), it is difficult for fungal organisms are to use carbon compounds in very complex forms such as the crystal and aromatic structures. The result of gas chromatography-mass spectrometry (GCMS) analysis showed that oil palm empty fruit bunch biochar contained organic compound with functional group of C = C aromatic ring structure. This was marked by the appearance of peak on the wave number of 1500–1600 cm^{-1} which might indicate the presence of C = C aromatic ring group. This result was also supported by the analysis of oil palm empty fruit bunch biochar with GCMS that obtained organic compounds suspected as cyclopropane (15.51%) and cycloheptatriene (12.04%). Therefore, carrier media combined with oil palm empty fruit bunch compost showed a higher phosphate-solubilizing fungi population, since compost contained simple compounds that could be used as nutrients by phosphate-solubilizing fungi. *Acremonium*-TB1 seemed to be more suited to the composition of biocom media of 80:20 (B2), whereas *Aspergillus*-TB7 had the highest population with 60:40 (B4) biocom compositions with slower population decline than other com-

a)

b)

c)

d)

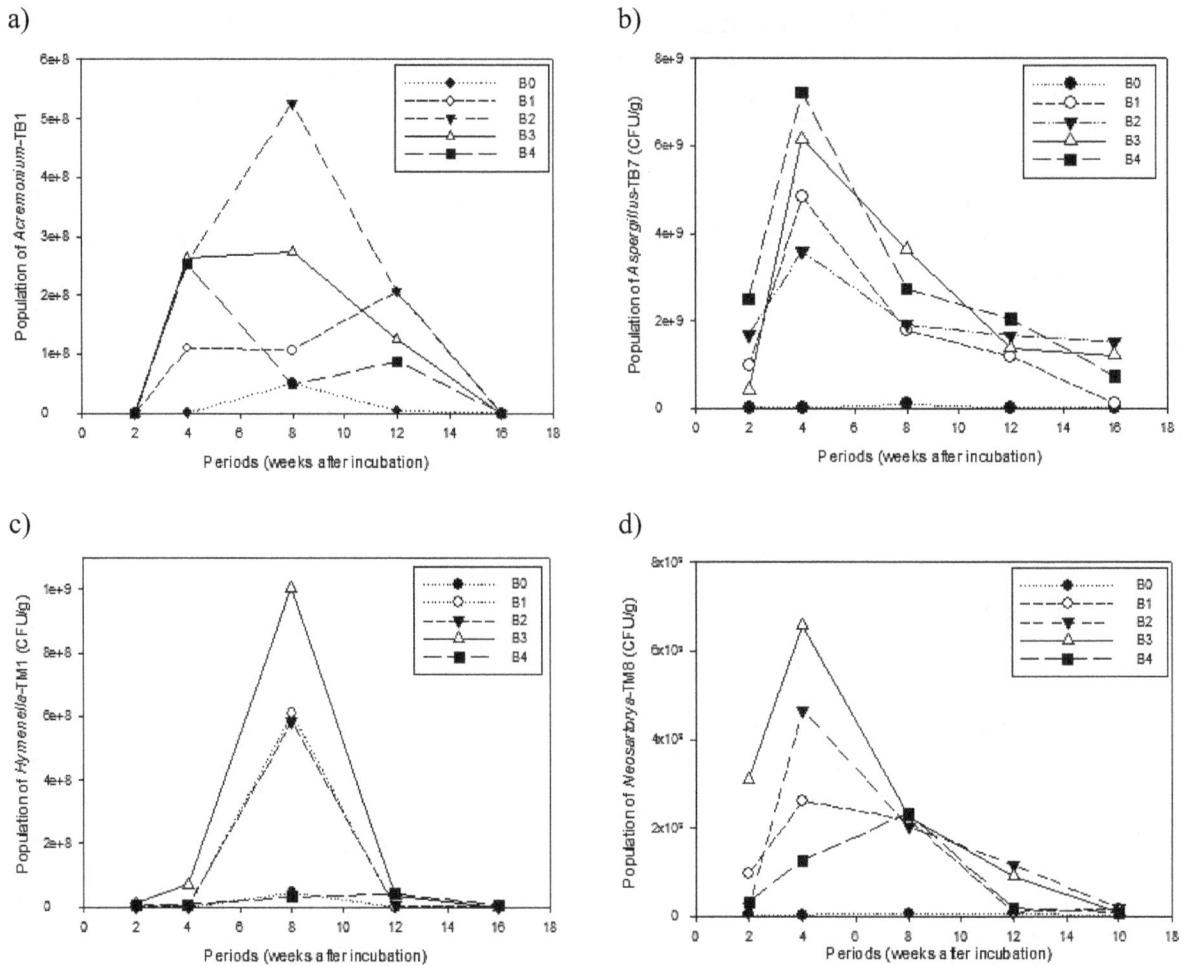

Figure 1. The survival of *Acremonium*-TB1 (A), *Aspergillus*-TB7 (B), Hymenella-TM1 (C), and *Neosartorya*-TM8, (D) for 2–6 weeks in several biocom formulations

positions. *Hymenella*-TM1 and *Neosartorya*-TM 8 were more suited to the 70:30 biocom media composition (B3) with the highest population in each isolate. The population decline in the 70:30 biocom media composition was also slower than the composition of other biocom media. Compost contains microaggregate with micropore structure (Somarathne et al., 2013). The materials having a micropore structure such as charcoal will be good carriers for soil inoculums (Senoo et al., 2002). Carrier media of compost and biochar can maintain the vitality of microorganisms (Senoo et al., 2002). Therefore, biochar and compost are good carriers for microbial inoculums (Somarathne et al., 2013). Figure 2 shows the results of microstructure analysis of oil palm empty fruit bunch biochar inoculated with *Aspergillus*-TB7 and *Neosartorya*-TM8.

The effect of adding oil palm empty fruit bunch compost onto oil palm empty fruit bunch biochar did not linearly increase the population of inoculated isolates of phosphate-solubilizing

fungi. For *Aspergillus*-TB7, the increase in the proportion of oil palm empty fruit bunch compost was accompanied by an elevated phosphate-solubilizing fungi population. However, for *Hymenella*-TM1 and *Neosartorya*-TM8, the increase of phosphate solubilizing fungi population was only up to the B3 (biochar 70%: 30% compost) and in the proportion of 40% compost, the population of both isolates decreased. The results of analyzes pertaining to pH and organic C of biocom media at 4 and 8 weeks showed that the decrease in the proportion of oil palm empty fruit bunch biochar in the media (decreasing the proportion of oil palm empty fruit bunch compost) has decreased the pH of the biocom medium (Table 1). In addition, the pH value of each biocom media formulation decreased 8 weeks after inoculation, compared with the pH value of biocom medium 4 weeks after inoculation.

The addition of oil palm empty fruit bunch compost to oil palm empty fruit bunch biochar helped in the adaptation of phosphate-solubiliz-

a)

b)

c)

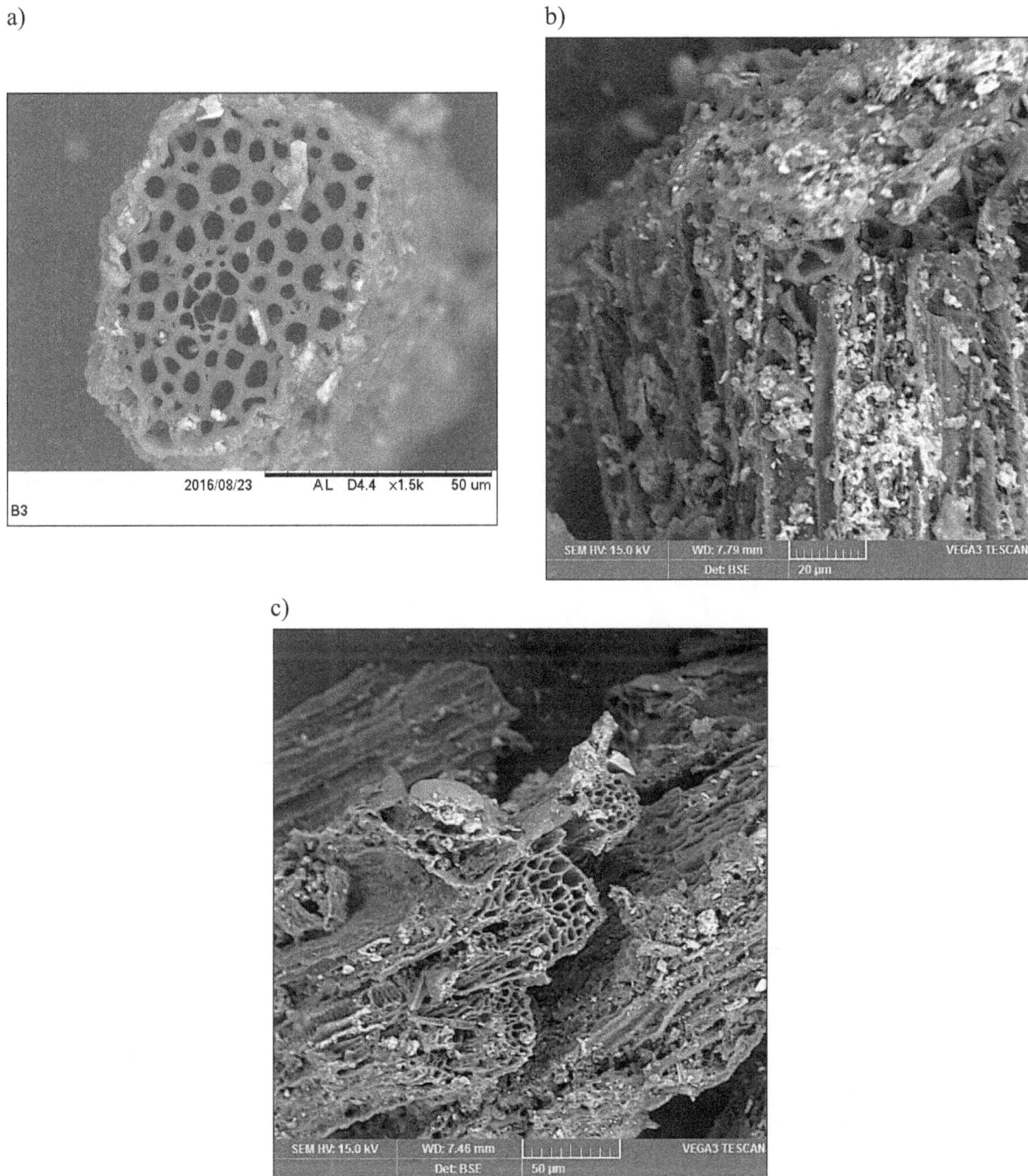

Figure 2. Microstructures (SEM EDX analysis) (A) biochar of oil palm fruit empty bunches; (B) biocom 60:40 (B4) plus *Aspergillus*-TB7; and (C) biocom 70:30 (B3) plus *Neosartorya*-TM8

ing fungi to oil palm empty fruit bunch biochar media. Although the content of organic-C media of biocom media was relatively unchanged.

Effect of biocom + phosphate solubilizing-fungi application on maize growth and P uptake efficiency

The application of oil palm empty fruit bunch biocom plus phosphate-solubilizing fungi significantly affected the growth and yield of maize as well as the P uptake by maize (Table 2). The significant effect of biochar (B) treatment on plant vegetative component (plant height and number of leaves) was seen faster 2 weeks after planting. However, after 2 weeks, the treatment of oil palm empty fruit bunch biocom plus phosphate-solubilizing fungi showed a better influence on the plant growth. The results of Duncan's Multiple Range Test at 5% level showed that when biochar (B) and biocom plus phosphate-solubilizing fungi (BTB and BTM)

Table 1. pH and organic C of biocom media 4 and 8 weeks after inoculation

Biocom	pH		Organic C (%)	
	4 weeks	8 weeks	4 weeks	8 weeks
Acremonium-TB1				
B0+TB1	10.13	10.12	8.13	8.72
B1+TB1	10.07	10.03	8.36	7.83
B2+TB1	10.08	9.87	8.77	7.87
B3+TB1	9.90	9.84	8.96	8.40
B4+TB1	9.83	9.70	10.56	10.56
Hymenella-TM1				
B0+TM1	10.19	10.17	9.54	8.46
B1+TM1	10.07	9.98	10.96	7.98
B2+TM1	10.01	9.85	9.39	9.27
B3+TM1	9.97	9.81	8.28	8.52
B4+TM1	9.83	9.65	9.61	9.68
Aspergillus-TB7				
B0+TB7	10.54	10.08	10.40	6.82
B1+TB7	10.49	10.00	10.68	7.90
B2+TB7	10.42	9.96	8.20	8.11
B3+TB7	10.34	9.83	10.97	9.99
B4+TB7	10.13	9.65	9.62	9.87
Neosartorya-TM8				
B0+TM8	10.00	10.00	10.10	9.27
B1+TM8	10.07	10.05	9.77	7.68
B2+TM8	10.04	9.92	9.44	9.44
B3+TM8	10.04	9.87	10.59	7.75
B4+TM8	9.77	9.80	11.35	10.46

Table 2. Effect of application of biocom + phosphate-solubilizing fungi on maize growth

Treatments	Plant Height (cm)			
	2 weeks	4 weeks	6 weeks	8 weeks
T	33.34	48.36 [a]	102.52 [a]	164.58 [a]
B	34.86	77.22 [b]	147.60 [b]	198.06 [b]
BTB	37.68	82.98 [d]	161.70 [d]	207.10 [c]
BTM	35.22	79.94 [c]	154.86 [c]	211.88 [d]
Treatments	Number of Leaves			
	2 weeks	4 weeks	6 weeks	8 weeks
T	4.00	4.80 [a]	5.40 [a]	9.20 [a]
B	4.00	5.60 [b]	7.40 [b]	11.60 [b]
BTB	4.00	6.40 [c]	8.01c	12.60 [c]
BTM	4.00	6.20 [c]	8.00 [c]	12.80 [c]
Treatments	Stem Diameter (mm)			
	2 weeks	4 weeks	6 weeks	8 weeks
T	1.52 [a]	3.62 [a]	7.54 [a]	8.92 [a]
B	2.38 [c]	6.82 [b]	11.48 [b]	14.94 [b]
BTB	2.22 [bc]	7.22 [c]	13.12 [d]	15.56 [c]
BTM	2.08 [b]	7.26 [c]	11.82 [c]	15.80 [d]

Remarks: numbers followed by the same letters in each column indicate no significant different at 5% Duncan's Multiple Range Test.

treatments were compared with control (T), the above-mentioned ameliorant materials had very significant effects on vegetative, generative, biomass and P uptake by maize. At the begin- ning of the growth of maize (2 weeks), the ap- plication of biochar alone (B) resulted in good and fast growth of maize. At 4–6 weeks, the biocom plus *Aspergillus*-TB7 (BTB) treatment gave the best growth performance and biomass of maize even though its P uptake was lower than that of P in biocom plus *Neosartorya*-TM8 (BTM) treatment. However, at 8 weeks, the performance of maize (growth, biomass, and P uptake) as well as the yield components with the BTM treatment was the best. At 4 weeks, the biocom plus phosphate-solubilizing fungi provided a 100–200% increase in the P uptake compared with the control treatment. Generally, biological fertilizer in carrier media increases the plant growth more effectively than free cell biofertilizer; this is because carriers protect the functional microbes from soil or climate stress (Jain et al., 2010). Mechanisms such as the production of phytohormones, vitamins or amino acids may be involved in the effect of microorganisms on phosphate dissolution (Chakkaravarthy et al., 2010).

A number of theories explain the mechanism of inorganic phosphate dissolution i.e. the production of mineral dissolving compounds such as organic acids, siderophores, protons, hydroxyl ions, and CO_2 (Rodríguez and Fraga, 1999; Sharma et al., 2013). The organic acids produced together with carboxyl and hydroxyl ion cations then chelate or reduce the pH to release P (Seshachala and Tallapragada, 2012). Organic acids are produced in the periplasmic chamber by direct oxidation pathway (Zhao et al., 2014). The excretion of these organic acids is accompanied by a decrease in pH, resulting in acidification of microbial cells and its surroundings, so that the P ions are released by substitution of H^+ to Ca^{2+} (Goldstein, 1994). Another mechanism of dissolving phosphate minerals by microorganisms is the production of inorganic acids (such as sulfuric acid, nitrate, and carbonate) and the production of chelating agents (Alori et al., 2017). The increase in the percentage of P uptake is in line with the elevated plant biomass and supports the increase of crop production. The fertilization efficiency of P fertilizer is based on the efficiency of P nutrient uptake. In this study, 32.6% increase of the P uptake efficiency was observed for biocom plus *Aspergillus*-TB7 (BTB) and 42.5% increase of the P uptake efficiency was observed for biocom plus *Neosartorya*-TM8 (BTM), while the application of biochar alone (B) only resulted in 2.5% increase of the P uptake efficiency. This suggests that the application of biochar plus phosphate-solubilizing fungi, either *Aspergillus*-TB7 or *Neosartorya*-TM8, could improve the P uptake efficiency to meet the P nutrient needs for maize. In addition, the use of biochar in fungi carrier positively contributed to the improved availability of micronutrients for plants and water balance in the soil, because biochar has a high porosity (Ścisłowska et al., 2015).

Table 3. Effect of application of biocom + phosphate-solubilizing fungi on yield components

Treatments	Shoot Dry Weight (g)		Root Dry Weight (g)		Fresh Weight of Cob with Husk (g)	Fresh Weight (g)	Dry Weight (g)
	4 weeks	8 weeks	4 weeks	8 weeks		Cob without husk	
T	8.76 [a]	57.58 [a]	5.07 [a]	10.03 [a]	153.58 [a]	97.48 [a]	31.09 [a]
B	11.73 [b]	98.54 [b]	6.68 [b]	16.37 [b]	221.56 [b]	154.60 [b]	46.07 [b]
BTB	13.59 [c]	133.08 [c]	7.91 [d]	19.36 [d]	266.92 [c]	200.58 [c]	60.58 [c]
BTM	11.95 [b]	150.94 [d]	6.76 [c]	17.82 [c]	277.12 [d]	210.44 [d]	65.42 [d]

Remarks: numbers followed by the same letters in each column indicate no significant different at 5% Duncan's Multiple Range Test.

Table 4. Effect of application of biocom + phosphate-solubilizing fungi on P uptake and efficiency of P uptake by maize

Treatments	P uptake by Maize Shoot (g plant^{-1})		Efficiency of P uptake P (%)	
	4 weeks	8 weeks	4 weeks	8 weeks
T	0.06 [a]	0.64 [a]	0.0	0.0
B	0.13 [b]	1.44 [b]	1.3	2.5
BTB	0.22 [c]	2.34 [c]	3.1	32.6
BTM	0.24 [d]	2.85 [d]	3.5	42.5

Remarks: numbers followed by the same letters in each column indicate no significant different at 5% Duncan's Multiple Range Test.

CONCLUSION

The phosphate-solubilizing fungi that were able to adapt to biocom carrier media were *Aspergillus*-TB7 with 60:40 biocom proportions and *Neosartorya*-TM8 with 70:30 biocom proportions. The performance of maize growth and P fertilization efficiency of maize could be improved by applying biocom plus phosphate-solubilizing fungi. Biocom plus *Neosartorya*-TM8 (BTM) application on an Ultisol of Central Kalimantan provided the best maize growth, maize yield, P uptake by maize, and P fertilization efficiency.

Acknowledgements

The authors thank the Ministry of Research, Technology and Higher Education, financial support to carry out this study. The authors also gratefully acknowledge PT. Surya Inti Sawit Kahuripan, Central Kalimantan for providing oil palm empty fruit bunch for this study. Very valuable technical assistance and support from staff of Soil Laboratory of Brawijaya University and Lambung Mangkurat University are highly appreciated.

REFERENCES

1. Alori E.T., Glick B.R., Babalola O.O. 2017 Microbial phosphorus solubilization and its potential for use in sustainable agriculture. Frontiers in Microbiology, 8, 971.
2. Ariani E. 2009. A test of NPK Mutiara fertilizer of 16:16:16 and various types of mulch on yield of pepper plant (Capsicum annum L.). Sagu, 8(1), 5–9 (in Indonesian).
3. Baldock J.A., Smernik R.J. 2002. Chemical composition and bioavailability of thermally altered Pinus resinosa (Red Pine) wood. Organic Geochemistry, 33, 1093–1109.
4. Central Bureau of Statistics. 2015. Statistics of Indonesian Oil Palm 2014. Jakarta (in Indonesian).
5. Chakkaravarthy V.M., Arunachalam R., Vincent S., Paulkumar K., Annadurai G. 2010. Biodegradation of tricalcium phosphate by phosphate solubilizing bacteria. Journal of Biological Sciences, 10(6), 531–535.
6. Chang C.H., Yang S. 2009. Thermo-tolerant phosphate-solubilizing microbes for multi-functional biofertilizer preparation. BioresourceTechnology, 100, 1648–1658.
7. Cheng C.H., Lehmann J., Engelhard M.H. 2007. Natural oxidation of black carbon in soils: changes in molecular form and surface charge along a climosequence. Geochimica et Cosmochimica Acta, 72, 1598–1610.
8. Coutinho F., Yano-Melo A.M., Felix W. 2012. Solubilization of phosphates in vitro by Aspergillus spp. and Penicillium spp. Ecological Engineering, 42, 85–89.
9. Crombie K., Mašek O., Cross A., Sohi S. 2015. Biochar-synergies and trade-offs between soil enhancing properties and C sequestration potential. GCB Bioenergy, 7(5), 1161–1175.
10. Delvasto P., Valverde A., Ballester A., Igual J.M., Muñoz J.A., González F., Blázquez M.L., García C. 2006. Characterization of brushite as a re-crystallization product formed during bacterial solubilization of hydroxyapatite in batch cultures. Soil Biology and Biochemistry, 38, 2645–2654.
11. Dobermann, A. 2007. Nutrient use efficiency. Measurement and management. In: Kraus A., Isherwood K., Heffer P. (Eds.), Fertilizers Best Management Practices. Proc. International fertilizer Industry Association, Brussels, Belgium, 7–9 March 2007, 1–22.
12. Douds Jr D.D., Lee J., Uknalis J., Boateng A.A., Ziegler-Ulsh C. 2014. Pelletized Biochar as a carrier for AM Fungi in the on-farm system of inoculums production in compost and vermiculite mixtures. Compost Science & Utilization, 22, 253–262
13. Duponnois R., Colombet A., Hien V., Thioulouse J. 2005. The mycorrhizal fungus Glomus intraradices and rock phosphate amendment influence plant growth and microbial activity in the rhizosphere of Acacia holosericea. Soil Biology and Biochemistry, 37, 1460–1468.
14. El-Azouni I.M. 2008. Effect of phosphate solubilizing fungi on growth and nutrient uptake of soybean (Glycine max L.) plants. Journal of Applied Sciences Research, 4(6), 592–598.
15. Fitriatin B.N., Yuniarti, A., Turmuktini T., Ruswandi F.K. 2014. The effect of phosphate solubilizing microbe producing growth regulators on soil phosphate, growth and yield of maize and fertilizer effiency on Ultisol. Eurasian Journal of Soil Science, 3(2), 101–107.
16. Goldstein A.H. 1994. Involvement of the quinoprotein glucose dehydrogenase in the solubilization of exogenous phosphates by gram-negative bacteria. In Phosphate in Microorganisms: Cellular and Molecular Biology, eds Torriani-Gorini A., Yagil E. and Silver S. ASM Press, Washington, DC.
17. Graber E.R., Harel Y.M., Kolton M., Cytryn E., Silber A., David D.R., Tsechansky L., Borenshtein M., Elad Y. 2010. Biochar impact on development and productivity of pepper and tomato grown in fertigated soilless media. Plant and Soil, 337, 481–496.

18. Gyaneshwar P., Naresh K.G., Parekh L.J., Poole P.S. 2002. Role of soil microorganisms in improving P nutrition of plants. Plant and Soil, 245, 83–93.

19. Hadi A., Gafur, A., Udiantoro, Mukhlis. 2014. Design of Pyrolysis Installation of Agricultural Waste in the Framework of Minimization of Greenhouse Gas Emissions from Wetlands. Proc. the 5th SNST of 2014. Faculty of Engineering, Wahid Hasyim University, Semarang. ISBN 978–602–999334–3–7. pp 1–9 (in Indonesian).

20. Hale L., Luth M., Kenney R., Crowley D. 2014. Evaluation of pinewood biochar as a carrier of bacterial strain Enterobacter cloacae UW5 for soil inoculation. Applied Soil Ecology, 84, 192–199.

21. Han X., Boateng A.A, Qi P.X., Lima I.M., Chang J. 2013. Heavy metal and phenol adsorptive properties of biochars from pyrolyzed switch grass and woody biomass in correlation with surface properties. Journal of Environmental Management, 118, 196–204.

22. Ichriani G.I., Nion Y.A., Chotimah H.E.N.C., Jemi R. 2016. Utilization of oil palm empty bunches waste as biochar-microbes for improving availibity of soil nutrients. Journal of Degraded and Mining Lands Management, 3(2), 517–520.

23. Ichriani G.I., Syekhfani, Nuraini Y., Handayanto E. 2017. Solubilization of inorganic phosphate solubilizing fungi isolated from oil palm empty fruit bunches of Central Kalimantan. Bioscience Research, 14(3), 705–712

24. Isgitani M., Kabirun S., Siradz S.A. 2005. The effect of bacterial solvent inoculation of phosphate on the growth of shorgum on various P content of soil. Journal of Soil Science and Environment, 5, 48–54 (in Indonesian).

25. Jain R., Saxena J., Sharma V. 2010. The evaluation of free and encapsulated Aspergillus awamori for phosphate solubilization in fermentation and soil-plant system. Applied Soil Ecology, 46, 90–94.

26. Jain R., Saxena J., Sharma V. 2012. Effect of phosphate-solubilizing fungi Aspergillus awamori S29 on mungbean (Vigna radiata cv. RMG 492) growth. Folia Microbiologica, 57, 533–540.

27. Joseph S.D., Camps-Arbestain M., Lin Y., Munroe P., Chia C.H., Hook J., Zwieten L., Kimer S., Cowie A., Singh B.P., Lehmann J., Foidl N., Semrnik R.J., Amonette J.E. 2010. An investigation into the reactions of biochar in soil. Soil Research, 48(7), 501–515.

28. Kang J., Amoozegar A., Hesterberg D., Osmond D.L. 2011. Phosphorus leaching in a sandy soil as affected by organic and incomposted cattle manure. Geoderma, 161, 194–201.

29. Khan M.S., Zaidi A., Wani P.A. 2007. Role of phosphate-solubilizing microorganisms in sustainable agriculture-A review. Agronomy for Sustainable Development, 27, 29–43.

30. Kresnawaty I., Budiani A., Darmono T.W. 2012. Population dinamic of Trichoderma harzianum DT38 on mixture of empty fruit bunches of oil palm (EFBOP) biochar and peat. Menara Perkebunan, 80(1), 17–24 (in Indonesian).

31. Kuppusamy S., Krishnan P.S., Kumutha K., French J., Carlos G.E., Toefield B. 2011. Suitability of UK and Indian source Acacia wood based biochar as a best carrier material for the preparation of Azospirillum inoculum. International Journal of Biotechnology, 4, 582–88.

32. Mao J.D., Johnson, R.L., Lehman, J., Olk D.C., Neves E.G., Thompson M.L., Schmidt-Rohr K. 2012. Abundant and stable char residues in soils: implications for soil fertility and carbon sequestration. Environmental Science and Technology, 46, 9571–9576.

33. Mittal V., Singh O., Nayyar H., Kaur J., Tewari R. 2008. Stimulatory effect of phosphate-solubilizing fungal strains (Aspergillus awamori and Penicillium citrinum) on the yield of chickpea (Cicer arietinum L. cv. GPF2). Soil Biology and Biochemistry, 40, 718–727.

34. Mukome F.N.D., Zhang X., Silva L.C.R., Six J., Parikh. S.J 2013. Use of chemical and physical characteristics to investigate trends in biochar feedstocks. Journal of Agriculture and Food Chemistry, 61, 2196– 2204.

35. Narsian V., Patel H.H. 2000. Aspergillus aculeatus as a rock phosphate solubilizer. Soil Biology and Biochemistry, 32, 559–565.

36. Novak J.M., Busscher W.J., Laird D.L., Ahmedna M., Watts D.W., Niandou M.A.S. 2009. Impact of biochar amendment on fertility of a Southeastern Coastal Plain soil. Soil Science, 174, 105–112.

37. Nurbaity A., Herdiyantoro D., Mulyani O. 2009. The use of organic matter as a carrier of innoculant of arbuscular mycorrhizal fungi. Journal of Biology, Padjadjaran University, 13(1), 17- 11 (in Indonesian).

38. Ogbo F.C. 2010. Conversion of cassava wastes for biofertilizer production using phosphate solubilizing fungi. Bioresource Technology,101, 4120–4124.

39. Osorio N.W., Habte M. 2014. Soil phosphate desorption induced by a phosphate solubilizing fungus. Communications in Soil Science and Plant Analysis, 45(4), 451–460

40. Prasetyo B.H., Suriadikarta D.A. 2006. The characteristics, potential, and technology of Ultisol management for agricultural development in Indonesia. Journal of Agricultural Research and Development, 25(2), 39–47 (in Indonesian).

41. Rebah F.B., Tyagi R.D., Prevost D. 2002. Wastewater sludge as a substrate for growth and carrier for rhizobia, the effect of storage conditions on sur-

vival of Sinorhizobium meliloti. Bioresource Technology, 831, 45–51.

42. Rivera-Cruz M.C., Narcía A.T., Ballona G.C., Kohler J., Caravaca F., Roldán A. 2008. Poultry manure and banana waste are effective biofertilizer carriers for promoting plant growth and soil sustainability in banana crops. Soil Biology and Biochemistry, 40, 3092–3095.

43. Rodríguez H., Fraga R. 1999. Phosphate solubilizing bacteria and their role in plant growth promotion. Biotechnology Advances, 17, 319–339.

44. Santi L.P., Goenadi D.H. 2012. Utilization of biochar from oil palm shells as an aggregate feeding microbial material. Buana Sains, 12(1), 7–14 (in Indonesian).

45. Ścisłowska M., Włodarczyk R., Kobyłecki R., Bis Z. 2015. Biochar to improve the quality and productivity of soils. Journal of Ecological Engineering, 16(3), 31–35

46. Senoo K., Keneko M., Taguchi R., Murata J., Santasup C., Tanaka A., Obata H. 2002. Enhanced growth and nodule occupancy of red kidney bean and soybean inoculated with soil aggregate-based inoculant. Soil Science and Plant Nutrition, 48(2), 251–259.

47. Seshachala U., Tallapragada P. 2012. Phosphate solubilizers from the rhizosphere of Piper nigrum L. in Karnataka, India. Chilean Journal of Agricultural Research.72, 397–403.

48. Sharma S.B., Sayyed R.Z., Trivedi M.H., Gobi T.A. 2013. Phosphate solubilizing microbes: sustainable approach for managing phosphorus deficiency in agricultural soils. Springerplus 2, 587–600.

49. Šimanský V., Klimaj A. 2017. How does biochar and biochar with nitrogen fertilization influence soil reaction?. Journal of Ecological Engineering, 18(5), 50–54.

50. Somarathne R., Yapa P., Yapa N. 2013. Use of Different Carrier Materials for Culture and Storage of Native Forest Soil Microorganisms. 3rd International Conference on Ecological, Environmental and Biological Sciences (ICEEBS'2013) April 29–30, 2013, Singapore.

51. Spokas K.A., Cantrell K.B., Novak J.M, Archer D.W., Ippolito J.A., Collins H.P., Boateng A.A.,

Lima I.M., Lamb M.C., McAloon A.J., Lentz R.D., Nichols K.A. 2011. Biochar: a synthesis of its agronomic impact beyond carbon sequestration. Journal of Environmental Quality, 41, 973–989.

52. Stephens J.H.G., Rask H.M. 2000. Inoculant production and formulation. Field Crops Research, 65, 249–258.

53. Subagyo H., Suharta N., Siswanto A.B. 2004. Agricultural Soils in Indonesia. p. 21–66. In Adimihardja A., Amien L.I., Agus F., Djaenudin D. (Eds.). Indonesia's Land Resources and Its Management. Center for Soil and Agroclimate Research and Development, Bogor (in Indonesian).

54. Sukartono, Utomo W.H., Kusuma, Z., Nugroho W.H. 2011. Soil fertility status, nutrient uptake, and maize (Zea mays L.) yield following biochar and cattle manure application on sandy soil of Lombok, Indonesia. Journal of Tropical Agriculture, 49(1–2), 47–52.

55. Sundara B., Natarajan V., Hari K. 2002. Influence of phosphorus solubilizing bacteria on the changes in soil available phosphorus and sugarcane and sugar yields. Field Crops Research, 77, 43–49.

56. Uzoma K.C., Inoue M., Andry H., Fujimaki H., Zahoor A., Nishihara E. 2011. Effect of cow manure biochar on maize productivity under sandy soil condition. Soil Use and Management. Wiley Online Library, USA.

57. Warnock D.D., Lehmann J., Kuyper T.W., Rillig M.C. 2007. Mycorrhizal responses to biochar in soil-concepts and mechanisms. Plant and Soil, 300, 9–20.

58. Widowati, Utomo W.H., Soehono L.A., Guritno B. 2011. Effect of biochar on the release and loss of nitrogen from urea fertilization. Journal of Agriculture and Food Technology, 1(7),127–132

59. Zhao K., Penttinen P., Zhang X., Ao X., Liu M., Yu X., Chen Q. 2014. Maize rhizosphere in Sichuan, China, hosts plant growth promoting Burkholderia cepacia with phosphate solubilizing and antifungal abilities. Microbiological Research, 169,76–82.

60. Zhu H.J., Sun L.F., Zhang Y.F., Zhang X.L., Qiao J.J. 2012. Conversion of spent mushroom substrate to biofertilizer using a stress-tolerant phosphate-solubilizing Pichia farinose FL7. Bioresource Technology, 11, 410–416.

18

Hydrogeological Conditions and Natural Factors Forming the Regime of Groundwater Levels in the Ivano-Frankivsk Region (Ukraine)

Lidiia Davybida[1*], Dmytro Kasiyanchuk[1], Liudmyla Shtohryn[1],
Eduard Kuzmenko[1], Mariia Tymkiv[1]

[1] Department of Geotechnogenic Safety and Geoinformatics, Ivano-Frankivsk National Technical University of Oil and Gas, 15 Karpatska Street, Ivano-Frankivsk, 76019, Ukraine

[*] Corresponding author's e-mail: davybida61085@gmail.com

ABSTRACT

The purpose of the article is to carry out a complex analysis of conditions and natural factors forming the hydrogeological regime as well as to integrate the existing information and groundwater monitoring data in Ivano-Frankivsk region for further creating the unified automated system to collect, process, analyze and store the monitoring observation data. The authors offer methodology, which is based on the geoinformation approach, for organizing and monitoring of the groundwater. The structure of spatial and attributive data that afford to systemize the existing schemes of the hydrogeological zoning as well as the results of long-term observations over variability of the groundwater levels and factors forming the hydrogeological regime were developed. The new approaches for improving the techniques to forecast the groundwater level with due consideration of temporal patterns for changes in water level, the Moon phases and seismic activity, are proposed. The obtained results are considered as an informational basis for reorganization of the state hydrogeological monitoring on the example of Ivano-Frankivsk region.

Keywords: groundwater, hydrogeological zoning, geodatabase, the Moon phases, earthquakes

INTRODUCTION

The groundwater, as the most dynamic component of the geologic environment, responds quickly to changes in their forming conditions and, hence, to the impact of anthropogenic factors. It means that the condition peculiar to the underground hydrosphere, and primarily, to the groundwater, as the first constant water-bearing formation, may serve as an indicator of the ecological state of the geological environment as a whole. Nowadays, it is believed that the functioning of a scientifically grounded stationary system of hydro-geological monitoring supplies the most complete information about the groundwater condition [Koshliakov et al. 2016, Shestopalov et al. 2016]. However, since such system is practically absent in the most territories of Ukraine, and in the studied Ivano-Frankivsk region in particular, and its creation and effective functioning is problematic because of purely economic reasons, we have to develop the new approaches to obtain and process information on the groundwater conditions, as well as on the dynamics of their levels first of all.

Upon analyzing the current experience in the application of the geoinformation approach to optimization of groundwater monitoring [Hudak et al. 1992, Ben-Jema et al. 1994, Berezko et al. 2012, Brovko et al. 2015, Pivovarova 2016, Tymkiv et al. 2016], the authors identify the priority tasks to renovate the state network of hydrogeological monitoring and to adapt it to the European standards in the shortest possible period. These tasks are as follows: inventory of the observation points, evaluation of their representativeness, and development of the concept for monitoring system reformation as well as formation of a single database with the involvement of GIS-technologies.

It is well known that the regime of underwater is affected by the exogenous factors (meteorological, hydrological characteristics, lunar tides), the endogenous factors (tectonic movements, volcanism, and earthquakes) and the technological factors (water intake). It is clear that the climatic factors (atmospheric precipitations, air temperature) and the hydrological factors (a regime of surface stream flows) have effect on the hydrogeological regime formation [Ruban et al. 2005] and, in particular, on the examined area, which is characterized by natural or slightly degraded regime undergoing the influence of local technological factors. First, the earthquakes should be enumerated among the geological factors, the effect of which may be fixed by hydrogeological observations. The impact of earthquakes on the groundwater regime is usually episodic and it is the most often seen in sharp changes in the level of pressure water. The study of peculiarities in changes of the natural regime of groundwater, in turn, gives the possibility to monitor the process of earthquake emerging. It is noted in scientific papers [Adushkin et al. 2017; Maréchal et al 2002] that the lunar and solar tides are clearly distinguished in the variations of level of underground water. It is confirmed by the spectral analysis. It is believed that the monthly lunar tides within the pressure water have an influence on the water-bearing horizon by periodically increasing the load, changing the elastic state of the water-bearing horizon. The influence of tidal force is fixed in water-table wells, because the gravity forces stretch the

Earth's crust, thus increasing the level of groundwater. These, so called ground, tidal oscillations are measured only by few centimeters and they have clear periodicity in 12 and 24 hours, and less clear periodicity in 14 and 28 days.

It should be noted that there has been no purposeful study of the influence produced by the seismic activity and lunar cycles on the hydrogeological regime formation in the Ivano-Frankivsk region. At the same time, it has been studied for nowadays that in the Western Ukraine, abnormal rains (one of the direct factors of influence over variability of the groundwater levels) usually occur in the Last or the First Moon Quarter. As a rule, the maximum precipitations are recorded on the first – third day of the Moon Phase [Shtohryn et al. 2015]. The cycles of extreme precipitations in 5.5–6.9.4 and 11 years are present in the spectra of all specified phases and in the whole period, as well as they are multiple of the solar activity periods and the Moon cycle (Fig. 1).

The purpose of the study is to integrate the existing information and groundwater monitoring data as well as the factors forming the hydrogeological regime in the Ivano-Frankivsk region for further creation of the unified automated system to collect, process, analyse and store the monitoring observation data. The Ivano-Frankivsk region was chosen due to three reasons: 1) its location on the border of mountainous areas and flat terrains; 2) the availability of output data; 3) possibility to spread the findings obtained to other territories of Ukraine and abroad regions.

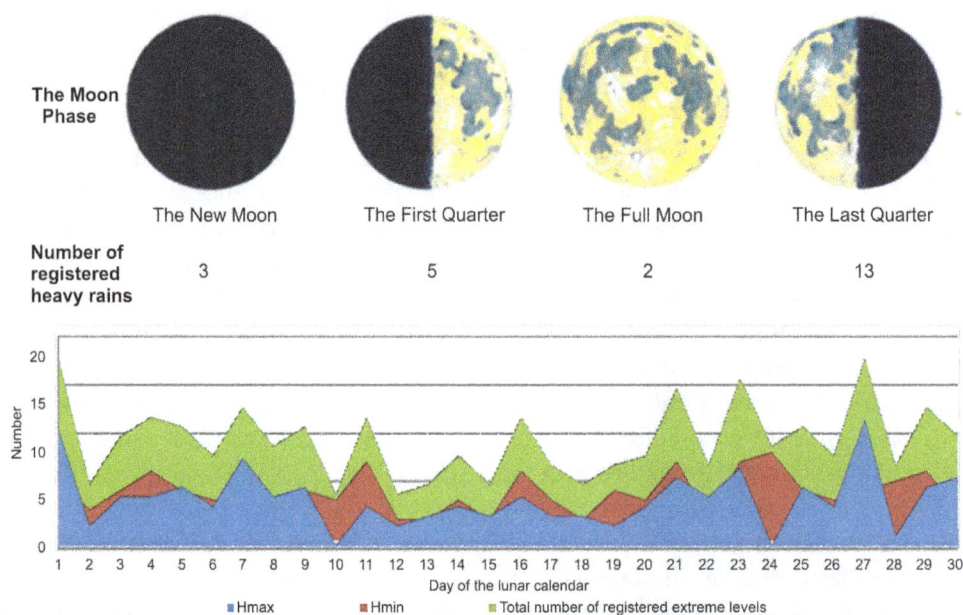

Figure 1. Diagram of interrelations between the Moon phases and extreme levels of groundwater

MATERIALS AND METHODS

The studied Ivano-Frankivsk region is characterised by the well-developed river network and belongs to the area, which is one of the richest areas in Ukraine in terms of surface water resources. However, this area takes one of last places with regard to the reserves of the drinking groundwater. Despite this fact, the individual water supply to the population and enterprises of the region is mainly carried out by the groundwater. The first water-bearing horizon is used the most often.

According to the zoning of Ukraine under the conditions for the groundwater formation (Fig. 2), Ivano-Frankivsk region is located in the area of such hydrogeological regions as Volyn-Podolian Artesian Basin and Precarpathian Artesian Basin as well as the Carpathian folded zone [Shestopalov et al. 2010]. The complex geological structure of the area has caused a great variety of its hydrogeological conditions, which are unequal for the groundwater using and its vulnerability to the industrial pollution. It should be stressed that the basic current network of the state hydrogeological monitoring, which in 2016 included only 5 observation wells and 2 wells within the water intake area in Ivano-Frankivsk region, did not fulfil the assigned functions and it requires serious reorganization and further development [Shestopalov et al. 2016].

The materials by 'Geoinform of Ukraine', State Research-and-Production Enterprise, Ukrainian State Geological Prospecting Institute, Institute of Geological Sciences of National Academy of Science of Ukraine, concerning the hydrological zoning pursuant to the water exchange principle, location of hydrogeological observation points as well as the results of long-term observations have been used for creating the geodatabase to provide the functioning of the hydrogeological monitoring system within the Ivano-Frankivsk region. The spatial and attributive information is supplemented by the topographical data and brought to a single scale (1:100 000) and format of digital cartographic models in GIS MapInfo Professional (Gauss-Krüger projection, Pulkovo, 1942, zone 6). The results of the monitoring observations are stored in the external relational database and they may be adjoined to the spatial objects (hydrogeological monitoring stations) using unique codes of the observations points.

The hydrogeological areas identified by various principles are considered as a territorial basis to conduct the groundwater monitoring as well as projecting and improving the current network of observation wells. In this paper, the zoning scheme based on groundwater formation conditions as well as the scheme of functional zoning pursuant to the basin principle are chosen as the initial cartographic materials that characterize conditions creating the regime on the area.

Zoning scheme (Fig. 2) has been developed pursuant to the conditions of the groundwater formation in the Ukrainian State Geological Prospecting Institute, which is detailed in the paper [Ruban et al. 2005]. The zoning scheme is formed in accordance with the geological and structural principle of hydrogeological zoning, which takes into account the heterogeneity of the Earth's crust composition and structure. The regions, which are covering the largest geo-structural units – the hydrogeological massifs (such as basins of produced-block water and basins of fracture-vein water) as well as artesian basins (produced water basins), are allocated at the core of the scheme. There are smaller taxonomic units that include the hydrogeological areas representing a part of the groundwater basin expressed on the certain territory, which is characterised by a specific combination of geological structure, relief, climatic, hydrological and, eventually, hydrogeological conditions. One may distinguish the sub-areas, which differ by some features of geological structure, relief and climatic conditions, within a majority of areas. The smallest taxonomic units of typing, which characterize the stratigraphic features of the conditions for the groundwater formation in the profile, are the landscape hydrogeological complexes (LHC). They include separate areas, where the first water bearing horizon is deposited having the defined stratigraphic sections, which reflect the structure of the aerated zone, water-bearing and waterproof layers and, to a certain extent, stipulate the peculiarities of the groundwater recharge and discharge as well as the reservoir properties of the upper layer of rocks, where their formation takes place. LHC are azonal units of zoning and they are characterized by common features of the most hydrogeological areas, mainly within the groundwater basins. In general, 26 varieties of plain LHC and 6 varieties of mountainous LHC are identified in Ukraine. There are 10 varieties (7 plain LHC and 3 mountainous LHC) in the Ivano-Frankivsk region. The indexation of LHC is specified in accordance with the hydrogeological zoning scheme of Ukraine [Ruban et al. 2005].

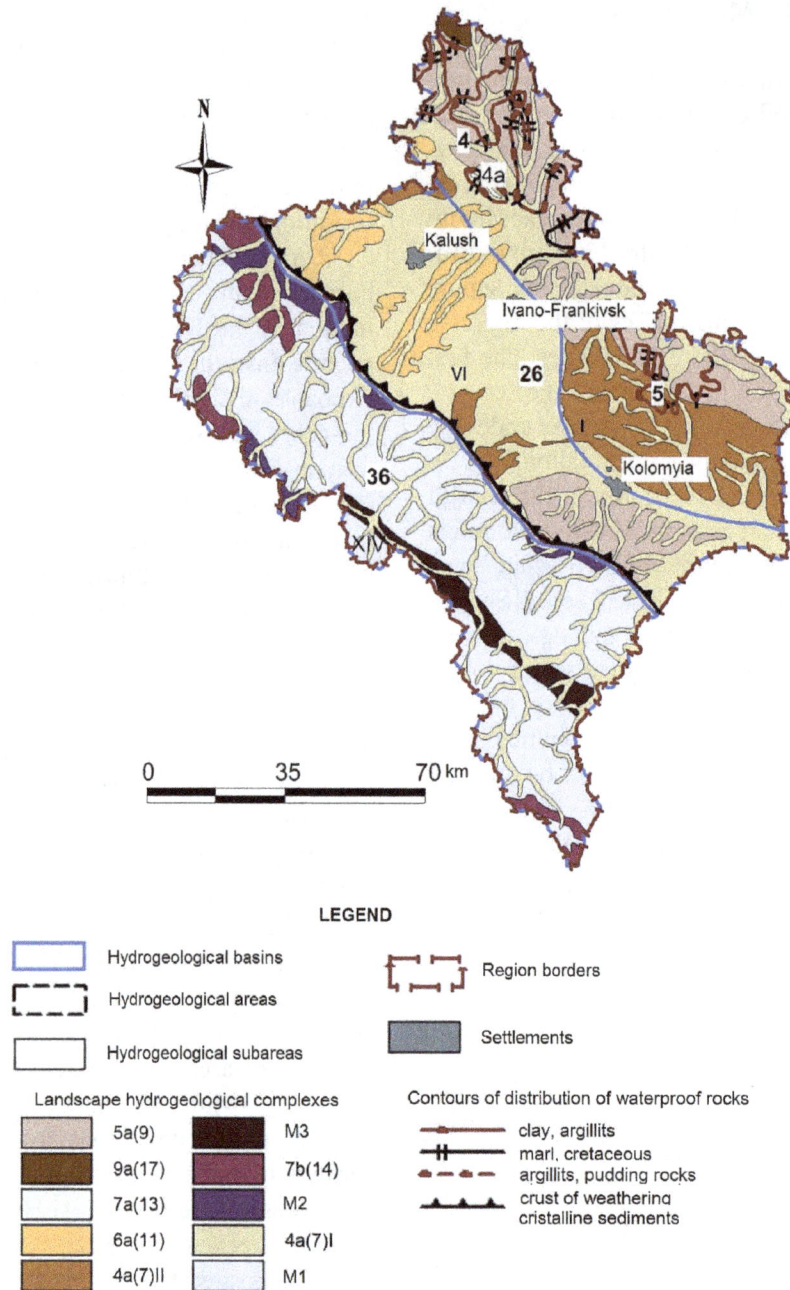

Figure 2. Zoning of Ivano-Frankivsk region according to the formation conditions and groundwater occurrence: I – Volyn-Podolian Artesian Basin (4 – Podolian area (4a – Roztochya-Opilian suberea), 5 – Prut-Dniester area); VI – Precarpathian Artesian Basin (26 – Precarpathian area); XIV – Carpathian Groundwater Basin of (36 – Mountain Carpathian area)

The scholars of the Institute of Geological Sciences of National Academy of Science of Ukraine have developed a functional zoning scheme, according to the basin principle (under the conditions of water exchange in the upper hydrogeological layer) [Shestopalov et al. 2010] (Fig. 3). The given scheme may be considered as the basic one for reorganization of the state system for the groundwater monitoring, in accordance with the principles of the EU Water Framework Directive, and for transition from the administrative-territorial approach to creation of the integrated basin model to control the water resources. This model envisages coordination of water protection measures as to the surface water and groundwater, which belong to the common ecologic, hydrologic, and hydrogeological system. The system for studying the groundwater drainage is based on the dynamic features of objects allocated according to the principle of groundwater runoff unity from the areas of drainage formation to its main discharge, which is controlled by a single closed balance of the groundwater. Within the hydrogeological structures, one may allocate the water exchange basins of the first order, which are characterized by single or close trends of the regional water exchange. The hierarchy of this zoning is as follows: there are basins of sea drainage (regions) and main rivers that flow into the seas (provinces). The next zoning level takes into account the peculiarities of the hydrogeological conditions as well as the close trends of lateral water exchange; thus, the basins of rivers' groundwater flow are united within the bounds of the single zoning units (provinces), where the subareas corresponding to water exchange basins of different main rivers tributaries are allocated. The water exchange basins, sizes and boundaries of which are determined by sizes and configuration of the river basins as a rule, should be considered as characteristic systems with homogenous regime in forecasting the changes in level of the ground water, which is a part of the underground hydrosphere that is most closely related to the external surface factors forming the groundwater recharge and discharge. The expediency of monitoring study

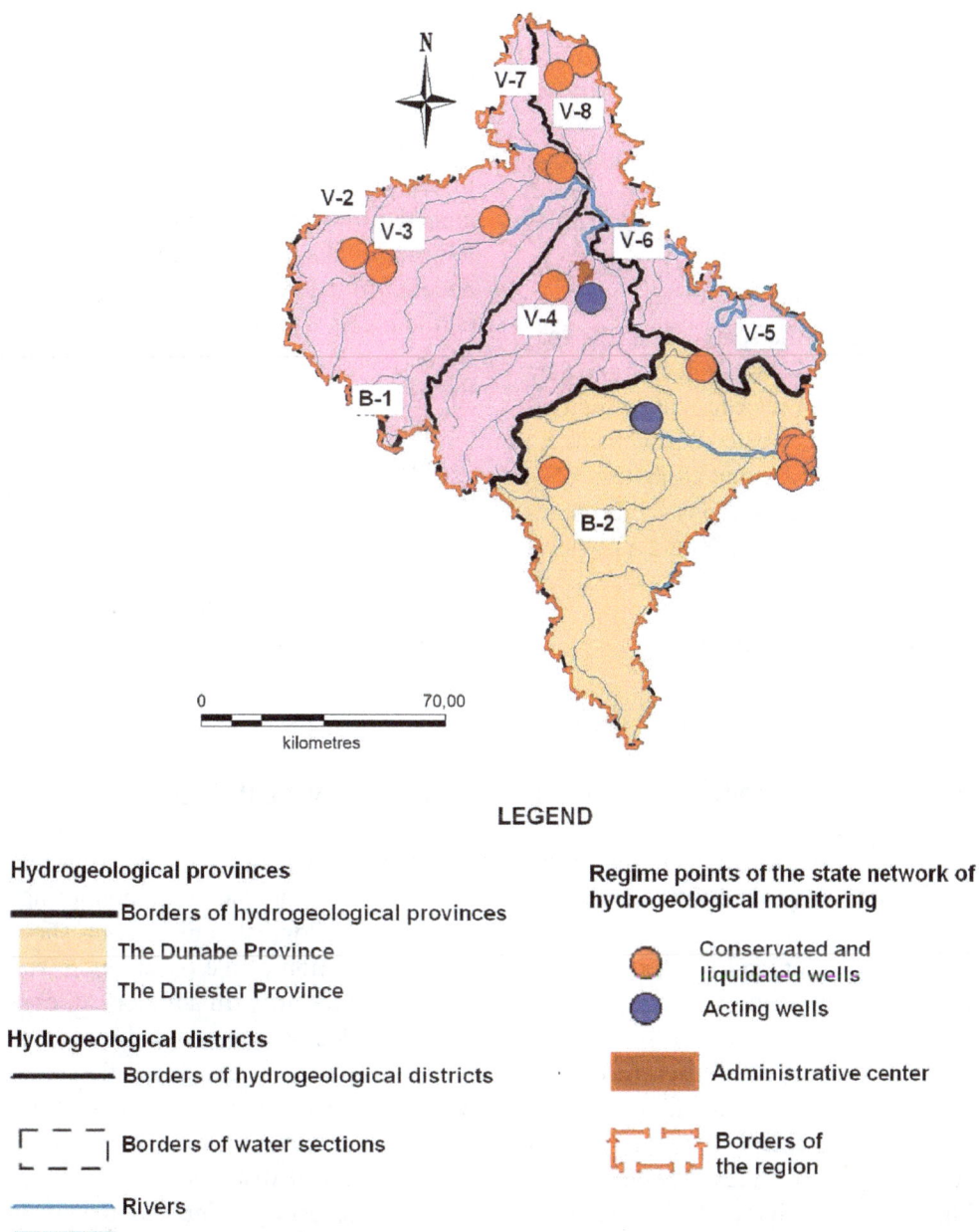

Figure 3. Zoning of the Ivano-Frankivsk region under the conditions for water exchange formation in the hydrogeological structures

for both surface and underground water is also confirmed by the synchronicity of the long-term variability of the mid-annual surface water flow and the annual average depth of the groundwater level. The synchronicity is observed for both the actual values of the regime elements and for long-term trends established for the hydrological stations and hydrogeological wells located in the same hydrogeological district [Davybida 2017]. The studied territory of Ivano-Frankivsk region belongs to the hydrogeological provinces of the Dniester River and the Dunabe River, which are divided within the studied region in 7 hydrological districts (V-2 – the Strypa River, V-3 – the Svich River and the Limnytsia River, V-4 – the Bystrytsia Nadvirnianska River and the Bystrytsia Solotvynska River, V-5 – the right bank of the Dniester River from the mouth of the Bystrytsia River to the mouth of the Liadova River, V-6 – the Stryvigora River and the Vereshchytsia River, V-7 – the Davydivka River and Svira River, V-8 – the Gnyla Lypa River and the Zolota Lypa River) and 2 hydrogeological districts (B-1 – the Tysa River, B-2 – the Prut River).

In order to integrate the spatial data, the spatial binding of raster maps was conducted and they were transferred into the vector format by digitalization. A geodatabase was created that includes the following layers with the relevant attributes:

1. hydrological monitoring stations (code, number, type, depth, coordinates X and Y, type of water-bearing horizon, regime type, disturbance of regime);
2. hydrological provinces (code, index, water exchange basin);
3. hydrological districts (code, index, water exchange basin);
4. water-resource regions (code, index, departmental subordination);
5. hydrological basins, allocated under the conditions of formation and groundwater occurrence (code, index, basin);
6. hydrogeological district (code, index, district);
7. hydrogeological sub-district (code, index, sub-district);
8. landscape hydrological complexes (code, index, age of rocks of the aeration zone, rocks of the aeration zone, chemical composition of the aeration zone, limit values of aeration zone depth, limit values peculiar to the coefficient of the aeration zone filtration, age of rocks contained in the water-bearing formation; rocks of the water-bearing formation; chemical compo-

sition of waters contained in the ground water-bearing horizon, limit values peculiar to the filtration coefficient of water-bearing horizon, age of watercourse rocks, filtration coefficient of watercourse rocks, limit values of mineralization, water hardness and pH index;
9. vector layers of the basic information including administrative boundaries, settlements, roads, rivers, vegetation, relief.

The factual input database contains the information on the structure of the state hydrogeological monitoring network (spatial coordinates and attributive data), location of hydrogeological points, meteorological points as well as the results of long-term regime observations over the groundwater levels, presented in the form of average monthly values of groundwater depth levels calculated under the data of timing observations, results of observations on the variability of weather and climate characteristics (average annual air temperature, amount of precipitation), long-term change of numerical characteristics peculiar to the geological features (the energy of earthquakes) and the cosmogenic features (the Moon phases, solar activity) required to form the hydrogeological regime.

It should be noted that due to the absence of observation wells for intermediate waters as well as of the results of observations over the hydrochemical and hydrothermal regime in the Ivano-Frankivsk region, the present informative elements were not entered into the developed database and they are considered as an object of separate study.

RESULTS AND DISCUSSION

The spatial analysis concerning the subject layers of hydrogeological monitoring stations and hydrogeological zoning taxon shows the inefficient location of the observation wells within the studied territory. In addition, all observations are conducted over the groundwater regime only. However, given the characteristic features of the conditions peculiar to the groundwater regime formation, number, configuration and location of hydrogeological districts in Ivano-Frankivsk region, it is necessary to increase the number of hydrogeological monitoring stations up to 18–20 (3 wells per each hydrogeological district allocated under the water exchange principle; they should be located in such a way so as to be representative for riverside, terraced and water-dividing types of the groundwater regimes). Another criterion for choosing the location of the observation point is the LHC occurrence (Fig. 2). The location of the monitoring stations with due allowance for conditions of the hydrogeological regime formation will permit to obtain more complete information and to organize a random placement of wells to provide the uniformity and independence of tests. The next stage of the study is determination of optimal coordinates to locate the monitoring wells on the basis of a formal statistical estimation of the points' density, taking into account the given extrapolation error of the investigated parameters in a GIS environment.

An outcome analysis of long-termed observations over the groundwater level, the weather and climate characteristics as well as over the hydrological regime creating factors typical for the observation territory (Fig. 4) has established the presence of synchronically changed rhythm and significant relation, first of all, for the time series of the average annual groundwater levels and the series of average annual water consumption, the value of which is connected with the modulus of flow, including such factors as surface and groundwater flow. The regime of the surface water, in its turn, is also determined by climatic factor having an influence on the groundwater.

Synchronicity of the long-term variability of the annual average discharge of surface flows and the average annual depths of groundwater levels confirms the feasibility of joint monitoring studies for surface water and groundwater in the studied region, as provided by Water Framework Directive [Directive 2000]. The established regularities may be used for renovating the time series of observations over the groundwater levels, for bringing them to a single timeframe, and for drawing up the medium and long-term hydrogeological forecasts.

In order to improve the reliability of forecasting the annual regime of variability peculiar to the groundwater levels as well as the examination of its interconnection with the seismic activity in the studied Ivano-Frankivsk region, the extreme values of groundwater levels in five wells from 1973 to 2003 were analyzed. These wells were drilled into the water-bearing low upper quaternary alluvial horizon AQ_{1-3} within Volyn-Podolian Artesian Basin. The location of the wells may be set by the boundary coordinates East Longitude

$25^{\circ}54' - 25^{\circ}57'$, North Latitude $48^{\circ}44' - 48^{\circ}48'$. The considered earthquakes were registered within the region and adjacent areas with coordinates East Longitude $24^{\circ} - 24^{\circ}6'$, North Latitude $47^{\circ}7' - 48^{\circ}8'9$ (Table 1).

The extreme values of groundwater levels, which were calculated according to the periodic measurements including season levels (pre-spring minimum level (H_1), spring maximum level (H_3), summer-autumn minimum level (H_5)) and annual levels (high level (H_{max}), low level (H_{min})), were considered in particular. The lowest groundwater levels are H_1, H_5, which mostly coincide with the extreme level H_{min}, and, respectively, are the highest groundwater levels are H_3 or H_{max}.

In Table 1 shows the coordinates (geographic latitude φ° N and longitude λ° E) of the hydrogeological points, for which were conducted the following measurements of the groundwater levels are summarized as well as coordinates of the registered earthquakes, their energies (E) and energy classes (K) and the Moon Phases (NL – the New Moon, PQ – the First Quarter of the Moon, PL – the Full-Moon, DQ – the Last Quarter of the Moon) observed at the fixation extremums and earthquakes.

While analyzing the Table 1, one may notice that the average value of the groundwater level equals 2.3–3.6 meters. The extreme levels, as a rule, were observed in the same months when the earthquake was registered. When the water levels are recorded before the earthquakes, the water rises closer to the surface in relation to the average value.

The authors may give the data under observation well No. 95p, as an ex ample, where the average level H_{min} equals 3.94 m, and the registered index that is fixed a day before the earthquake of 28.05.96 with the energy class K=7.9 amounts to 3.16 m. This is closer to the surface by 0.78 m in comparison with the average value. It should be noted that in the Ivano-Frankivsk region, the abnormal quantity of the extreme precipitations was not observed in May of 1996. Such information may indicate the series of the Earth's break ups and rifts, along which the water rises before the earthquakes.

While analyzing the graphs of the extreme groundwater levels (Fig. 5), we can see that the minimum and maximum groundwater levels are in-phase during the second part of the New Moon Phase (NL), from the first to the third day within the First Quarter of the Moon Phase (PQ) and the

a)

Q, m3/h

350,00	0
300,00	10
250,00	20
200,00	30
150,00	40
100,00	50
50,00	60
0,00	70

P, %

1968 1973 1978 1983 1988 1993 1998 2003 2008

— Average water discharge (Dniester)   — Probability of grondwater annual level

Series I: Groundwater annual level
Series II: Water discharge (Dniester)

Year	Corr.	StDv
-14	,2586	,1925
-12	,3572	,1857
-10	-,105	,1796
-8	-,032	,1741
-6	-,068	,1690
-4	-,305	,1644
-2	-,045	,1601
0	-,191	,1562
2	,1160	,1601
4	,2723	,1644
6	,1089	,1690
8	,2289	,1741
10	-,134	,1796
12	-,167	,1857
14	-,208	,1925

b)

W, mm

1200	0
1000	10
800	20
600	30
400	40
200	50
0	60

P, %

1968 1973 1978 1983 1988 1993 1998 2003 2008

— Annual precipitation (Ivano-Frankivsk)   — Probability of grondwater annual level

Series I: Groundwater annual level
Series II:Annual precipitation

Year	Corr.	StDV
-14	-,003	,1925
-12	,1033	,1857
-10	-,035	,1796
-8	,0400	,1741
-6	,0773	,1690
-4	-,151	,1644
-2	,0734	,1601
0	-,270	,1562
2	-,139	,1601
4	,0242	,1644
6	,1135	,1690
8	,2972	,1741
10	-,127	,1796
12	-,027	,1857
14	-,079	,1925

c)

T, grad C

10	0
9	10
8	20
7	30
6	40
5	50
4	60
3	70
2	80
1	90
0	100

P, %

1968 1973 1978 1983 1988 1993 1998 2003 2008

— Average annual temperature (Ivano-Frankivsk)   — Probability of grondwater annual level

Series I: Groundwater annual level
Series II: Average annual temperature

Year	Corr.	StDv
-14	-,118	,1925
-12	-,225	,1857
-10	-,299	,1796
-8	,0809	,1741
-6	,3162	,1690
-4	,0167	,1644
-2	,0870	,1601
0	,0462	,1562
2	,0271	,1601
4	-,138	,1644
6	,0347	,1690
8	-,011	,1741
10	,1269	,1796
12	,0094	,1857
14	-,176	,1925

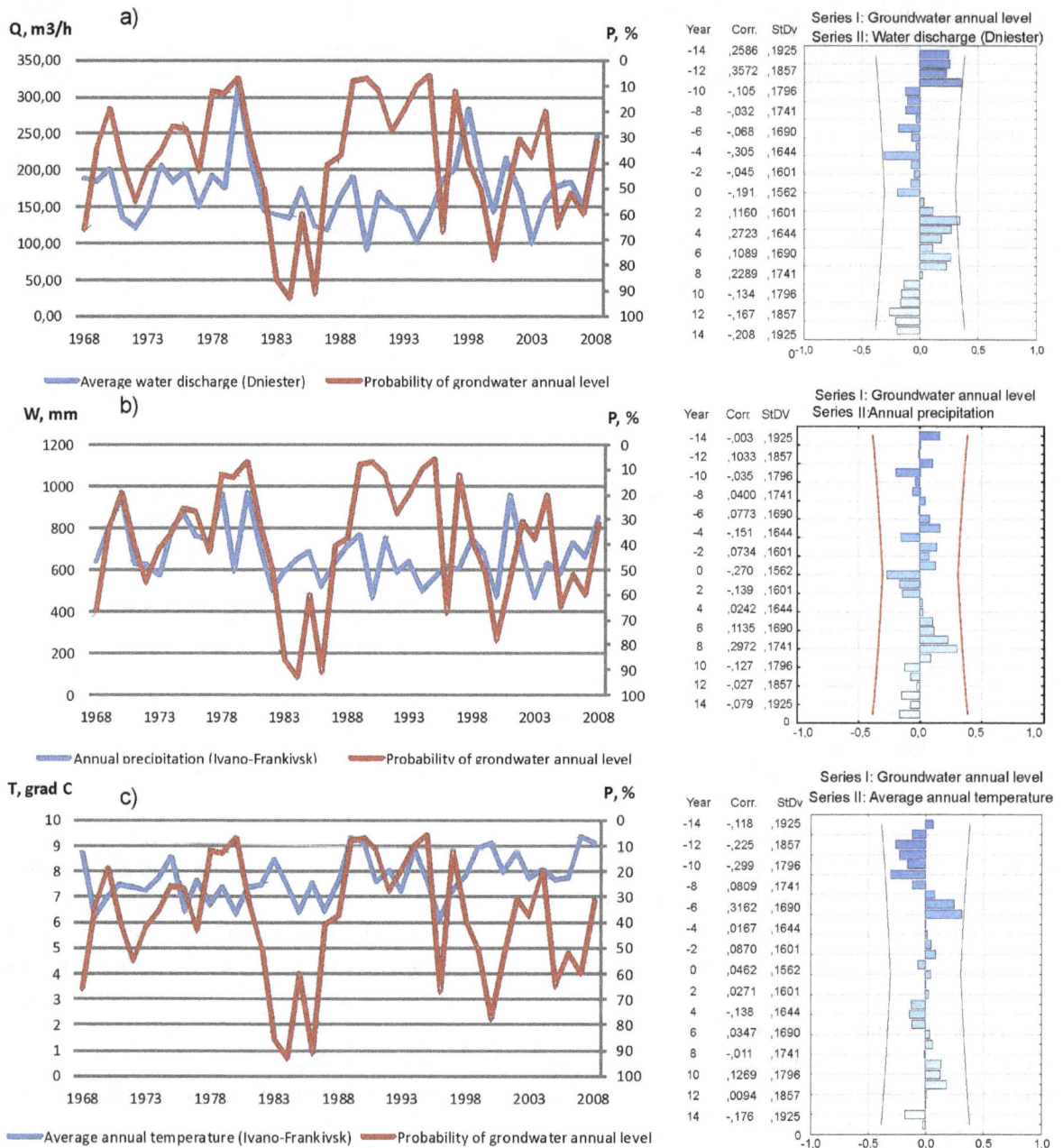

Figure 4. Estimation of interconnections peculiar to changing in groundwater levels and numerical characteristics of the regime creating factors applied for the monitoring stations in the Ivano-Frankivsk region: a) graphs and cross-correlation function for the groundwater level and surface water discharge; b) graphs and cross-correlation function for the groundwater level and annual amount of precipitation; c) graphs and cross-correlation function for the level of groundwater and average annual air temperature

Full Moon (PL) of the Moon cycle. The groundwater levels are in the antiphase during the second part of the Last Quarter Phase (DQ) during the twenty-fourth-twenty-sixth days as well as from the beginning of the New Moon Phase (NL) from the twenty-eighth to the thirties day.

As it seen from Table 2, the maximum groundwater levels are registered during the New Moon Phase (NL) (8 cases of 22 values were registered within the phase) and 2 days before the First Quarter of the Moon Phase (PQ) (7 cases of 21 values registered within the phase), that constitutes 36% and 33%, respectively.

There is a characteristic feature as to the registration of the minimum groundwater levels, which is conducted the next day of the phase. Such cases are registered during the New Moon (NL), the Last Quarter of the Moon Phase (DQ), the Full Moon (PL) (6–7 cases), which constitutes from 28% to 37 % of the registered values within the phase.

Table 1. Interconnection between the Moon Phases, earthquakes, and groundwater level

Earthquaekes		Date of earthquake registration	K	E×10⁸, J	Well's cadastral number	φ° N of well	λ° E of well	Date of level registration	Day of the Lunar Cycle and the Moon Phase		Level of ground-water (H)	$H_{registere}$, m	$H_{average}$, m
φ° N	λ° E												
47.90	24.60	14.11.74	9.5	31.60	113p/e	48.44	25.56	06.11.74	23	/DQ	H_5	3.63	3.69
47.97	24.51	14.11.74	7.2	0.16	113p/e	48.44	25.56	06.11.74	23	/DQ	H_5	3.63	3.69
49.03	24.00	18.11.74	7.6	0.40	113p/e	48.44	25.56	06.11.74	23	/DQ	H_5	3.63	3.69
49.05	24.02	14.01.76	11.0	1000.0	114p	48.44	25.57	03.01.76	3	NL//	H_1	2.98	3.16
49.05	24.02	14.01.76	11.0	1000.0	3	48.48	25.54	03.01.76	3	NL//	H_{min}	3.61	4.53
49.05	24.02	14.01.76	11.0	1000.0	95p	48.47	25.54	03.01.76	3	NL//	H_{min}	3.48	3.94
47.83	24.75	11.06.76	7.0	0.10	113p/e	48.44	25.56	06.06.76	9	PQ/	H_{max}	2.81	3.08
47.83	24.75	11.06.76	7.0	0.10	114p	48.44	25.57	06.06.76	9	PQ/	H_{max}	1.93	2.25
47.83	24.75	11.06.76	7.0	0.10	95p	48.47	25.54	06.06.76	9	PQ/	H_{max}	2.58	2.92
48.95	24.07	31.10.79	9.5	31.60	114p	48.44	25.57	30.10.79	13	//PL	H_5	2.80	3.24
48.05	25.00	13.07.85	7.7	0.50	113p/e	48.44	25.56	27.07.85	11		H_{min}	4.56	4.06
48.05	25.00	13.07.85	7.7	0.50	114p	48.44	25.57	06.07.85	19		H_{max}	1.90	2.25
47.70	24.50	28.08.90	7.8	0.63	102p/e			27.08.90	8	/PQ	H_5	1.80	2.02
48.20	24.70	08.01.93	7.6	0.40	102p/e			03.01.93	11	PQ//	H_{min}	1.68	2.09
48.80	24.29	16.08.95	8.2	1.58	114p	48.44	25.57	09.08.95	14	/PL	H_5	2.97	3.24
48.69	24.40	28.05.96	7.9	0.79	95p	48.47	25.54	27.05.96	11	PQ//	H_{min}	3.16	3.94

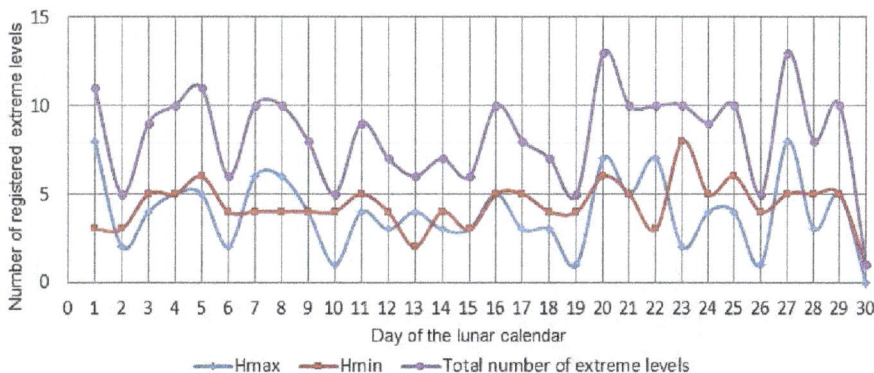

Figure 5. Dependency graph for a number of the registered extreme groundwater levels (by the lunar calendar

While considering the graphs of the extreme groundwater levels and number of the earthquakes both in the Ivano-Frankivsk region and in the adjacent territories with the energy class K> 6 (Fig. 6) for 1962–2003, we may notice an inverse relation. It is especially traced for maximum groundwater levels. The same trend is observed for the minimum groundwater levels from the first to third day of New Moon Phase (NL), during the First Quarter (PQ) and the full Moon (PL). However, since the second half of Full Moon Phase (PL) and during the Last Quarter Phase (DQ) we can notice a significant correlation between changing of minimum groundwater levels and registration of strong earthquakes with energy class K> 6 (correlation class from the fifteenth to thirties day of the Moon cycle equals 0.47).

We must also note the increase in seismic activity during the Full Moon Phase (PL), when it coincided with Perigee (the distance between the Moon and the Earth is the closest). Thus, eight cases of the registered earthquakes with energy class K=7.1–9.1 occurred during Perigee or within it (≈ 2 days).

In the days of the New Moon Phase (NL) (from the first to seventh day) and the Full Moon (from the thirteenth to seventeenth day), we may observe the inverse relation, which is primarily agreed with Hmax, between the number of measurements of the extreme groundwater levels and number of the registered earthquakes. Only a graph of changing measurement of Hmin repeats the trend of the earthquake registration.

The maximum values peculiar to occurrence of the groundwater levels (Hmax) are most often registered during the First and the Last Moon Quarter (PQ and DQ). The coefficients of correlation equal 0.98 and 0.8, respectively. It should

Table 2. Interconnection between the Moon Phases, earthquakes, and groundwater levels

The Moon Phase	Day of the Lunar Cycle	Number of registered extreme groundwater levels			Number of earthquakes with K>6
		H_{max}	H_{min}	Total number	
//NL	28, 29	3	4	7	2
/NL	29, 30	5	5	10	1
NL	1	8	2	10	5
NL/	2	2	6	8	3
NL//	3	4	4	8	1
//PQ	6, 7	7	3	10	2
/PQ	7, 8	4	3	7	6
PQ	9	2	3	5	3
PQ/	8, 9	5	4	9	1
PQ//	11, 12	2	5	7	4
//PL	13, 14	4	1	5	5
/PL	14, 15, 16	4	5	9	2
PL	15, 16	4	3	7	6
PL/	16, 17, 18	5	7	12	3
PL//	18, 19	3	3	6	1
//DQ	21, 22	5	4	9	2
/DQ	22, 23	5	5	10	3
DQ	23	2	5	7	3
DQ/	23, 25	2	6	8	3
DQ//	25, 26	4	3	7	4
Total number of registered extreme groundwater levels during the Moon Phases / Total number		80/118	81/131	161/249	60/94

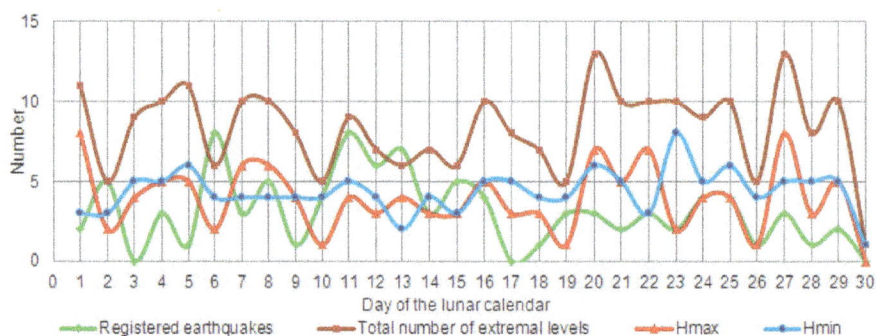

Figure 6. Graph of the earthquake numbers in the Ivano-Frankivsk region and the extreme groundwater levels (according to the lunar calendar)

be noted that 35 earthquakes were registered during PQ Phase; it constitutes 37% of the total number of the studied earthquakes. The minimal extreme values Hmin were also registered at that time and the coefficient of correlation is positive and equals 0.51.

Thus, one may state that the tides occurring on the Earth during the Full Moon (PL) are one of the factors, which not only influence the changing in the occurrence of the groundwater levels but also cause the earthquakes (at that time 17 cases were registered, which equaled to 28% of the total number of earthquakes registered during the Moon Phases).

The registration of the extreme underground water levels Hmax and Hmin is in-phase in the days of the New Moon (NL), the coefficient of correlation equals 0.8–0.9. During the Full Moon (PL), the registration of Hmin decreases with the increasing of the earthquake number (the coefficient of correlation is negative (-0.64). This shows a rise of the water level up to the surface. Such regularities may indicate an increase in probability of the earthquake arising in the days of the First Quarter (PQ), the Full Moon (PL), and the New Moon (NL) in seismic-active territories.

CONCLUSION

It is necessary to review and expand the network of the observation wells as well as to provide the development and improvement of the automated database on the grounds of contemporary geo-informative technologies in order to ensure the functioning of the groundwater monitoring system. The geo-informative approach to decision of the hydrogeological monitoring tasks envisages the systematic use of geo-informative analysis for collection, process and storage of information. In addition, the approach aims at the creation of permanent hydrogeological models based on GIS and allows the efficient analysis of the spatially organized information obtained because of the regime observations. Moreover, the introduction of GIS-technologies permits to implement the basin principle for processing of the groundwater monitoring data while the single geodatabase within the groundwater monitoring system will allow the maximum preservation of the existing hydrogeological survey results, ensure the integration of heterogeneous data and their operational analysis. A complex analysis of the periodical observation data in the Ivano-Frankivsk region, in particular, confirms the interconnections of the recording time of the seasonal and annual extremes peculiar to the groundwater levels with seismic activity and the lunar calendar. The results obtained may be used not only for clarification of the hydrogeological forecasts but also for the risk assessment of the earthquake arising in the adjacent territories.

REFERENCES

1. Adushkin V., Ryabova A., Spivak A. 2017. Effects of the moon-solar tide in the Earth crust and Earth atmosphere. Physics of the Earth, 4, 76–92.

2. Berezko O., Vasneva O. 2012. Groundwater monitoring in Belarus: implication and future prospects. Transboundary Aquifers in the Eastern Borders of the European Union, Springer Science+Business Media. Dordrecht, 115–120. DOI: 10.1007/978–94–007–3949–9_10.

3. Ben-Jema F., Marino M.A., Loaiciga H.A. 1994. Multivariate geostatistical design of groundwater monitoring networks. Journal of water resources planning and management-ASCE, 120, 505–522. DOI: 10.1061/(ASCE)0733–9496(1994)120:4(505).

4. Brovko A., Brovko G., Koshliakov O. 2015. Estimation of geosystem stability as a methodological approach for determination of the technogenic impact on groundwater (case study of quaternary aquifer on the territory of Rivne NPP). Visnyk of Taras Shevchenko National University of Kyiv: Geology, 69, 75–78. DOI: 10.17721/1728–2713.69.12.75–78.

5. Directive 2000/60/EC of the European Parliament and of the Council of 23 October 2000 establishing a framework for Community action in the field of water policy [official website]. http://eur-lex.europa.eu/resource.html?uri=cellar:5c835afb-2ec64577bdf8756d3d694eeb.0004.02/DOC_1&format=PDF [access: 29.01.2018].

6. Hudak P., Loaiciga H. 1992. A location modeling approach for groundwater monitoring network augmentation. Water Resources Research, 28, 643–649. DOI: 10.1029/91WR02851

7. Koshliakov O., Dyniak O., Koshliakova I. 2016. Hydrogeological and geoinformatical aspects of European standards implementation in Ukraine in the area of natural waters quality and water management. Proc. «Geoinformatics 2016» – XVth International Conference on Geoinformatics – Theoretical and Applied Aspects. http://www.earthdoc.org/publication/publicationdetails/?publication=84608 [access: 18.02.2018]. DOI: 10.3997/2214–4609.201600507.

8. Maréchal J.-C. et al. 2002. Establishment of earth tides effect on water level fluctuations in an unconfined hard rock aquifer using spectral analysis. Current Science, 83 (1), 101–104.

9. Pivovarova I. 2016. Optimization methods for hydroecological monitoring system. Journal of Ecological Engineering, 17(4), 30–34. DOI: 10.12911/22998993/64503.

10. Ruban S., Shinkarevsky M., Nikolishina A. 2005. Groundwater of Ukraine, Kyiv.

11. Shestopalov V. et al. 2010. Modern principles of hydrogeological zoning. Zbirnyk of scientific works of UkrSGEI, 3–4, 147 – 157.

12. Shestopalov V., Liuta N. 2016. State and ways of reforming the state groundwater monitoring system taking into account international experience and the requirements of the European Union Water Framework Directive. Mineral Resources of Ukraine, 2, 3–4.

13. Shtohryn L., Kasiyanchuk D. 2015. About the possible connection between periodicity of precipitation, activation of landslides and phases of the moon, Zbirnyk of scientific works of UkrSGEI, 4, 93–102.

14. Tymkiv M., Davybida L. 2016. Analysis of the state of hydrogeological monitoring network within the territory of Ukraine and the possibilities of its optimization on the basis of the geoinformation approach. Proc. International scientific and technical conference of young scientists «GeoTerrace-2016», 157–160.

19

Estimating Environmental Impact Potential of Small Scale Fish Processing using Life Cycle Assessment

Rahayu Siwi Dwi Astuti[1], Hady Hadiyanto[1,2*]

[1] Master Program of Environmental Studies, School of Postgraduate Studies Diponegoro University, Semarang, Indonesia

[2] Chemical Engineering Departement, Faculty of Engineering Diponegoro University, Semarang, Indonesia

* Corresponding author's e-mail: hadiyanto@live.undip.ac.id

ABSTRACT

Post-harvest handling / processing of fishery commodities requires large amounts of water and energy to overcome their perishable properties. Water is needed as raw/auxiliary material and to ensure that the production process and its environment meet the sanitary and hygiene principles. Meanwhile, large amount of energy is required for the transportation of raw materials and products, cold chain system during the process and operations of processing machines. They contribute towards the environmental impact of fish processing. This study used life cycle assessment to estimate the potential environmental impact of small scale mackerel fish processing. The results showed that the fish processing has contributed to 0.079 kg SO_2 eq acidification potential, 9.66 kg CO_2 eq climate chang-GWP 100, 0.02 kg PO_4 eq Eutrophication-generic, 0.17 kg 1.4 DCB eq human toxicity-HTP inf, and 0.0015 kg ethylene eq photochemical oxidation-high NO_x. Wastewater treatment implementation simulation showed elimination of direct emissions that contribute to eutrophication and increasing the potential of other process associated with energy consumption.

Keyword: wastewater treatment, fish processing, life cycle assessment

INTRODUCTION

The life cycle of a fishery product begins with the capture. When fish are removed from the ocean, cold chain systems and sanitation guarantees are required. The perishable nature of fishery commodities requires hygienic and cold chain conditions for their handling. As a consequence, the handling process need large amount of water and energy, and in turn produce large volumes of wastewater as well (Hall & Kose 2014). Hall and Kose (2014) describe four common problems of sustainability in fisheries processing technology, which include energy consumption, water consumption, waste control and by-product development. Water usage in the pretreatment stage includes the process of handling fresh fish storage, weeding process as well as cleaning equipment and work area (Doorn et al. 2006, Duangpaseuth et al., 2010). The generated wastewater contains organic matter (fat, protein and suspended solids),

phosphate and high amount of nitrates (Duangpaseuth et al., 2010). Energy is used to operate machinery, produce ice, heating, cooling and drying (Arvanitoyannis & Kassaveti 2008).

Considering the high amount of organic material, an effective method for treating the industrial waste is a biological treatment, but it must be conducted under optimum conditions (Sunny & Mathai 2013). One of the wastewater treatment methods of food industry that contain high amount of organic matter is anaerobic-aerobic biofilter method. This method is believed to have high waste degradation efficiency, reach up to 95%. Integration of these two methods, in addition to providing better results, can also reduce the energy consumption and sludge production (Sunny & Mathai 2013, Said 2017). At the anaerobic stage, the organic pollutants in the waste water are decomposed into CO_2 gas without energy use, but ammonia and H_2S are not removed. Furthermore, the unorganized organic material remains, described in the aerobic stage, in which the organic material is decomposed into CO_2 and water, ammonia to nitrite and subsequently nitrates, and H_2S into sulfate (Said 2017).

The decomposition process at the anaerobic stage involves 4 groups of bacteria, namely hydrolytic bacteria, fermentative acidogen bacteria, acetogenic bacteria and methanogenic bacteria. These four bacteria convert organic matter into CH_4 and CO_2, as well as a little NH_3, H_2 and H_2S (Said 2017). At the aerobic stage, the bacterial metabolism process breaks down organic materials into a simple form of CO_2, H_2O, oxide compounds such as nitrates, sulfates, phosphates and the formation of new cell mass. In general, the reaction is (Said 2017):

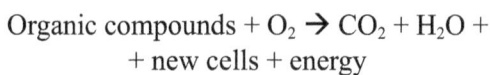

$$\text{Organic compounds} + O_2 \rightarrow CO_2 + H_2O + \text{new cells} + \text{energy}$$

The main processes used in wastewater treatment to reduce nitrogen are nitrification and denitrification, involving autotrophic bacteria. During nitrification, ammonia is converted to nitrites by nitrosomonas bacteria; then, nitrites are converted to nitrate by nitrobacterial bacteria. In this process, N_2O can be produced even if it is not an intermediate product. The nitrification process requires a considerable amount of oxygen – 3.43 g to oxidize nitrogen to nitrite, and 1.14 g to oxidize nitrogen to nitrate (Said 2017, Snip 2010).

Denitrification occurs under anoxic conditions, where heterotrophic bacteria use nitrate, nitrite, nitric oxide and nitrous oxide as electron acceptor. In this process N_2O becomes an intermediate material. Thus, N_2O can be produced and released by the imperfect denitrification process (Snip 2010, Said 2017).

$$NO_3^- \rightarrow NO_3^- \rightarrow NO \rightarrow N_2O \rightarrow N_2 \qquad (1)$$

A method used to assess potential environmental impact of a fishery product system is Life Cycle Assessment (LCA). ISO 14040 1997 explains that LCA is a technique to assess the environmental aspects and potential impacts of a product. LCA is carried out throughout the product life cycle, including raw material extraction, production process, usage to waste disposal (cradle to grave). The categories of environmental impact common to LCA are the use of resources, human health and ecological consequences. LCA can be utilized in (International assessment – Principles and framework, 1997):

1. Identification of the opportunities to improve the environmental aspects of a product throughout its lifecycle,
2. Decision making in industry or organization
3. Marketing (e.g. providing support in the form of claims for environmental performance)

The scope of the LCA study on a system of fishery products includes pre-manufacture, manufacture, packaging and distribution, and end use. The assessment can be based on the energy and water consumption and waste production. The six common categories of environmental impacts in LCA of the fisheries industries are global warming, acidification, eutrophication, ozone depletion, land use and photochemical smog (Hall & Kose 2014).

Energy consumption in pre-manufacturing stage is the use of fuel in the process of fish catching and transporting from the landing site to a fish processing plant. The formulas to estimate emission of a fishing vessel energy consumption are provided in Boer, et al. (Rizaldi Boer et al., 2012). Meanwhile, the consumption of electrical energy is related to the indirect emissions generated. The emission was estimated using Widiyanto, et al. (Widiyanto, Kato, & Maruyana, 2003) models.

Other environmental impacts connected to fisheries products system that deserve attention is the waste produced, both solid and wastewater. The proportion of solid waste depends on the proportion of the body of each species of fish and the type of product. The wastewater depends on the water volume used and fish processing method. Generally, water is mostly used in fish washing and the process of maintaining sanitation and hygiene of machinery and equipment, as well as rooms and employees. The fish processing wastewater contains contaminants in soluble, colloid and particulate forms, high BOD content, fat, and mineral (Tay et al. 2004). Fish filleting can produce 1–3 m^3 of waste water with COD content of 4–15 kg (Arvanitoyannis & Kassaveti 2008). The high organic matter released into the environment has the potential of causing eutrophication.

Life cycle processes of wastewater treatment facility, as described previously, produce emissions either directly due to biological processes or indirectly as a result of energy consumption. Assessment of the potential environmental impacts of the life cycle of some wastewater treatment plants had been done in several places using the

LCA (Foley et al. 2010; Glick et al. 2005; Kalbar et al. 2013; Snip 2010). Glick dan Guggemos (2013) used the LCA to estimate the environmental impact of the manufacturing, construction, usage and maintenance phase of the facility.

As previously described, the major emissions generated in biological waste processing are CH_4, CO_2 and N_2O. CH_4 and N_2O have high GWP values, while the resulting CO_2 is considered safe because it is biogenic origin. IPCC, 2001, noted that CH_4 has 23 value of GWP and N_2O has 296 values of GWP (Midgley et al. 2001).

The aims of this study are to assess the environmental impacts of amplang processing, and what are the potential environmental impacts when a sewage treatment facility is implemented in an amplang product system.

RESEARCH METHODOLOGY

System boundary and functional unit

The study was conducted at a fish processing SME in Kumai Sub-district in July 2017, using the secondary data obtained from SMEs and the primary data for liquid waste analysis. In addition to the data to sourced from SME financial reports, it was also acquired from interviews with business owners of processing and fishing.

The data analysis conducted in this study using the LCA method "cradle to gate", i.e. from the stage of arrest until production (Figure 1). The functional unit used was 1 kg amplang.

Life Cycle Inventory

The analysis was restricted to energy, water and raw fish consumption, excluding environmental loads of other raw materials, packaging plastics and transportation. The formulas to estimate emissions from fuel consumption for sea transport, i.e. (Sunny et al. 2015) are as follows:

$$\text{Tier-2: } Emission =$$

$$\sum_{ab} fuel\ consumption_{ab} \times NK_a \times EF_{ab} \quad (2)$$

where: $Emission$ = emission of CO_2, CH_4 or N_2O
$Fuel\ consumption_{ab}$ = vessel fuel consumption
NK_a = fuel caloric value a
EF_a = Emission factor of CO_2, CH_4 or N_2O according to fuel type (kg/TJ) (Table 1)
a = fuel type (solar, IDO etc.)
b = vessel / motor type

Emissions per 1 kW of electrical energy generated were estimated using a model developed by Widiyanto et al. (2003) and shown in Table 2.

Table 1. Typical emission factor for river/sea vessel in Indonesian cities using Tier-**2** (Suhadi & Febrina 2013)

Parameter	Unit	Emission
NO_x	kg/ton of fuel	57.1
CO	kg/ton of fuel	19.8
CO_{2+}	kg/GJ	74.1
HC	kg/ton of fuel	7.45
SO_x	kg/ton of fuel	2 x %S x FCm
PM_{10}	kg/ton of fuel	4.6

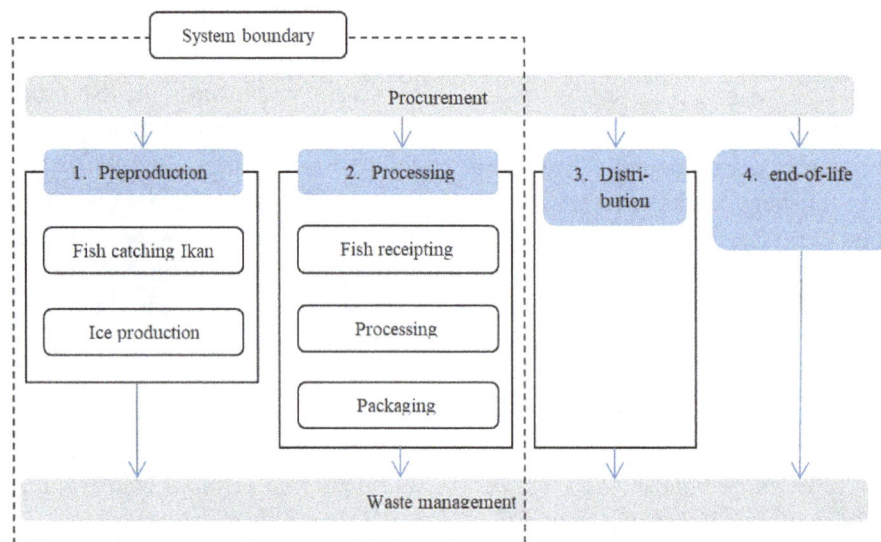

Fig. 1. Amplang production system

Table 2. Direct Emissions Measurement of Fuel Burning at Power Plant (Widiyanto et al. 2003)

Emission	Eletricity generation (kg/kWh)		
	1	2	3
CO_2	9.22×10^{-1}	9.22×10^{-1}	7.72×10^{-1}
SO_2	4.36×10^{-3}	3.99×10^{-3}	2.01×10^{-3}
NO_x	4.39×10^{-3}	4.19×10^{-3}	8.64×10^{-3}
SPM	6.70×10^{-4}	6.12×10^{-4}	3.24×10^{-4}
N_2O	4.25×10^{-5}	3.64×10^{-5}	2.19×10^{-5}
NMHC	3.20×10^{-5}	3.20×10^{-5}	4.68×10^{-4}
CH_4	1.13×10^{-5}	1.03×10^{-5}	3.83×10^{-5}
CO	1.47×10^{-4}	1.34×10^{-4}	1.64×10^{-4}

Information:
1 =   Coal steam turbine; using Bukit Asam coal (Sumatra) (in a 600 MW power plant)
2 =   Coal steam turbine; using Kalimantan coal (in a 600 MW power plant)
3 =   Indonesia's diesel-fueled power plant.
The wastewater content was obtained through the waste sample test, and the sample was taken by grab sampling.

Life cycle impact assessment of amplang poduction system

The input and output data of amplang production system were analyzed using an OpenLCA application developed by Green Delta with the methodology of CML (baseline) [v4.4, January 2015].

Predesigning and conducting life cycle inventory of small wastewater treatment plant

The second step in this study was to conduct a theoretical wastewater treatment plant predesign. On the basis of this predesign, the emissions produced were estimated either directly or indirectly. The CH_4 and N_2O emissions were calculated using the Tier 2 formula in IPCC 2006 and Guidelines for National Greenhouse Gas Inventory Book II. On the other hand, the emissions from electricity consumption were calculated using the model Widianto et al. (2003).

Emission of CH_4 and N_2O in a wastewater plant was estimated using the formula adapted from IPCC 2006 and the inventory guidelines (Doorn et al. 2006; Kementerian Lingkungan Hidup 2012):

$$CH_4 emission = \sum_t [(TOW_i - S_i)EF_i - R_i] \quad (3)$$

where: TOW_i = total organically degradable material in wastewater from industry i in inventory year, kg COD
i = industrial sector
S_i = organic component removed as sludge in inventory year, kg COD
EF_i = emission factor for industry i, kg CH_4/kg COD for treatment/discharge pathway or system(s) used in inventory

R_i = amount of CH_4 recovered in inventory year, kg CH_4

Choosing emission factor (Doorn et al., 2006; Kementerian Lingkungan Hidup, 2012):

$$EF_j = B_o \times MCF_j \quad (4)$$

where: EF_j = emission factor for each treatment/ discharge pathway or system, kg CH_4/kg COD
j = each treatment/discharge pathway or system
B_o = maximum CH_4 producing capacity, kg CH_4/kg COD MCF_j
MCF_j = methane correction factor (fraction) = 0.8 (Doorn et al., 2006; Kementerian Lingkungan Hidup, 2012)

Organically degradable material in industrial wastewater (Doorn et al., 2006; Kementerian Lingkungan Hidup, 2012) :

$$TOW_i = P_i \times W_i \times COD_i \quad (5)$$

where: TOW_i = total biodegradable material in wastewater for industry i, kg COD
i = industrial sector
P_i = total industrial product for industrial sector i,
W_i = wastewater generated, m³/t product
COD_i = chemical oxygen demand (industrial degradable organic component in wastewater), kg COD/m³

N_2O emissions from wastewater effluent:

$$N_2O \; emission = $$
$$N_{effluent} \times EF_{effluent} \times 44/28 \quad (6)$$

where: *N₂O emissions* = N₂O emissions in inventory, kg N₂O

$N_{effluent}$ = nitrogen in the effluent discharged to aquatic environments, kg N

$EF_{effluent}$ = emission factor for N₂O emissions from discharged to wastewater, kg N₂O-N/kg N, 0.005 kg N₂O-N/kg-N

The factor 44/28 is the conversion of kg N₂O-N into kg N₂O

The estimated emissions obtained from step 2 were analyzed using OpenLCA application with methodology CML (baseline) [v4.4, January 2015] to assess the environmental impact potential with boundary, as shown in Figure 2.

The final step was to implement the theoretical preliminary WWTP into amplang production system and assess its potential environmental impact using LCA.

RESULT AND DISCUSSION

Data Inventory

The fishing data was obtained from fishermen and fish collectors' financial statements during the June – August 2017 fishing period, as well as using the information from boat owners and fish collectors. The data obtained include the inputs of fuel and ice and the output of fish catch per species during the period of June – August 2017 (Table 3). The production process of amplang, fish cracker and sticks in outline consists of 4 stages, including fish receiving, storing, processing and packaging. Mass balance of those 4 stages is presented in Table 4.

Life Cycle Impact Assessment

The output of amplang production was analyzed using CML methodology (baseline) [v4.4, January 2015]. The results of the analysis show that the processing has potential environmental impacts in the categories of acidification, climate change, eutrophication and photochemical oxidation. This potential is the impact of indirect emissions on the consumption of electrical energy, the use of diesel fuel in the process of catching, the use of gas and frozen fish storage, and also wastewater generated.

Using the result of LCA analyses (Table 5), we also can see contribution of each process to impact categories, as shown in Figure 3.

Acidification potential

In the production system studied, the highest contributor of the acidification was fish catching and electricity consumption due to fossil fuels consumption of fishing vessels and power plants. The emissions generated by fishing vessel that contribute to acidification potential were NO$_x$ and SO$_x$, while power plants produced NO$_x$ and SO₂. The fuels used in power plant were diesel and coal. According to Widiyanto et al. (2003), in order to generate 1 kWh of electric energy, the SO₂ and NO$_x$ emissions produced by burning coal in power plants are greater than in the case of diesel. The NO$_x$ contribution in 1 kg amplang production system was 6.826E^{-02} kg SO₂ eq, while SO₂ was 1.035E^{-02} kg SO₂ eq.

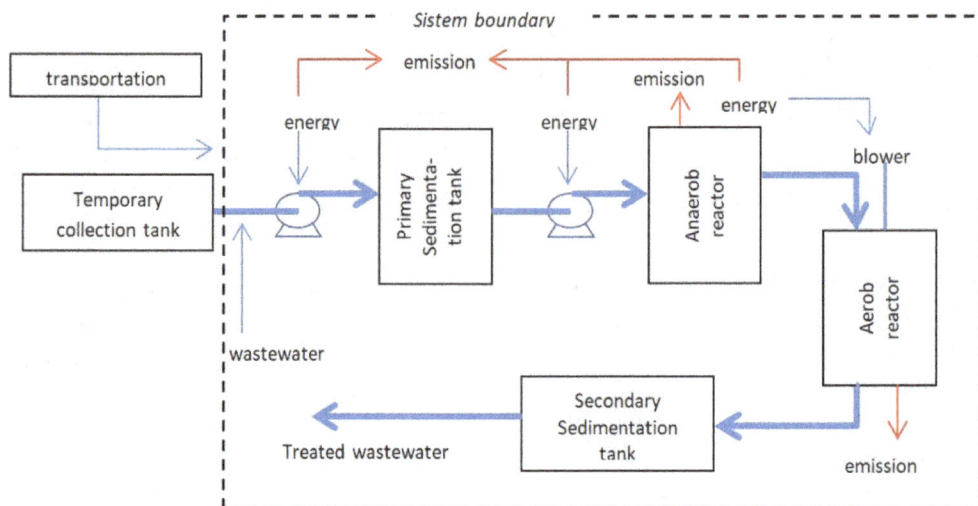

Fig. 2. WWTP's LCA boundary

Table 3. Fishing yield of 7 studying object vessel in June – August 2017 periods

Object	Vessel's motor and capacity	
	PS 120 4 Tak, 15 GT	Dongfeng 4 Tak, 15 GT
Trips	33	27
Diesel consumption per trip (liter)	250	250
Ice consumption per trip (kg)	1250	1250
HTSU		
Tenggiri (kg)	4,708.77	4,592
Otek (kg)	3,718	4,269
Telang (kg)	2,375	2,473
Kakap merah (kg)	111.5	104
Senangin (kg)	620.5	713.12
Other fish (kg)	3,013.8	3,493
Total (kg)	14,547.57	15,644.12

Tabel 4. Mass balance of amplang processing

Input		Processing phase	Output	
- Surimi - Eggs - Seasoning - Energy	4,320 kg 1,728 kg 432 kg 108 kWh	*Mixing I*	- Dough 1 - Eggshell - Energy	6,315.84 kg 164.16 kg 108 kWh
- Dough 1 - Tapioca - Energy	6,315.84 kg 7,200 kg 243 kWh	*Mixing II*	- Dough 2 - Energy	13,515.84 kg 243 kWh
- Dough 2	13,515.84 kg	*Shaping*	- Raw amplang	13,515.84 kg
- Raw amplang - Cooking oil - LPG	13,515.84 kg 5,760 liter 1,728 kg	*Frying*	- Amplang - Steam - Used cooking oil - Energy	8,640 kg 4,875.84 kg 2,160 liter 8.17E-02 TJ
- Amplang - Packaging Plastic - Energy	8,640 kg 188 kg 173 kWh	*Packaging*	- Amplang packaged - Energy	8,828 kg 173 h

Table 5. Impact category of amplang production process

Impact category	Reference unit	Result amplang
Acidification potential – average Europe	kg SO2 eq.	0.079
Climate change – GWP100	kg CO2 eq.	9.66
Eutrophication – generic	kg PO4--- eq.	0.02
Human toxicity – HTP inf	kg 1,4-dichlorobenzene eq.	0.17
Photochemical oxidation – high Nox	kg ethylene eq.	0.0015

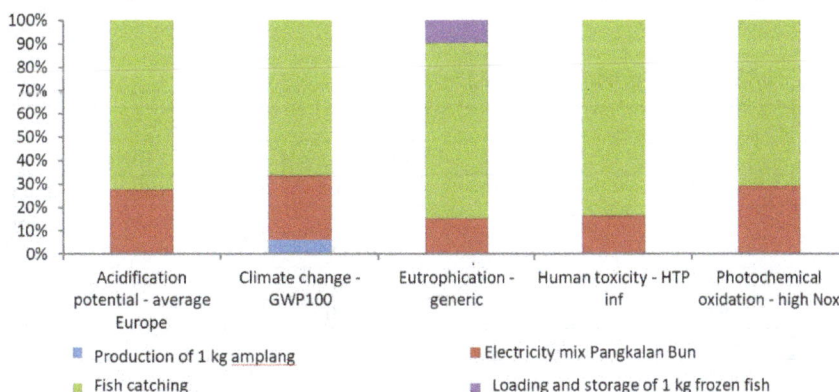

Fig. 3. The contribution of process to impact

Climate change – GWP100

In the system studied, CO_2, CH_4 and N_2O emissions were emitted by fishing vessel, power generation and little in the production process. In these three stages, emissions were released by burning the fuel of fishing vessels, power plants and LPG burning. The greenhouse gas emissions produced from fuel consumption on fishing vessels amounted to 6.41 kg CO_2 eq, power plants 2.65 kg CO_2 eq and cooking processes using LPG gas stoves $5.98E^{-01}$ kg CO_2 eq. The contribution of CO_2 to climate change potential reached as much as 9.555 kg, CH_4 $2.245E^{-03}$ kg CO_2 eq, and N_2O $2.561E^{-02}$ kg CO_2 eq.

Eutrophication potential

Eutrophication factors emitted in the process of fish catching amounted to about $1.48E^{-02}$ kg PO_4 eq, fish receipting $1.92E^{-03}$ kg PO_4eq, amplang frying $7.01E^{-05}$ kg PO_4eq and power plants $2.93E^{-03}$ kg PO_4 eq. The emission factors included N_2O and NO_x emitted to air, emissions of ammonia to water, BOD, COD, nitrate, nitrite, nitrogen, phosphate and phosphorus (Table 6). N_2O in the SME production system was only emitted by the electricity generation process. On the other hand, NO_x was generated from LPG burning on gas stoves, diesel consumption as a source of energy for fishing vessels and fossil fuels consumption (diesel and coal) in power plants. Meanwhile, all emissions to water are generated from the washing stage of the fish receiving and storage process.

Human toxicity potential

NO_x was the highest factor of the human toxicity impact category. NO_x in the human toxicity category has respiratory distress potential (Najjar 2011; WHO 2005). As a result of the study, NO_x

was mostly generated in the fish catching process due to the diesel fuel combustion in ship engines. In the production chain of 1 kg of amplang, at the fishing stage, NO_x of $1.14E^{-01}$ kg, $1.14E^{-01}$ kg, and $1.36E^{-01}$ kg were produced, respectively. As previously mentioned, NO_x contributes to three impact categories: acidification, eutrophication and human toxicity. On the basis of the analysis results obtained from this study, NO_x contribute to the category of human toxicity impact the most.

Photochemical oxidation potential

The contributors to photochemical oxidation impact potential of 1 kg amplang production system include: CO about $1.084E^{-03}$ kg ethylene eq, methane $5.389E^{-07}$ kg ethylene eq and SO_2 $4.139E^{-04}$. Out of the three compounds, carbon monoxide was the highest contributor.

Biofilter Anaerobic-aerobic processes

Wastewater treatment with anaerobic-aerobic biofilter process is a combination of anaerobic and aerobic processes. This method has high decomposition efficiency of organic material but requires relatively low amounts of energy. In this process, CO_2 and CH_4 gases are produced in the anaerobic phase, whereas in the aerobic phase, NH_3 breaks down into nitrite and nitrate, and H_2S becomes sulfate. In this study, LCA analyses only consider the energy use and emissions generated that are CH_4 and N_2O. The CO_2 also generated in this process is not considered due to it is biogenic origin (Kementerian Lingkungan Hidup 2012).

The SME studied is located in the middle of a fairly dense settlement, similarly to several other fish processing SMEs. The problems faced by these SMEs are the lack of wastewater treatment and direct discharge of the wastewater into the river. Considering the high organic matter content

Table 6. Characteristic of Mackerel processing wastewater

No	Parameter	Result	
		Amount (mg/l)	Environmental burdon (kg/ton)
1	BOD	1,051.5	2.52E+00
2	COD	1,741	4.18E+00
3	NO_2	0.065	1.56E-04
4	NO_3	10.0	2.40E-02
5	NH_3	0,170	4.08E-04
6	P total	37.1	8.90E-02
7	PO_4-P	36.8	8.83E-02
8	N total	1,741	4.18E+00
9	Wastewater (m^3/ton)		2.4

characteristic of the fish processing wastewater, location and IPAL requirement for community or SMEs, this study selected the anaerobic-aerobic biofilter method to be analyzed.

In the amplang industry, water was used for clean fish but not as one of the raw materials in the processing. Thus, the volume of wastewater is considered equal to the volume of water used. On the basis of volume and composition of wastewater, wastewater treatment plant (WWTP) was preliminary designed as follows (Table 7):

Comparison of amplang product system with and without wastewater treatment

Figure 4 shows that there is a little difference within four impact categories due to implementation of BWRO system water treatment and waste treatment. A relatively high decrease of potential impact is only seen in the eutrophication category. The reason of this condition is an increasing of electricity consumption due to water treatment purposes and reduction of emissions to water in the presence of WWTP (Table 8).

Table 8 above shows that if a WWTP applied, the process of fish receiving no longer contributes to eutrophication. However, potential impacts still arise in the waste treatment process, even though the value is much smaller. Due to waste treatment process, the emissions of ammonia, BOD,

COD, nitrate and nitrogen to water contributed to the impact of eutrophication, while ammonia and ammonium contributed from the water treatment process . The highest contribution to the eutrophication impact category actually comes from the process of fish catching, as a consequence of fossil fuel consumption. This process contributes to eutrophication with emissions to air of NO_x by 0.0280 kg PO_4^- eq.

The process of energy generation and fish catching was the largest contributor to the five impact categories, as a result of fossil fuel consumption to meet energy needs. The electricity demand on fish receipting process was estimated to increase from 0.173 kWh/kg of frozen fish to 0.1778 kWh/kg of frozen fish. On the other hand, the average requirement of diesel fuel in the catching process equals 1.29 liters per kg of mackerel fish. One of the efforts to decrease the specific energy needs of the fishing process is by increasing the catch of fishermen per trip, as an attempt to optimize the utilization of energy resources, in addition to substituting the transfer of fuel with a more environmentally friendly energy source.

Sensitivity analysis

As described previously, the implementation of water and wastewater treatment has a major effect on the fish receiving process; thus, the sen-

Table 7. Portable WWTP design input output per day

Parameter *Input*	Unit	Value	Parameter *Output*	Unit	Value
Influent rate	m³/d	3.21	*Product:*		
BOD_{in}	kg	3.375315	Treated wastewater	m³/d	3.21
COD_{in}	kg	5.58861	Recovered CH_4	kg	0.51
NO_{3in}	kg	10	*Emission/waste:*		
N-total$_{in\ total}$	kg	1741	NO_{3out}	Kg	0.024075
NH_{3in}	kg	0.0005457	N-total$_{out\ l}$	Kg	3.406261514
Energy	kWh	1.147	$NH_{3\ out}$	Kg	0.000544495
			$BOD_{\ out}$	kg	0.335347438
			$COD_{\ out}$	kg	0.717337114
			N_2O	kg	0.0373

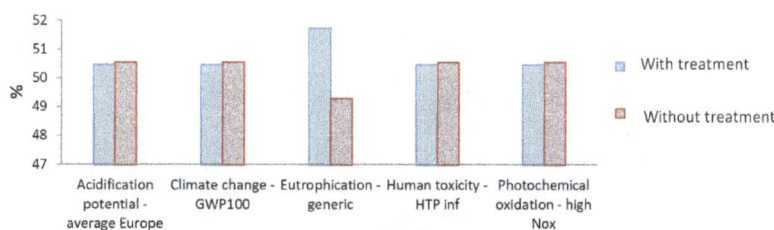

Fig. 4. Comparison of amplang product system with and without wastewater treatment

Table 8. Process contribution to impact category

Impact category	Referency unit	Receipting and storing 1 kg of frozen fish	Electricity mix Pangkalan Bun	Fish catching	Wastewater treatment	BWRO 1.050 m³
Without water and wastewater treatment						
Acidification potential – average Europe	kg SO$_2$ eq.	-	3.33E-02	9.04E-02	-	-
Climate change – GWP100	kg CO$_2$ eq.	-	4.09E+00	1.02E+01	-	-
Eutrophication – generic	kg PO$_4^{-2}$ eq.	3,05,E-03	4.53E-03	2.35E-02	-	-
Human toxicity – HTP inf	kg 1,4-DCB eq.	-	4.28E-02	2.29E-01	-	-
Photochemical oxidation – high NO$_x$	kg ethylene eq.	-	6.60E-04	1.69E-03	-	-
With water and wastewater treatment						
Acidification potential – average Europe	kg SO$_2$ eq.	-	3.34E-02	9.05E-02	-	-
Climate change – GWP100	kg CO2 eq.	-	4.11E+00	1.02E+01	-	-
Eutrophication – generic	kg PO$_4^{-2}$ eq.	-	4.55E-03	2.35E-02	1.49E-03	2.88E-05
Human toxicity – HTP inf	kg 1,4-DCB eq.	-	4.29E-02	2.29E-01	-	-
Photochemical oxidation – high NO$_x$		-	6.63E-04	1.69E-03	-	-

sitivity analysis is limited to this process. The sensitivity analysis aims to test the differences that may occur when the input value of fish raw materials is changed in the process of receiving fish. This sensitivity analysis is based on high fluctuations in the fish availability. This fluctuation results in variable power consumption, especially in the frozen fish storage stage. In addition, fluctuations in raw materials also impact the water consumption.

Figure 5 shows that the categories of climate change impacts are the most sensitive parameters due to the increase in the electricity consumption along with the growing amount of raw materials that must be stored in a frozen state. In the SME, the amount of production depends on the order received, while the duration of storage of raw materials depends on the order and stock of the product. In addition, the acceptance of raw materials can be done when stocks of frozen fish still exist. With the principle of first in first out, the later fish

must be kept frozen for longer until the stock of the earlier fish were used up.

More specifically, the sensitivity of the five categories of impacts to changes in raw inputs of fish is shown in Figure 5 below.

The second sensitive parameter is human toxicity, which is contributed by NO$_x$, PM$_{10}$, and SO$_2$. They are the result of burning fossil fuels. NO$_x$ comes from power plants and fishing vessels, PM$_{10}$ from fishing vessels and SO$_2$ from power plants. The largest contributor towards the human toxicity in fish receipt 1500 kg is NO$_x$ that is equal to 281.3885807 kg 1,4-dichlorobenzene eq.

The impact category of eutrophication was less sensitive to the fluctuation in the amount of frozen fish it receives. This can be due to portable waste treatment. In this processing, the volume of wastewater input can be adjusted according to its capacity, then it can be expected that the quality of treated wastewater would not fluctuate much.

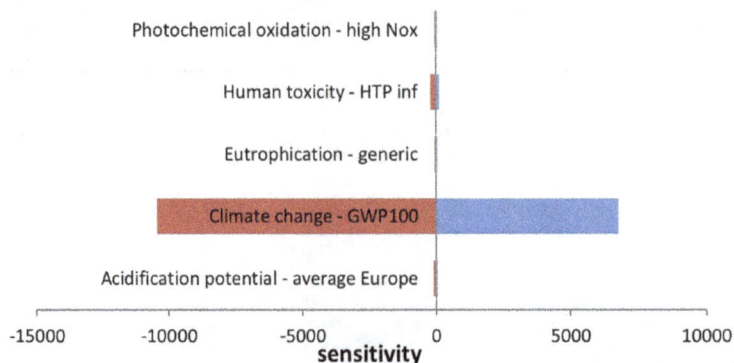

Fig. 5. Sensitivity of impact category potential by raw fish input fluctuation

CONCLUSION

The system product of fish-based amplang, in its life cycle, as well as fish processing in general, requires considerable amounts of water and energy. These water and energy requirements have implications for the generation of waste and gas emissions. This study showed that the life cycle production of the amplang from cradle to gate has the environmental impact on the acidification potential, climate change, eutrophication, human toxicity and photochemical oxidation categories. The processes that contribute the most to the impact category are power generation and fishing as a consequence of using fossil fuels (diesel and coal) as an energy source.

The amplang production process itself produces direct emissions from wastewater discharged into rivers. This wastewater contains COD, BOD and nitrogen in the amounts that are high enough to contribute to eutrophication impact categories.

The results of analysis in this study indicate that the implementation of waste treatment of anaerobic-aerobic biofilter method can decrease the eutrophication impact category. As methane produced by WWTP can be recovered, only N_2O increases the climate change potential, in addition to N_2O still contributing to eutrophication with the remaining COD.

REFERENCES

1. Arvanitoyannis, I.S., & Kassaveti, A. 2008. Fish industry waste: Treatments, environmental impacts, current and potential uses. International Journal of Food Science and Technology, 43(4), 726–745. https://doi.org/10.1111/j.1365–2621.2006.01513.x

2. Doorn, M.R.J., Towprayoon, S., Vieira, S.M.M., Irving, W., Palmer, C., Pipatti, R., & Wang, C. 2006. Wastewater Treatment and Discharge. IPCC Guidelines for National Greenhouse Gas Inventories. Prepared by the National Greenhouse Gas Inventories Programme. IGES, Japan.

3. Duangpaseuth, SDas, Q., Chotchamlong, N., Ariunbaatar, J., Khunchornyakong, A., & Prashanthini, V. 2010. Seafood Processing.

4. Foley, J., de Haas, D., Hartley, K., & Lant, P. 2010. Comprehensive life cycle inventories of alternative wastewater treatment systems. Water Research, 44(5), 1654–1666. https://doi.org/10.1016/j.watres.2009.11.031

5. Glick, S., Guggemos, A.A., & Asce, A.M. 2005. Rethinking Wastewater-Treatment Infrastructure : Case Study Using Life-Cycle Cost and Life-Cycle Assessment to Highlight Sustainability Considerations, (Coldham 1996), 1–8. https://doi.org/10.1061/(ASCE)CO.1943–7862.0000762.

6. Hall, G.M., & Kose, S. 2014. Fish Processing Installations: Sustainable Operation. In I.S. Boziaris (Ed.), Seafood Processing: Technology, Quality and Safety (pp. 1–488). West Sussex: John Wiley & Sons. https://doi.org/10.1002/9781118346174

7. International assessment – Principles and framework 1997. International Organization for Standardization.

8. Kalbar, P.P., Karmakar, S., & Asolekar, S.R. 2013. Assessment of wastewater treatment technologies : life cycle approach, 27(3), 261–268. https://doi.org/10.1111/wej.12006

9. Kementerian Lingkungan Hidup 2012. Pedoman Penyelenggaraan Inventarisasi Gas Rumah Kaca Nasional Buku II. Jakarta: Kementerian Lingkungan Hidup (in Indonesian).

10. Midgley, P., Wang, M., Berntsen, T., Bey, I., Brasseur, G., Buja, L., Yantosca, R. 2001. Atmospheric Chemistry and Grenhouse Gases. In Climate Change 2001: The Scientific Basis. IPCC. Retrieved from https://www.ipcc.ch/ipccreports/tar/wg1/pdf/TAR-04.PDF

11. Rizaldi Boer, Dewi, R. G., Siagian, U. W., Ardiansyah, M., Surmaini, E., Ridha, D. M., et al. 2012. Pedoman Penyelenggaraan Inventarisasi Gas Rumah Kaca Nasional Buku Ii. Metodologi Penghitungan Tingkat Emisi Gas Rumah Kaca Kegiatan Pengadaan Dan Penggunaan Energi (Vol. 1). Kementerian Lingkungan Hidup (In indonesian).

12. Said, N.I. 2017. Teknologi Pengolahan Air Limbah. Jakarta: Penerbit Erlangga (In indonesian).

13. Snip, L.J.P. 2010. Quantifying the greenhouse gas emissions of wastewater treatment plants. Wageningen University. Retrieved from http://modeleau.fsg.ulaval.ca/fileadmin/modeleau/documents/Publications/MSc_s/sniplaura_msc.pdf

14. Suhadi, D.R., & Febrina, A.S. 2013. Pedoman Teknis Penyusunan Inventarisasi Emisi Pencemar Udara Di Perkotaan, 153 (In indonesian).

15. Sunny, N., & Mathai P.L. 2013. Physicochemical process for fish processing wastewater. International Journal of Innovative Research in Science, Engineering and Technology, 2(4), 901–905.

16. Tay, J.-H., Show, K.-Y., & Hung, Y.-T. 2004. Seafood Processing Wastewater Treatment. In: L.K. Wang, Y. Hung, H.H. Lo, & C. Yapijakise (Eds.), Handbook of industrial and hazardous wastes treatment (2nd ed.). New York, Basel: Marcel Dekker, Inc., 29–66.

17. Widiyanto, A., Kato, S., & Maruyana, N. 2003. Environmental Impact Analysis of Indonesian Electric Generation Systems. JSME International Journal Series B, 46(4), 650–659. https://doi.org/10.1299/jsmeb.46.650.

20

Bioactive Membranes from Cellulose with a Graphene Oxide Admixture

Alicja Machnicka[1*], Beata Fryczkowska[2]

[1] Institute of Environmental Protection and Engineering, University of Bielsko-Biała, ul. Willowa 2, 43-309 Bielsko-Biała, Poland

[2] Institute of Textile Engineering and Polymer Materials, University of Bielsko-Biała, ul. Willowa 2, 43-309 Bielsko-Biała, Poland

* Corresponding author's e-mail: amachnicka@ath.bielsko.pl

ABSTRACT

The paper presents the results of microbiological tests of composite membranes made of cellulose (CEL) with graphene oxide (GO) admixture. At the beginning, the antibacterial properties of the GO in aqueous solutions of various concentrations (0.001; 0.01; 0.1% w/w) were studied, and the obtained results allowed to use GO as an additive to cellulose membranes. The solution used to prepare the membranes was a 5% cellulose solution (CEL) in 1-ethyl-3-methylimidazolium acetate (EMIMAc), into which various amounts of graphene oxide (GO) dispersed in N,N-dimethylformamide (DMF) were added (0.5÷28.6% of GO). From this solution, composite membranes were formed using phase inversion method. It was observed that the GO addition influences the process of membrane formation and their physicochemical properties. The obtained membranes were subjected to microbiological tests using the Gram-negative bacteria (*Escherichia coli*), Gram-positive bacteria (*Staphylococcuc aureus*) and fungi (*Candida albicans*). It was observed that the GO addition to the cellulose membrane (GO/CEL) inhibited-the growth of bacteria and fungi, and the biological activity as dependent on the type of living organism and the size of GO particles.

Keywords: cellulose, graphene oxide, membranes, bactericidal properties, fungicidal properties

INTRODUCTION

Cellulose is one of the most widespread, inexpensive and biodegradable polymers, which is widely used in many industries [Ramamoorthy et al 2015, Yang et al. 2016]. Chemically, this polysaccharide is a polymer in which the chains of cellulose are linked with hydrogen bonds (intramolecular and intermolecular), which hinder the dissolution of this biopolymer in classical solvents [Fink et al. 2001, Lindman et al. 2010].

An interesting group of solvents which dissolve polysaccharides are ionic liquids, which due to their biodegradability and low toxicity [Novoselov et al. 2007, Pinkert et al. 2009, Zhu et al. 2006] are called "green" solvents. Cellulose dissolved in ionic liquids can be precipitated with polar solvents to obtain "flocs," fibres or membranes [Kuo & Hong 2005, Rambo et al. 2008].

Graphene oxide (GO) is a modern material, which due to the presence of oxygen groups (epoxide, hydroxyl, carbonyl, carboxyl) [Guerrero-Contreras & Caballero-Briones 2015] shows hydrophilic properties. GO can be easily dispersed both in water [Texter 2014, Yoon et al. 2013] and in classic organic solvents such as N,N-dimethylformamide (DMF), N-methyl-2-pyrrolidone, tetrahydrofuran and ethylene glycol [Parades et al. 2008].

Graphene oxide is used as a component of composite materials. Combining GO with cellulose, a hydrophilic composite can be obtained. Zhang et al. (2015) described the method to obtain GO-containing microbeads in NaOH and urea solution, by coagulation in mineral acid. Aerogels were obtained from bamboo fibres dissolved in a NaOH/polyethylene glycol (PEG) mixture to which water-dispersed GO was added [Wan & Li 2016]. The hydrogels were prepared from an aqueous dispersion of GO, NaOH and urea and cellulose, which were combined with a polyvinyl alcohol (PVA) solution [Rui-Hong et al. 2016]. Liu et al. (2016) obtained cellulose composite membranes in the simplest way possible – by filtration of GO solution on pure cellulose membrane. The paper-making method was used to obtain paper from cellulose, with GO and polyacrylamide addition [Huang et al. 2016] or by mixing the suspension of hydrolysed microcrystalline cellulose with a GO dispersion [Kafy et al. 2016]. Kim et al. (2011), on the other hand, obtained a membrane by dissolving GO and cellulose in N-methylmorpholine N-oxide (NMMO). Tang et al. (2012) received composite membranes using a layer-by-layer (LbL) method. For this purpose, a cellulose solution in an ionic liquid 1-butyl-3-methylimidazolium chloride ([Bmim]Cl) was applied to the glass plate, dried and then coated with a GO dispersion. Yang et al. (2016) mixed the grinded bacterial cellulose with GO dispersion and formed a composite film. Another research team obtained GO/CEL composite granulate during cellulose synthesis by *Acetobacter xylinum* [Zhu et al. 2015]. The team of Luo used the *Komagataebolacter xylinus X-2* bactcria to which a suspension of GO was added, for the synthesis of the hydrogel [Luo et al. 2017].

Graphene oxide also has bactericidal properties, both for Gram-positive and Gram-negative bacteria such as *Escherichia coli* (E. coli) [Liu et al. 2012, Tu et al. 2013], *Pseudomonas aeruginosa* [Shahnawaz Khan et al. 2015, Singh & Singh, 2017], *Staphylococcus aureus* (S. aureus) [Akhavan & Ghaderi 2010, Liu et al. 2017, Singh 2016], *Bacillius subtilis* [Musico et al. 2014]. In addition, studies of GO fungicidal properties against *Candida utilis* and *Saccharomyces cerevisiae* [Shahnawaz Khan et al. 2015] as well as *Mucor racemosus* [Li et al. 2017] were also carried out.

GO's bactericidal properties in combination with other materials allowed to obtain products that could be used as packaging [Hu et al. 2010], pharmaceutical carriers [Luo et al. 2017], scaffolds [Kanayama et al. 2014, Mahmoudi et al., 2017, Pal et al. 2017]. Owing to its biocidal properties, GO can be used to design membranes not susceptible to biofouling. Yang et al. (2018) described a method for obtaining membranes of enhanced biofouling resistance, by creating a GO laminate with silver nanoparticle and polydopamine. These membranes can be used for water treatment and ion separation. Lim et al. (2017) developed antibacterial graphene oxide membranes functionalised with acid and polyethyleneimine, which can be used for ion separation. Other researchers have obtained biofouling resistant membranes based on cellulose acetate coated with graphene oxide-silver nanoparticles [Sun et al. 2015]. Musico et al. (2014) used the modification of commercial membranes using poly(N-vinylcarbazole)-graphene oxide to obtain antibacterial membranes that can be used for treating water and wastewater.

This paper presents the results of bactericidal properties of graphene oxide (obtained from graphite with two different particle sizes <20 μm and <150 μm) and the effect of the concentration of these nanoparticles and their size on the growth of *E. coli and S. aureus bacteria*. GO with the best bactericidal properties was used to prepare cellulose membranes with graphene oxide addition. 1-ethyl-3-methylimidazole acetate (EMIMAc) was used to prepare the cellulose solution. GO was dispersed in DMF. Composite membranes were prepared from a homogenous CEL/EMIMAc and GO/DMF solutions by phase inversion and then subjected to microbiological tests. *E. coli* and *S. aureus* bacteria as well as *Candida albicans* fungus were used for testing the membranes. It was observed that the bactericidal and fungicidal properties of membranes depend on the type of microorganism and the concentration of GO in the sample.

MATERIALS AND TEST METHODS

Reagents

Cellulose (long fibres), ionic fluid: 1-ethyl-3-methylimidazolium acetate (EMIMAc), graphite powder <20 μm and graphite < 150 μm, were purchased from Sigma-Aldrich. $NaNO_3$, 98% H_2SO_4, $KMnO_4$, 30% H_2O_2, N,N-dimetyloformamide (DMF), NaCl were purchased from Avantor Performance Materials Poland S.A. The *Staphylococcus aureus* (ATCC 33741-B1), *Escherichia coli* (ATCC 35925-B2), and *Candida albicans* (ATCC BAA-473) were purchased from ATCC (American Type Culture Collection).

Blood agar, Chapman medium, MacConkey medium, Candida agar were purchased from BTL Ltd. Department of enzymes and peptones, Łódź.

Graphene oxide

Graphene oxide was obtained according to modified Hummers method [Hummers & Offeman 1958], described in our previous paper [Fryczkowska et al. 2015].

At the beginning, 1 g of $NaNO_3$ and 46 cm^3 of H_2SO_4 and 2 g of graphite powder were placed in the flask in ice bath. The reaction mixture was stirred intensively for 30 minutes until the temperature was reduced to approx. 5 °C. Thereafter, 6 g of $KMnO_4$ was slowly added in portions, taking care to prevent the exothermic reaction not to increase the temperature above 20 °C. After the entire $KMnO_4$ was added, stirring was continued for another 5 minutes, after which the reaction mixture was warmed to 35 °C. Stirring was continued for 4 h, after which 92 cm^3 of distilled water was slowly added to dilute the acid. Excess $KMnO_4$ was removed by introducing a solution containing: 80 cm^3 of distilled water and 50 cm^3 of 3% H_2O_2. Finally, the graphene oxide obtained in the reaction was centrifuged and washed several times with distilled water until pH 7. Wet graphene oxide was dried in a drying oven at 60 °C turning into a brown-coloured solid.

The synthesis of graphene oxide was carried out for two types of graphite: graphite powder <20 µm, obtaining GO1 and graphite with grain size <150 µm, obtaining GO2.

Then in the volumetric flasks, solutions of GO1 and GO2 in distilled water were prepared with concentrations of: 0.001; 0.01; 0.1%.

GO/CEL composite membranes

The membrane-forming solutions were prepared as described in our earlier article [Fryczkowska & Wiechniak 2017]. Initially, a 5% solution of cellulose in the ionic fluid – 1-ethyl-3-methylimidazolium acetate (EMIMAc) was prepared. The mixture of cellulose and EMIMAc was thoroughly mixed and then heated in a laboratory microwave oven, taking care that the temperature of the mixture did not exceed approx. 40 °C. The resulting cellulose solutions were left for 24 hours to deaerate.

In order to prepare solutions for forming GO/CEL composite membranes, adequate amounts of cellulose and ionic fluid were first weighed (Table 1) and cellulose solutions were prepared as described above. Then, a GO1 dispersion was prepared. To do this, dry GO1 was dispersed in DMF in an ultrasonic bath, resulting in a dispersion with a concentration of 3.7% GO1/DMF.

The appropriate amounts of GO1/DMF dispersion were then added to the cellulose solutions (Table 1) and mixed intensively using a laboratory stirrer for 1 week.

Cellulose membranes were prepared using wet phase inversion method. For this purpose, the cellulose-forming solution was poured onto a leveled, clean glass plate. Then, a polymer film was formed using casting knife with an adjustable thickness fixed at 0.2 mm and coagulated in distilled water. The precipitated membranes were dried.

As a result of the experiment, a pure cellulose membrane ("0") and composite GO1/CEL cellulose membranes with different amounts of GO (A, B, C, D, E, F) were obtained. The physicochemical properties of the membranes were described in an earlier publication (Fryczkowska & Wiechniak 2017).

Microbiological analysis

The samples were exposed to bacteria and fungi capable of causing infections in humans, i.e. the Gram-positive *Staphylococcus aureus* and Gram-negative *Escherichia coli*, and *Candida albicans*. The microorganisms were growing on blood agar. Microorganisms were incubated

Table 1. The composition of solutions to prepare the membranes

Membrane designation	„0"	A	B	C	D	E	F
The amount of 3.7% GO/DMF solution [g]	0	0.135	0.34	0.67	1.35	6.76	13.5
The amount of CEL [g]	2.5	2.5	2.5	2.5	2.5	2.5	2.5
The amount of EMIMAC [g]	47.5	47.4	47.2	46.8	46.2	40.1	34.0
W/w conc. of GO1 [%]	0	0.5	1.0	2.0	3.8	16.7	28.6
W/w conc. of CEL [%]	100	99.5	99.0	98.0	96.2	83.3	71.4

at 36±2°C, for 24 hours. Grown cultures were washed out with 1 ml of physiological salt solution, and 0,1 ml added to the sterile selective agar. The following mediums for cultivation of microorganisms were used: Chapman agar – *S. aureus*, MacConkey agar – *E.coli* and Candida agar – *Candida albicans*. Seeding of "grated tiles" was using. A samples were placed in the centre of the Petri plate. The Petri plates with samples were subsequently placed into a laboratory heater and then kept heated at 36±2°C for 24 hours. Sterile paper discs (diameter 1.0 cm) impregnated (two drops) with GO solution (graphene oxide – 0.001; 0.01; 0.1% w/w) and composite membranes GO1/CEL were samples for testing. The cellulose foils were of a size 1 x 1 cm. A control samples (cellulose foil – membrane "0"). were done. The experiment was performed three times for each sample of membrane. Growth inhibition zones were read by stereoscopic microscope equipped with Olympus CCD ARTCAM camera.

RESULTS AND ANALYSIS

GO characterization

Graphene oxide, which was used to obtain composite GO1/CEL membranes, was studied using X-ray diffraction, DSC thermal gravimetric analysis and FTIR spectroscopy. The obtained results were similar to the results described in our earlier work [Fryczkowska et al., 2015].

Bactericidal properties of the aqueous GO solution

The GO obtained from graphite with different particle size was used to study the antimicrobial properties of graphene oxide, resulting in GO1 with an area <20 µm and GO2 with an area <120 µm.

Studies shown that GO1 had bactericidal properties against *E. coli*, regardless of the graphene oxide concentration. In all cases, a marked inhibition of the growth of this bacteria was observed. For cellulose discs with a diameter of 1 cm, the width of the inhibition zone was ~ 2mm (Fig. 1). *S. aureus*, on the other hand, is more resistant to the GO1 particles dispersion. The highest rate of bacterial growth inhibition was observed for the highest concentration of graphene oxide (0.1% w/w) and it was ~ 0.5 mm wide. For lower concentrations of GO1 (0.01 and 0.001% w/w) it was observed that the *S. aureus* growth inhibition width zone was even smaller: ~ 0.2 and ~ 0.1 mm (Fig. 1).

When conducting microbiological tests using GO2, it was observed that *S. aureus* is completely resistant to large flakes of graphene oxide. Regardless of the concentration of nanoparticles in the aqueous solution, no zones of *S. aureus* growth inhibition were observed (Fig. 2). However, studies conducted on *E. coli* shown that the GO2 bactericidal properties depend on the concentration of graphene oxide in the aqueous solution. At the highest concentration of GO2 (0.1% w/w), the largest zone of inhibition of ~ 2 mm was recorded.

GO1 concentration in water solution[% w/w]

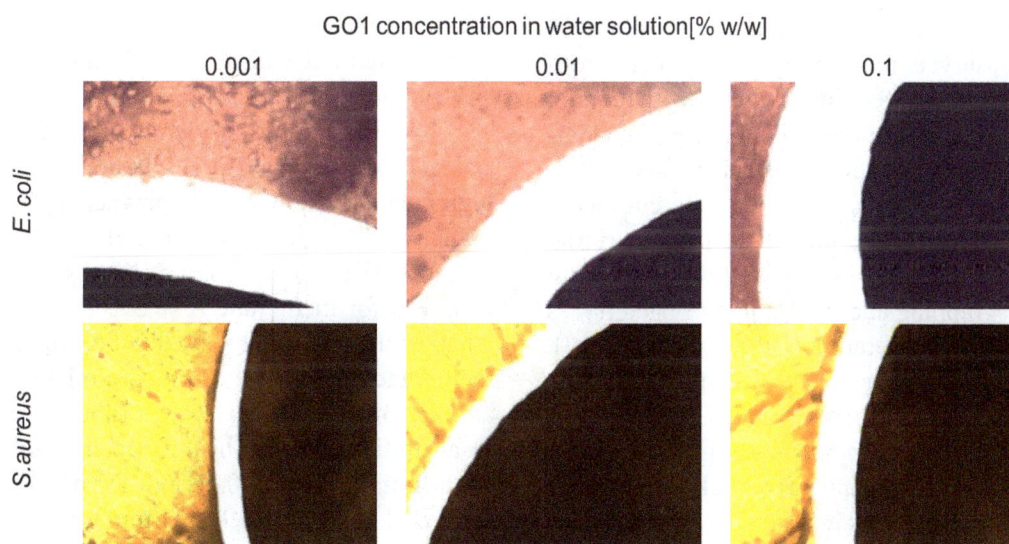

Fig. 1. Microphotographs of cellulose discs (dark box) impregnated with a GO1 solution after 24 h incubation of *E. coli* and *S. aureus*

GO2 concentration in water solution[% w/w]

0.001   0.01   0.1

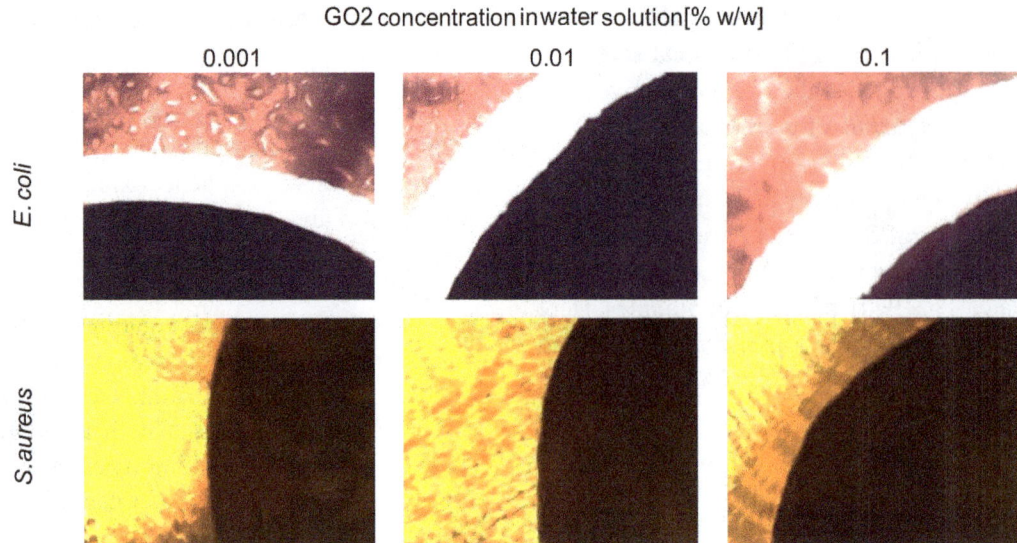

Fig. 2. Microphotographs of cellulose discs (dark box) impregnated with a GO2 solution after 24 h incubation of *E. coli* and *S. aureus*

Then, along with the decrease in GO2 concentration, the *E. coli* inhibition zones were decreased, successively, to ~ 1.5 mm for 0.01% w/w of GO and to ~ 1.0 mm for 0.001% w/w GO2 (Fig. 2).

Analysing the results of the experiment, it was observed that *E. coli*, which belongs to the Gram negative bacteria, was less resistant to GO molecules in aqueous solution (Fig. 3). In the case of GO1 particles with a size of <20 μm, the growth of small cells of bacteria (2 × 0.8 μm) were inhibited, which results in a relatively large area of *E. coli* growth inhibition observed in the pictures. One of the reasons could be breaking of the thin cell membrane, resulting in a leakage of cytoplasm, as described in the literature [Palmieri et al. 2017].

In the studies conducted on GO2 solutions, it was observed that strong bactericidal properties were observed in solutions with a high (0.1% w/w) concentration of GO2 (Fig. 3). Comparing the size of the *E. coli* cell to the size of GO2 particles, one can suspect a different mechanism of stopping the growth of these bacteria. In this case, one may assume that we are dealing with the wrapping and/or trapping of bacteria, as described in the literature [Palmieri et al. 2017], resulting in membrane stress and/or oxidative stress, leading to the bacteria death.

Gram positive *S. aureus* has a spherical shape. Its size is 0.8 ÷ 1 μm and has a thick, single-layer cell wall, which makes it difficult for external factors to penetrate the interior of the bacteria. The size and structure of the bacteria makes it completely resistant to large GO2 molecules (Fig. 3).

On the other hand, in the case of GO1, it was observed, that the toxicity of graphene oxide increases with its concentration in solution. The observed phenomenon could be explained by the fact that in low concentration solutions the probability of encountering the GO1 particle and damage to the bacterial cell wall was smaller than in high concentration solutions. Therefore it could be assumed that the destruction of *S. aureus* occurs as a result of cutting the membrane and leakage of cytoplasm [Palmieri et al. 2017].

Bactericidal and fungicidal properties of GO/CEL membranes

Microbiological investigations for pure cellulose membrane ("0") and GO/CEL membranes were carried out with *E. coli*, *S. aureus* bacteria and additionally with *Candida albicans fungi*.

GO/CEL composite membranes have bactericidal properties against *E. coli* both for low and high concentrations of nanopowder in the membrane (Fig. 4). The largest area of growth inhibition of 0.43 mm wide for 1 × 1 cm samples was observed for membrane A (0.5% w/w of GO1), what indicated the bioavailability of GO1 in membrane A. The use of GO1 admixture in the amount of 1% w/w or more (Table 1) resulted in an initial decrease in the bacterial inhibition zone from a width of 0.16 mm for membrane B, through 0.24; 0.25; 0.29; 0.31 mm for C, D, E and F membranes, respectively. The obtained results allowed to conclude, that a fairly good resistance to *E. coli* was ensured by the lowest GO1 addi-

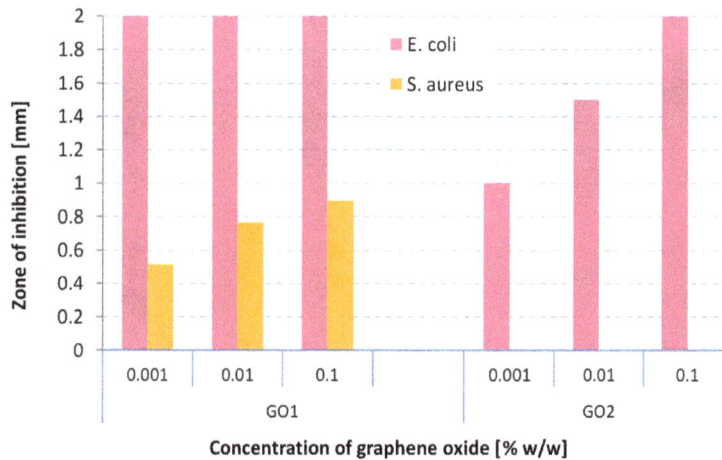

Fig. 3. Zone of bacterial growth inhibition (of *E. coli* and *S. aureus*) around cellulose discs impregnated with GO1 and GO2 solutions of appropriate concentrations

tion in the GO/CEL composite membranes. The introduction of graphene oxide above 0.5% w/w into composite cellulose membranes did not significantly improve the antibacterial properties of these membranes against *E. coli*.

Other results were obtained for membrane studies using *S. aureus* (Fig. 4). Analysis of the obtained results showed that this bacteria was resistant to low concentrations of GO1 in composite membranes A and B. The largest inhibition zone was observed for membrane C – a width of 0.81 mm. In the case of consecutive membranes, this zones were: 0.57; 0.34; 0.23 mm wide for membranes D, E and F, repectively.

The obtained results with *Candida albicans* *fungus* clearly indicated that GO1 assured the fungicidal properties of GO/CEL composite membranes (Fig. 4). The fungicidal effect of the membranes was directly related to the concentration of graphene oxide in the membrane. The higher the concentration of GO1, the more resistant to the fungi it was. For membrane A, a 0.1 mm width of growth inhibition was observed. The following membranes had increasingly wider growth inhibition zones, starting from 0.14; 0.18; 0.25; 0.32 mm width for membranes B, C, D, E. The highest concentration of GO1 was for membrane F (28.6% w/w), and this membrane had the highest growth inhibition area for *Candida albicans* of 0.48 mm.

Adding GO1 to cellulose membrane "0" makes it resistant to selected bacteria and fungi (Fig. 5). Graphene oxide, as an additive to a cellulose-based composite, gave bactericidal properties against Gram negative *E. coli* and fungicidal properties against *Candida albicans*. It should also be noted that an admixture of graphene oxide

in the cellulose membrane as low as 0.5% w/w increased its resistance to microorganisms.

In the case of Gram positive *S. aureus*, the biocidal action occured only above 2% w/w of GO addition to the cellulose matrix. Membrane C had a very good bactericidal action against *S. aureus*, almost 2 times higher than the highest value obtained for *E. coli* and *Candida albicans*. The remaining membranes (D, E, F) were characterized by antibacterial action at the level similar to *E. coli*.

Comparison the bactericidal and fungicidal action of graphene oxide in the GO/CEL membranes prepared by us with the results described in the literature is difficult. The biological activity of a GO-containing composite depends on many factors, including the method of obtaining the composite, the size and degree of dispersion of the nanoaddition particles, and the microbiological availability of the bioactive components.

CONCLUSIONS

This paper presented the results of research on bioactive membranes made of cellulose (CEL) with graphene oxide (GO) admixture. The studies on the biocidal properties of the membranes were preceded with GO as the modifier. At the beginning, the influence of the size of nanoparticles and their concentration on the growth of *E. coli* and *S. aureus* were examined. For this purpose, graphene oxide was obtained from graphite with two different particle sizes, <20 μm and <150 μm, as a result of which GO1 and GO2 were formed. Both products were dispersed in water to con-

Fig. 4. Microscope photos of growth inhibition zones around cellulose membranes ("0") and GO/CEL composite membranes after 24 hours of incubation of *E. coli, S. aureus* and *Candida albicans fungi*

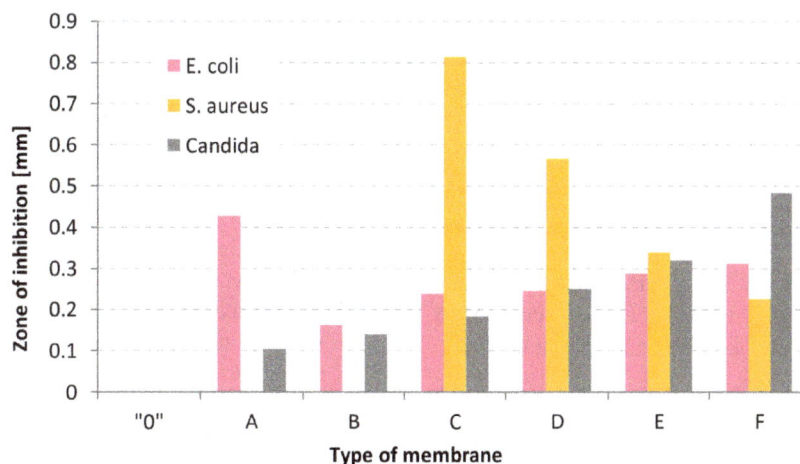

Fig. 5. Zones of bacterial (of *E. coli* and *S. aureus*) and *Candida albicans* fungi growth inhibition around the cellulose membrane (membrane "0") and GO/CEL composite membranes (membrane A ÷ F)

centrations of 0.001; 0.01; 0.1% w/w of GO. It was observed that *E. coli* was less resistant to the GO1 molecules, and the area of growth inhibition for this bacteria was high, regardless of the concentration of graphene oxide in the sample. The tests carried out with GO2 solutions showed that the strong bactericidal action against *E. coli* is only with the highest GO concentration. Studies conducted with *S. aureus* showed complete resistance to GO2 but that small particles of GO1 showed the antibacterial action.

The obtained GO/CEL membranes were subjected to microbiological analysis in which E. coli and S. aureus as well as Candida albicans fungi were used. The obtained results indicated that adding GO1 to GO/CEL cellulose membranes gave resistance to bacteria and fungus. Graphene oxide, introduced even in low concentrations (0.5% w/w) into the cellulose composite, gave bactericidal properties against E. coli and fungicidal properties against Candida albicans. In the case of S. aureus, on the other hand, the biocidal action occured only above 2% w/w of GO addition to the cellulose matrix.

The results of our studies indicate that GO/CEL composite membranes obtained in the process of GO1 addition to cellulose have a bioactive effect, both for Gram-negative and Gram-positive bacteria as well as fungi of *Candida albicans*. These properties enable the use of GO/CEL composite membranes as potential dressing and packaging materials. Graphene oxide enclosed in the structure of the GO/CEL membrane is not released into the environment, which allows its safe use, e.g. in membrane water treatment processes.

REFERENCES

1. Akhavan O., Ghaderi E. 2010. Toxicity of graphene and graphene oxide nanowalls against bacteria. ACS Nano, 4(10), 5731–5736.

2. Fink H. P., Weigel P., Purz H. J. Ganster J. 2001. Structure formation of regenerated cellulose materials from NMMO-solutions. Progress in Polymer Science (Oxford), 26(9), 1473–1524.

3. Fryczkowska B., Sieradzka M., Sarna E., Fryczkowski R., Janicki J. 2015. Influence of a graphene oxide additive and the conditions of membrane formation on the morphology and separative properties of poly(vinylidene fluoride) membranes. Journal of Applied Polymer Science, 132(46), 42789.

4. Fryczkowska B., Wiechniak K. 2017. Preparation and properties of cellulose membranes with graphene oxide addition. Polish Journal of Chemical Technology, 19(4), 41–49.

5. Guerrero-Contreras J., Caballero-Briones F. 2015. Graphene oxide powders with different oxidation degree, prepared by synthesis variations of the Hummers method. Materials Chemistry and Physics, 153, 209–220.

6. Hu W., Peng C., Luo W., Lv M., Li X., Li D., Huang Q, Fan C. 2010. Graphene-based antibacterial paper. ACS Nano, 4(7), 4317–4323.

7. Huang Q., Xu M., Sun R., Wang X. 2016. Large scale preparation of graphene oxide/cellulose paper with improved mechanical performance and gas barrier properties by conventional papermaking method. Industrial Crops and Products, 85, 198–203.

8. Hummers W. S., Offeman R. E. 1958. Preparation of Graphitic Oxide. Journal of the American Chemical Society, 80(6), 1339–1339.

9. Kafy A., Akther A., Shishir M. I. R., Kim H. C., Yun Y., Kim J. 2016. Cellulose nanocrystal/gra-

phene oxide composite film as humidity sensor. Sensors and Actuators, A: Physical, 247, 221–226.

10. Kanayama I., Miyaji H., Takita H., Nishida E., Tsuji M., Fugetsu B., Sun L., Inoue K, Ibara A., Akasaka T., SugayaT., Kawanami M. 2014. Comparative study of bioactivity of collagen scaffolds coated with graphene oxide and reduced graphene oxide, 3363–3373.

11. Kim C. J., Khan W., Kim D. H., Cho K. S., Park S. Y. 2011. Graphene oxide/cellulose composite using NMMO monohydrate. Carbohydrate Polymers, 86(2), 903–909.

12. Kuo Y. N., Hong J. 2005. A new method for cellulose membrane fabrication and the determination of its characteristics. Journal of Colloid and Interface Science, 285(1), 232–238.

13. Li G., Zhao H., Hong J., Quan K., Yuan Q., Wang X. 2017. Antifungal graphene oxide-borneol composite. Colloids and Surfaces B: Biointerfaces, 160, 220–227.

14. Lim M. Y., Choi Y. S., Kim J., Kim K., Shin H., Kim J. J., Shin D.M., Lee J. C. 2017. Cross-linked graphene oxide membrane having high ion selectivity and antibacterial activity prepared using tannic acid-functionalized graphene oxide and polyethyleneimine. Journal of Membrane Science, 521, 1–9.

15. Lindman B., Karlström G., Stigsson L. 2010. On the mechanism of dissolution of cellulose. Journal of Molecular Liquids, 156(1), 76–81.

16. Liu G., Ye H., Li A., Zhu C., Jiang H., Liu Y., Han K., Zhou Y. 2016. Graphene oxide for high-efficiency separation membranes: Role of electrostatic interactions. Carbon, 110, 56–61.

17. Liu S., Hu M., Zeng T. H., Wu R., Jiang R., Wei J., Wang L., Kong J., Chen Y. 2012. Lateral Dimension-Dependent Antibacterial Activity of Graphene Oxide Sheets. Langmuir, 28(33), 12364–12372.

18. Liu Y., Wen J., Gao Y., Li T., Wang H., Yan H., Niu B., Guo R. 2017. Antibacterial graphene oxide coatings on polymer substrate. Applied Surface Science, (2010).

19. Luo H., Ao H., Li G., Li W., Xiong G., Zhu Y., Wan Y. 2017. Bacterial cellulose/graphene oxide nanocomposite as a novel drug delivery system. Current Applied Physics, 17(2), 249–254.

20. Mahmoudi N., Eslahi N., Mehdipour A., Mohammadi M., Akbari M., Samadikuchaksaraei A., Simchi A. 2017. Temporary skin grafts based on hybrid graphene oxide-natural biopolymer nanofibers as effective wound healing substitutes: pre-clinical and pathological studies in animal models. Journal of Materials Science: Materials in Medicine, 28(5), 1–12.

21. Musico Y. L. F., Santos C. M., Dalida M. L. P., Rodrigues D. F. 2014. Surface Modification of Membrane Filters Using Graphene and Graphene Oxide-Based Nanomaterials for Bacterial Inactivation and Removal. ACS Sustainable Chem. Eng, 2, 1559–1565.

22. Novoselov N. P., Sashina E. S., Kuz'mina O. G., Troshenkova S. V. 2007. Ionic liquids and their use for the dissolution of natural polymers. Russian Journal of General Chemistry, 77(8), 1395–1405.

23. Pal N., Dubey P., Gopinath P., Pal K. 2017. Combined effect of cellulose nanocrystal and reduced graphene oxide into poly-lactic acid matrix nanocomposite as a scaffold and its anti-bacterial activity. International Journal of Biological Macromolecules, 95, 94–105.

24. Palmieri V., Carmela Lauriola M., Ciasca G., Conti C., De Spirito M., Papi M. 2017. The graphene oxide contradictory effects against human pathogens. Nanotechnology, 28(15), 152001

25. Parades J. I., Villar-Rodil S., Martínez-Alonso A., Tascón J. M. D. 2008. Graphene oxide dispersions in organic solvents. Langmuir, 24(19), 10560–10564.

26. Pinkert A., Marsh K. N., Pang S., Staiger M. P. 2009. Ionic liquids and their interaction with cellulose. Chemical Reviews, 109(12), 6712–6728.

27. Ramamoorthy S. K., Skrifvars M., Persson A. 2015. A Review of Natural Fibers Used in Biocomposites: Plant, Animal and Regenerated Cellulose Fibers. Polymer Reviews, 55(1), 107–162.

28. Rambo C. R., Rccouvreux D. O. S., Carminatti C. A., Pitlovanciv A. K., Antônio R. V., Porto L. M. 2008. Template assisted synthesis of porous nanofibrous cellulose membranes for tissue engineering. Materials Science and Engineering C, 28(4), 549–554.

29. Rui-Hong X., Peng-Gang R., Jian H., Fang R., Lian-Zhen R., Zhen-Feng S. 2016. Preparation and properties of graphene oxide-regenerated cellulose/polyvinyl alcohol hydrogel with pH-sensitive behavior. Carbohydrate Polymers, 138, 222–228.

30. Shahnawaz Khan M., Abdelhamid H. N., Wu H. F. 2015. Near infrared (NIR) laser mediated surface activation of graphene oxide nanoflakes for efficient antibacterial, antifungal and wound healing treatment. Colloids and Surfaces B: Biointerfaces, 127, 281–291.

31. Singh Z. 2016. Applications and toxicity of graphene family nanomaterials and their composites. Nanotechnology, Science and Applications, 9, 15–28.

32. Singh Z., Singh R. 2017. Toxicity of Graphene Based Nanomaterials Towards Different Bacterial Strains: A Comprehensive Review. American Journal of Life Sciences American Journal of Life Sciences. Special Issue: Environmental Toxicology, 5(5), 3–1.

33. Sun X. F., Qin J., Xia P. F., Guo B. B., Yang C. M., Song C., Wang S. G. 2015. Graphene oxide-silver

nanoparticle membrane for biofouling control and water purification. Chemical Engineering Journal, 281, 53–59.

34. Tang L., Li X., Du D., He C. 2012. Fabrication of multilayer films from regenerated cellulose and graphene oxide through layer-by-layer assembly. Progress in Natural Science: Materials International, 22(4), 341–346.

35. Texter J. 2014. Graphene dispersions. Current Opinion in Colloid and Interface Science. 19(2), 163–174.

36. Tu Y., Lv M., Xiu P., Huynh T., Zhang M., Castelli M., Liu Z., Huang Q., Fan C., Fang H., Zhou R. 2013. Destructive extraction of phospholipids from Escherichia coli membranes by graphene nanosheets. Nature Nanotechnology, 8(8), 594–601.

37. Wan C., Li J. 2016. Graphene oxide/cellulose aerogels nanocomposite: Preparation, pyrolysis, and application for electromagnetic interference shielding. Carbohydrate Polymers, 150, 172–179.

38. Yang E., Alayande A. B., Kim C.-M., Song J., Kim I. S. 2018. Laminar reduced graphene oxide membrane modified with silver nanoparticle-polydopamine for water/ion separation and biofouling resistance enhancement. Desalination, 426(October 2017).

39. Yang X. N., Xue D. D., Li J. Y., Liu M., Jia S. R., Chu L. Q., Wahid F., Zhang Y. M., Zhong C. 2016. Improvement of antimicrobial activity of graphene oxide/bacterial cellulose nanocomposites through the electrostatic modification. Carbohydrate Polymers, 136, 1152–1160.

40. Yoon K. Y., An S. J., Chen Y., Lee J. H., Bryant S. L., Ruoff R. S., Huh C., Johnston K. P. 2013. Graphene oxide nanoplatelet dispersions in concentrated NaCl and stabilization of oil/water emulsions. Journal of Colloid and Interface Science, 403, 1–6.

41. Zhang X., Yu H., Yang H., Wan Y., Hu H., Zhai Z., Qin J. 2015. Graphene oxide caged in cellulose microbeads for removal of malachite green dye from aqueous solution. Journal of Colloid and Interface Science, 437, 277–282.

42. Zhu S., Wu Y., Chen Q., Yu Z., Wang C., Jin S., Ding Y., Wu G. 2006. Dissolution of cellulose with ionic liquids and its application: a mini-review. Green Chemistry, 8(4), 325–327.

43. Zhu W., Li W., He Y., Duan T. 2015. In-situ biopreparation of biocompatible bacterial cellulose/graphene oxide composites pellets. Applied Surface Science, 338, 22–26.

21

Calibration of Activated Sludge Model with Scarce Data Sets

Dariusz Andraka[1*], Iwona Kinga Piszczatowska[2], Jacek Dawidowicz[1],
Wojciech Kruszyński[1]

[1] Białystok University of Technology; Faculty of Civil and Environmental Engineering; Wiejska 45E, 15-351 Bialystok; Poland
[2] Wodociągi Białostockie (Bialystok Water Supply) Sp. z o.o., Młynowa 52/1, 15-590 Białystok, Poland
* Corresponding author e-mail: d.andraka@pb.edu.pl

ABSTRACT

Mathematical models of activated sludge process are well recognised and widely implemented by researchers since 1980's. There is also numerous software available for modelling and simulation of activated sludge plants, but practical application of those tools is rather limited. One of the main reasons for such a situation is a difficult process of model calibration the requires extended data sets collected at investigated plant. Those data are usually not included in a standard plant monitoring plan. In the paper the problem of model calibration with the data sets derived from standard monitoring plan is discussed with a special regard to simulation objectives and data availability. The research was conducted with operational data from Białystok Wastewater Treatment Plant. The model of the plant was based on Activated Sludge Model No.3 developed by IWA Task Group and implemented in ASIM simulator. Calibration and validation of the model gave promising results, but further applications should be carefully considered, mainly due to uncertainties underlying input data.

Keywords: activated sludge models, modelling and simulation, model calibration

INTRODUCTION

Mathematical models of activated sludge process have been widely used by researchers and professionals for more than three decades. A task group formed in 1982, under the auspices of IWA (then the International Association on Water Pollution Research and Control) had a major contribution to the development of activated sludge models (ASM). The first model elaborated by the group came to be known as Activated Sludge Model No. 1 (ASM1) (Henze et al. 1987) and was followed by next generation models, including: ASM2 and ASM2d (Henze et al. 1995, 1999) and ASM3 (Gujer et al. 1999). Moreover, other researchers contributed to the development of activated sludge models, These include especially the Barker and Dold model (1997), as well as extension to ASM3 model developed by Riegger et al. (2001), both of which cover the biological phosphorus removal process.

With recent developments in IT (especially the popularization of powerful personal computers), the commercial software (simulators) that implements the above-mentioned mathematical models, became available. These simulators usually contain additional models for other unit processes (primary and secondary settlers, anaerobic digestion, thickening and dewatering), enabling simulation of the whole treatment facility (Rieger et al. 2013). In addition to commercial simulators, one can also find a few freeware tools, which are typically available for download (ASIM, STOAT®) or are Web-based and intended to be run through a Web-based application (JASS). The key aspects to consider while planning to use freeware tools are: limited functionality, less flexible user interface, and what can be deciding for less experienced modelers – the lack of support (WEF MOP31 2014).

As reported by Hauduc et al. (2009), Universities, public research centers and private consulting / engineering companies represent the majority of ASM users while only few of them are related with wastewater treatment plants (WWTPs). The main obstacles limiting modeling projects, expressed by the respondents of this survey, can be split into 4 topics: **cost and time demand**; **model structure** (complexity, reliability and non-adequacy of models); **model application** (for many potential users – models are not required to reach their objectives) and **modeling procedure** (data collection, calibration and validation, etc.). Particularly strong obstacles for the potential users from WWTPs are: costs and ASM complexity, related with large number of unit processes building the model, which are described by even more kinetic and stoichiometric parameters. These parameters can be evaluated from different information sources (Petersen et al., 2002):

- default parameter values from literature (usually used as defaults in built-in models of the simulators);
- full-scale facility data (average or dynamic data from collected samples, online data, measurements in reactors to characterize process dynamics);
- bioassays tests (laboratory-scale experiments with wastewater and activated sludge from the full-scale facility under study).

The parameter values obtained from defaults through fitting the model until simulation results agree sufficiently with the facility data are known as *calibrated parameters*, while those evaluated directly from measurements and experiments are referred to as *measured parameters*. In order to obtain reliable results, researchers use both types of parameters, which requires establishing special monitoring plan for the studied WWTP because routinely performed analyses of typical parameters characterizing influent and effluent (BOD, TSS, total nitrogen and phosphorus) are not consistent with the purpose of modeling and model requirements.

In most cases, the data available from historical records pertaining to monitoring results of the wastewater treatment facility include only the basic parameters (BOD, COD, TSS, TN, TP), which cannot be used directly for modeling purposes. As the result, there are few examples in the literature where plant operational data collected during standard monitoring plan were used as model input (Cinar et al., 1998; Sochacki et al., 2009) and practical applications of ASM are the few.

The main purpose of this study was to check the applicability of limited data sets obtained during routine monitoring of municipal WWTP in Białystok (Poland) for the calibration and validation of WWTP model under static conditions and to evaluate the possible application areas of such a simplified model.

MATERIALS AND METHODS

Bialystok WWTP characterization

Municipal wastewater treatment plant in Bialystok was constructed in 1974 for a design flow rate of 176.500 m^3/d. In 2002, the facility was significantly reconstructed in order to achieve higher efficiency of biogenic compounds elimination (to comply with compulsory regulations), and the capacity of the plant was reduced to 100.000 m^3/d (Simson, 2008). The technological layout of the facility consists of the following sections: preliminary mechanical treatment (screens, rectangular aerated grit chambers with sand separator, primary settlers with horizontal flow), biological reactors with activated sludge, comprising: predenitrification (PreDN) and anaerobic (DeP) sections organized in 4 parallel lines and anoxic (denitrification, DN) – aerobic (nitrification, N) sections organized into 8 parallel lines (with total volume $V_B = 63.200$ m^3) and six parallel secondary clarifiers (6.000 m^3 each). At present, PreDN and DeP (dephosphatation) basins (which are reconstructed from old primary settlers, with volume of 1.800 m^3 each) work only in 2 (out of 4) lines. They receive return activated sludge (RAS) from secondary clarifiers which can be split between PreDN and DeP with ratio 30/70%. The RAS flow is varying between 150–300% of daily inflow to the plant. Main activated sludge reactors form two technological blocs with different type of aeration (surface aerators and diffused air aerators). Each of 8 parallel lines consists of 3 sections: anoxic (DN, volume 1.375 m^3), alternative (either anoxic or aerobic, volume 1.125 m^3) and aerobic (N, volume 4500 m^3). Thus, the aerated volume makes up 60–75% of the total biological reactor's volume, depending on the state of the alternating section. The rate of internal recirculation of nitrate-rich mixture from aerobic to anoxic section is varying between 400–600% of daily inflow to the plant. In addition to biological phosphorus uptake, the facility is equipped with an installation for chemical precipitation of phosphates.

The Bialystok WWTP provides high efficiency of organic matter, solids and phosphorus removal, while nitrogen compounds elimination is unsteady (Table 1). For this reason, after a series of pilot studies (Simson, 2008; Ignatowicz et al., 2015), an installation for dosing external carbon source was introduced in 2009. Different agents are used for this purpose (with carbon content measured as COD no less than $1.000.000$ g/m^3) with the rate of 40–70 g per 1 m^3 of sewage inflow.

For the purpose of this study, only routine operational data, collected within standard monitoring plan of the facility, were used. In the Bialystok WWTP, the data, including raw and mechanically treated sewage, as well as effluent from the plant characteristics, are collected two times a month, which meets the requirements of applicable environmental regulations. The yearly averages estimated from the acquired data are presented in Table 1.

Modeling procedure

Together with the introduction of different simulators, several modeling protocols were published with the aim to guide model users through a series of defined steps and to obtain reliable results with less effort. However, the most popular protocols often presented different approach to modeling. STOWA protocol (Hulsbeek et al., 2002; Roeleveld and van Loosdrecht, 2002) was developed in order to help with modeling nitrogen removal using ASM No.1 model. On the other hand, the WERF guidelines (Melcer et al., 2003) were based on the experience with ASM from consulting companies, software developers and universities, mainly from North America) with targeted users from municipalities and consulting engineering companies. The BIOMATH protocol (Vanrolleghem et al., 2003) introduced a concept of step-wise calibration/validation of models, with a focus on the biokinetic model and sections on settling, hydraulics, and aeration. The HSG protocol (Langergraber et al., 2004) gathered the experience of researchers from German-speaking countries and encourages an objective-oriented approach. In order to bridge the gap between the existing protocols, a new IWA task group was formed – Good Modeling Practice (GMP) Task Group – with the aim to combine these proto-

Table 1. Wastewater characteristics of Białystok WWTP based on operational data

Parameter	Flow	BOD5	COD	TSS	TN	N-NH4	TP	temp.
	m³/d	mg/dm³	mg/dm³	mg/dm³	mg/dm³	mg/dm³	mg/dm³	⁰C
2016 – Influent (raw sewage)								
Average	-	443	1142	591	88	-	11.7	-
MIN	-	200	629	390	38.3	-	6.2	-
MAX	-	800	1600	930	147	-	26.8	-
2016 – After mechanical pretreatment								
Average	-	240	455.4	77.1	63	46.5	5.2	-
MIN	-	140	312	50	46.2	37.9	2.6	-
MAX	-	390	588	97	82.5	58.7	7.3	-
2016 – Final effluent								
Average	66.430	3.7	31	4	8.7	-	0.3	15.2
MIN	45.900	1.9	20	2	5.0	-	< 0.2	10.5
MAX	114.600	5.6	52	11	13.2	-	0.83	20.7
2017 – Influent (raw sewage)								
Average	-	485	1193	748	78.1	43.6	9.93	14.2
MIN	-	170	557	170	51.1	30.5	6.02	6.6
MAX	-	1020	2600	1860	113	55.6	16.8	18.5
2017 – After mechanical pretreatment								
Average	-	165.5	331.2	63.7	49.4	39.2	3.97	-
MIN	-	77	192	36	26.7	18.5	1.8	-
MAX	-	260	444	87	67	54.5	3.5	-
2017 – Final effluent								
Average	73.693	2.8	27.7	3.3	10.15	-	0.3	14.6
MIN	54.500	1.2	20	2.0	7.6	-	0.2	5.8
MAX	108.400	4.2	42	7.3	14.6	-	0.55	20.8

Symbols: BOD5 – 5-day Biochemical Oxygen Demand; COD – Chemical Oxygen Demand; TSS – Total Suspended Solids; TN – total nitrogen; N-NH4 – Ammonia Nitrogen; TP – Total Phosphorus

cols in one unified protocol intended mainly for practitioners. This unified protocol comprises following steps (Rieger et al., 2012): 1 – project definition; 2 – data collection and reconciliation; 3 – plant model setup; 4 – calibration and validation, 5 – simulation and results interpretation.

In the **project definition** step, the problem related with modeling task should be formulated and then – objectives of the project defined together with determination of requirements. In this study, after the analysis of the Bialystok WWTP performance it was decided that the main objective of modeling project will be the simulation of nitrogen removal processes in the plant.

Data collection and reconciliation aims at the preparation of reliable data sets for simulation projects, using dedicated methods based on statistical analysis, expert knowledge etc. According to the preliminary assumptions, only the data from routine plant monitoring were used in this study. The collected data were analyzed in order to eliminate outliers and detect possible faults in measurements and reports.

Plant model was created using ASIM simulator (Holinger, http://www.holinger.com). The main reasons for this choice were the software availability (it is free for noncommercial applications) and the ease of application for basic technological layouts of biological treatment units, which could promote its usage by less experienced modelers. Although ASIM allows only for the simulation of biological treatment systems (without preliminary treatment or sludge disposal processes) it has built-in IWA basic models: ASM No.1, ASM No.2d and ASM No.3 that may be freely edited, redefined and stored by the user. Since nitrification and denitrification processes were the main focus of this study, ASM1 and ASM3 were taken into account for the simulation. After preliminary investigations, ASM3 was selected for further simulations due to following premises:

- influent fractionation in ASM3 is relatively easier than in ASM1, which may be essential in the case of limited input information;
- although ASM3 includes significantly more unit processes than ASM1 (12 vs 7), as well as stoichiometric (7 vs 3) and kinetic (21 vs 14) parameters, the complexity of both models is comparable; furthermore, ASM3 is designed to be the core of many different models (for example modules on phosphorus removal can

be easily connected) and to satisfy primarily the requirements of practical model applications (Henze et al. 2000);

- initial simulations with default parameters showed better results for ASM3, especially in terms of the response of the model to temperature changes, which was essential for the examined WWTP, as the nitrogen removal efficiency strongly depends on the seasonal variations in wastewater temperature.

The plant model created with ASIM simulator is presented in Figure 1. In order to simplify the modeling, procedure only one technological line was imitated in the model. Assuming that wastewater after mechanical pretreatment is homogeneously mixed with RAS and then is evenly distributed between 8 parallel lines, the model represents the average conditions in the biological part of the plant. It is also important to note that due to the software limitations it was impossible to represent all bioreactors with their specifics in one model.

For the **calibration and validation** of the model, operational data were grouped in two data sets representing monthly averages of measured parameters: a) calibration data set from the period February – September, 2016, and b) validation data set for the period January – July, 2017. The sensitivity analysis was performed according to EPA guidelines (US EPA, 1987) to determine the parameters that may influence the model behavior significantly. The normalized sensitivity coefficients were evaluated with the following formula:

$$S_{i,j} = \frac{\Delta y_j / y_j}{\Delta x_i / x_i} \qquad (1)$$

where: Δy_j – increase in output variable (for example N-NH$_4$, TN etc.) relevant to Δx_i increase in input variable (for example stoichiometric or kinetic parameter of the model).

For the purpose of this research, a 10% increase in input variables was applied, as suggested by Liwarska-Bizukojc and Biernacki (2010). According to Petersen et al. (2003), the coefficients $S_{i,j} < 0.25$ have no significant influence on the model, while $1 < S_{i,j} < 2$ are very influential and $S_{i,j} > 2$ are extremely influential.

The calibrated values of influential parameters were obtained using a goodness-of-fit test, based

2.Influent =   0.25

Influent = 8180.00

O2 saturation   = 10.00
Sludge age, SRT 12.00

Secondary clarifier

| 1 | 2 | 3 | 4 | 5 | 6 | 7 |

Effluent

Rec1=42000.00

Return sludge = 12300.00

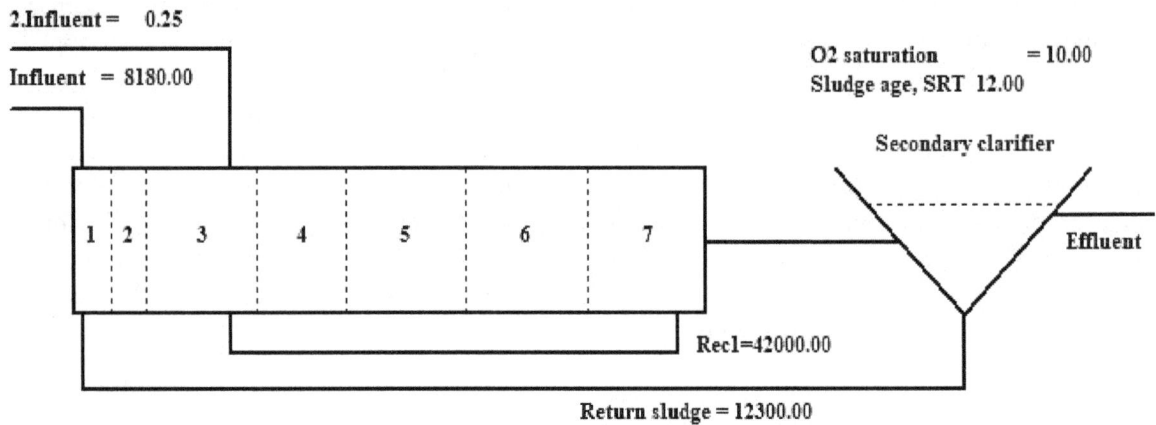

Reactor	1	2	3	4	5	6	7	Clar.1
Volume	450.00	450.00	1375.00	1125.00	1500.00	1500.00	1500.00	4500.00
O2 Conc.	-	-	-	1.00	1.00	1.00	1.00	
Kla Value	0.0	0.0	0.0	-	-	-	-	

Figure 1. Example of Bialystok WWTP model in ASIM simulator (data on the diagram – April, 2016)

on absolute criterion from residuals, calculated from following formula (WEF MOP31, 2014):

$$E_2 = \frac{1}{n} \sum_{i=1}^{n} (O_i - P_i)^2 \rightarrow min \qquad (2)$$

where: O_i – observed value;
P_i – simulated value;
n – number of simulations.

RESULTS AND DISCUSSION

The application of ASM requires influent fractionation according to input data structure for a given model. As the influent data available for this study did not include the information about COD fractions, it was necessary to estimate the input variables on the basis of preliminary simulations. The plant model was created using yearly average inflow characteristics and default parameters values. The simulation results were compared with the yearly average effluent quality and relevant parameters were adjusted to obtain acceptable agreement. The default and adjusted fractionation parameters are presented in Table 2.

The calibration procedure was performed with regard to the study goals. Since the target process of this research was nitrogen compounds removal, the calibration data set was prepared consisting of monthly averages for the period February-September, 2016. Moreover, the stop criterion

Table 2. Comparison of ASM3 model compounds for typical wastewater composition (Henze, 2000) and Bialystok WWTP (primary efluent)

Compounds	Dissolved compounds			Particulate compounds				
	S_I	S_S	$\frac{S_S}{S_{COD}}$	X_I	X_S	X_H	$\frac{X_S}{X_{COD}}$	$\frac{X_H}{X_{COD}}$
	gCOD/m³		-	gCOD/m³			-	-
Typical	30	60	0.60	25	115	30	0.69	0.10
	COD_{tot} = 260 gCOD/m³; TSS = 125 gSS/m³; TSS/X_{COD} = 0.75; TKN = 25 gN/m³; S_{NH4} = 16 gN/m³; S_{NH4}/TKN = 0.64							
Bialystok WWTP	29	234	0,89	40	133	19	0,69	0,1
	COD_{tot} = 455 gCOD/m³; TSS = 77 gSS/m³; TSS/X_{COD} = 0.40; TKN = 63 gN/m³; S_{NH4} = 46 gN/m³; S_{NH4}/TKN = 0.73							

Symbols: S_I – soluble inert organics, S_S – readily biodegradable substrates; S_{COD} – soluble COD; S_{NH4} – ammonium; X_I – inert particulate organics; X_S – slowly biodegradable substrates; X_H – heterotrophic biomass; X_{COD} – particulate COD; COD_{tot} – total COD; TSS – total suspended solids; TKN – total Kiejdahl nitrogen

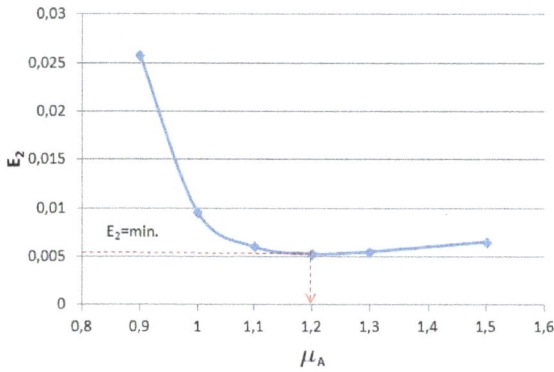

Figure 2. Calibration of autotrophic maximum growth rate (uA) by minimization of absolute criterion from residuals (E2)

(acceptable error range) was established, according to Rieger et al. (2012) at value of 1.0 gN/m³.

The initial run of the model was performed with default stoichiometric and kinetic parameters, built in ASIM simulator. The comparison of simulated and observed values indicated that the acceptable error range was exceeded in several points (compare Figure 3, series TN(1) and TN(2) for IV.2016, V.2016, VI.2016 and VIII.2016) and further parameters calibration is required. The analysis performed with Eq. (1) for the model parameters responsible for nitrogen removal allowed for determination of influential parameters, which were adjusted afterwards by minimization of average squared residuals (E_2) with Eq. (2). The graphical representation of the calibration process for autotrophic maximum growth rate (m_A) is shown in Figure 2 and the summary of calibration results for all influential parameters is presented in Table 3.

The data presented in Table 3 partly correspond with the results of Hauduc et al. studies (2011), presented later in Rieger et al. (2012) who examined several databases for ASM3 models of full scale WWTPs in Northern Europe and proposed new default parameter set, including autotrophic maximum growth rate (m_A) at the value of 1,3 d^{-1}.

The simulation results for ASM3 default and calibrated model, compared with the observed values of total nitrogen in the effluent from the plant are presented in Figure 3.

The obtained results show that in most cases, the calibrated ASM3 model has better accuracy of predictions than the default model and only in the case of June, 2016 simulation error is higher than the acceptable value (1.0 gN/m³).

At the last stage of this study, the calibrated ASM3 model of Bialystok WWTP was validated with the data set prepared for the period of January – July, 2017 (Figure 4).

The results of validation illustrated in Figure 4 show that the calibrated ASM3 model has an acceptable accuracy of predictions (all simulation points, except January 2017, have prediction error lower than 1.0 gN/m³), although it should be also noticed that default ASM3 model is able to predict effluent TN concentrations with similar or even better precision. This ambiguity may be explained by the uncertainty underlying modeling process based on scarce input data sets with limited informative value. For example, in this study the average monthly observations were estimated on the basis of two samples only, collected in different time intervals. In such a case, the input data used for the calibration and valida-

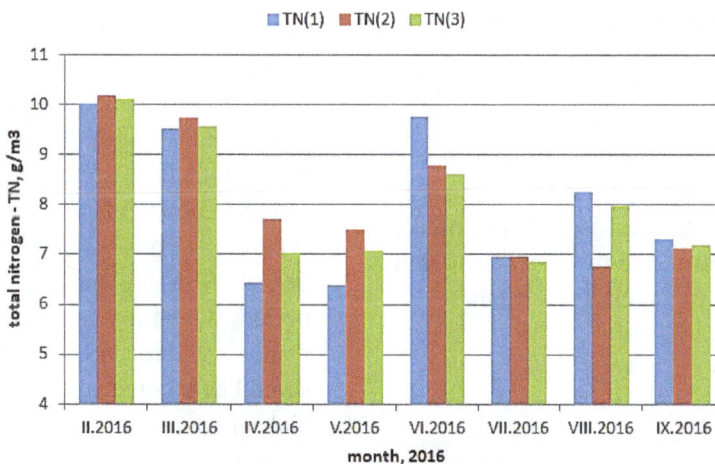

Figure 3. Comparison of observed and simulated effluent TN (total nitrogen) concentrations; TN(1) – observed, TN(2) – simulated with default ASM3 model, TN(3) – simulated with calibrated ASM3 model

Table 3. Sensitivity analysis and calibration results for ASM3 model parameters

Parameter	Default value	Sensitivity coefficient (S_{ij})		Calibrated value
		NO_x	$N-NH_4$	
(s) Anoxic storage of dissovled species. with regard to dinitrogen and nitrate (x_3); -	0.07	0.42	2.71	0.065
(k) Autotrophic maximum growth rate (m_A); d^{-1}	1.00	0.44	8.08	1.2
(k) Aerobic endogenous respiration rate ($b_{A,O2}$); d^{-1}	0.15	<0.25	2.17	0.15

(s) – stoichiometric parameter; (k) – kinetic parameter

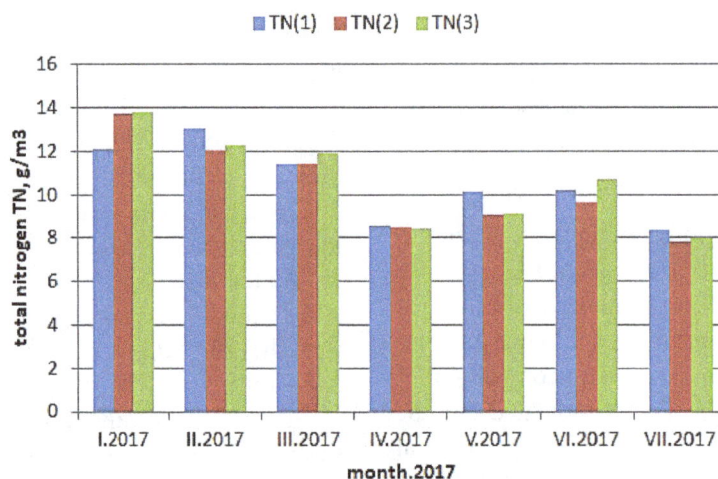

Figure 4. Results of ASM3 model validation for effluent total nitrogen (TN); TN(1) – observed, TN(2) – simulated with default ASM3 model, TN(3) – simulated with calibrated ASM3 model

tion of the model are very sensitive to the "noise" related with possible temporary disturbances in the process (like diurnal variations in hydraulic and contaminants load, operational errors, equipment failures etc.). The other factor that may influence the accuracy of model predictions in this work is related with specific mode of operation of Bialystok WWTP, which is focused on maximizing nitrogen removal by regulation of internal recirculation of nitrates rate, RAS rate and wasted sludge rate, depending on the current needs (in other words – sludge age is not a target operational parameter for the plant). Thus, the mass balance of microorganism in biological reactors, which is one of key components deciding about ASM quality, could not be verified during this study and sludge age values used in the model were not calculated from the measured data, but assumed on the basis of expert knowledge.

CONCLUSIONS

The research presented in this paper was performed on the Bialystok WWTP with the focus on applicability of limited data sets coming from standard plant monitoring program, for mathematical modeling of activated sludge process using the available ASM simulators. The obtained results allow for drawing the following conclusions:

1. Scarce data sets available from standard monitoring of WWTP performance may be used for setting up a facility model and for simulations of plant performance under steady state conditions;
2. Calibration of the Bialystok WWTP ASM3-based model with the available data represented by monthly averaged values and with regard to nitrogen compound removal process, significantly improved the accuracy of model predictions for a considered time period,
3. There is no significant difference between the accuracy of predictions for the calibrated and default ASM3 plant model for the validation period, which indicates that the created model is not reliable enough and modeling results should be studied and implemented with a special care and awareness of uncertainty underlying the whole modeling procedure;
4. Improvement of the model reliability is possible, but additional data allowing for mass balance completion should be available;

5. Despite the existing limitations and deficiencies of the model developed in this study, it can still be useful for various purposes, including: plant operators training (observation of plant response to the changes in basic operational parameters like recycle flows, anoxic to aerobic volume ratio; dissolved oxygen concentration etc., with regard to varying input characteristics), development of optimum control strategy, etc.

Acknowledgements

The authors highly appreciate the cooperation within this study with the Białystok Waterworks Ltd. (Wodociągi Białostockie Sp. z o.o.). We would like to especially thank the Chief Technologist of Bialystok WWTP, Mr Grzegorz Simson, who delivered invaluable information on the technological process.

The paper was accomplished under BUT Rector's grant S/WBIIS/2/14, supported by Polish Ministry of Science and Higher Education

REFERENCES

1. Barker P.S. and Dold P.L. 1997. General model for biological nutrient removal activated sludge systems: model presentation. Water Environment Research, 69(5), 969–984.

2. Çinar Ö., Daigger G.T., Graef S.P. 1998. Evaluation of IAWQ Activated Sludge Model No. 2 using steady-state data from four full-scale wastewater treatment plants. Water Environ. Res., 70 (6), 1216–1224.

3. Gujer W., Henze M., Mino T. and van Loosdrecht M.C.M. 1999. Activated Sludge Model No. 3. Water Science and Technology, 39(1), 183–193.

4. Hauduc H., Gillot S., Rieger L., Ohtsuki T., Shaw A., Takács I. and Winkler S. 2009. Activated sludge modelling in practice – An international survey. Water Science and Technology, 60(8), 1943–1951.

5. Hauduc, H., Rieger, L., Ohtsuki, T., Shaw, A., Takács, I., Winkler, S., Héduit, A., Vanrolleghem, P.A. and Gillot, S. 2011. Activated sludge modelling: Development and potential use of a practical applications database. Water Science and Technology, 63(10), 2164–2182.

6. Henze, M.; Grady, C.P.L.; Gujer, W.; Marais, G.V.R.; Matsuo, T. 1987. Activated Sludge Model No. 1; IAWPRC Scientific and Technical Report No. 1; International Association on Water Pollution Research and Control: London, U.K.

7. Henze, M.; Gujer, W.; Mino, T.; Matsuo, T.; Wentzel, M.C.; Marais, G.V.R. 1995. Activated Sludge Model No. 2; IAWQ Scientific and Technical Report No. 3; International Association on Water Quality: London, U.K.

8. Henze, M.; Gujer, W.; Mino, T.; Matsuo, T.; Wentzel, M.C.; Marais, G.V.R.; van Loosdrecht, M.C.M. 1999. Activated Sludge Model No. 2d, ASM2d. Water Sci. Technol., 39 (1), 165–182.

9. Henze, M., Gujer, W., Mino, T. and van Loosdrecht, M.C.M. (Eds), Activated sludge models ASM1, ASM2, ASM2d and ASM3. Scientific and Technical Report No. 9, IWA Publishing, London, UK

10. Hulsbeek, J.J.W.; Kruit, J.; Roeleveld, P.J.; van Loosdrecht, M.C.M. 2002. A Practical Protocol for Dynamic Modeling of Activated Sludge Systems. Water Sci. Technol., 45 (6), 127–136.

11. Ignatowicz, K., Piekarski, J., Kozłowski, D. 2015. Wspomaganie procesu denitryfikacji preparatem Brenntaplus VP1 jako zewnętrznym źródłem węgla. Rocznik Ochrona Środowiska, 17, 1178–1195.

12. Langergraber, G.; Rieger, L.; Winkler, S.; Alex, J.; Wiese, J.; Owerdieck, C.; Ahnert, M.; Simon, J.; Maurer, M. 2004. A Guideline for Simulation Studies of Wastewater Treatment Plants. Water Sci. Technol., 50 (7), 131–138.

13. Liwarska-Bizukojc, E.; Biernacki, R. 2010. Identification of the most sensitive parameters in the activated sludge model implemented in BioWin software. Bioresource Technology 101 (19): 7278–7205.

14. Melcer, H.; Dold, P. L.; Jones, R. M.; Bye, C. M.; Takacs, I.; Stensel, H. D.; Wilson, A. W.; Sun, P.; Bury, S. 2003. Methods for Wastewater Characterization in Activated Sludge Modeling; Water Environment Research Foundation: Alexandria, Virginia.

15. Petersen B., Gernaey K., Henze M. and Vanrolleghem P. A. 2002. Evaluation of an ASM1 model calibration procedure on a municipal-industrial wastewater treatment plant. Journal of Hydroinformatics, 4(1), 15–38.

16. Petersen, B., Gernaey, K., Henze, M., Vanrolleghem, P.A. 2003. Calibration of activated sludge models: a critical review of experimental designs. In: Agathos, S.N., Reineke, W. (Eds.), Biotechnology for the Environment: Wastewater Treatment and Modelling. Waste Gas Handling. Kluwer Academic Publishers, Dordrecht.

17. Rieger L., Koch G., Kühni M., Gujer W. and Siegrist H. 2001. The EAWAG Bio-P module for activated sludge model No. 3. Water Research, 35(16), 3887–3903.

18. Rieger, L.; Gillot, S.; Langergraber, G.; Ohtsuki, T.; Shaw, A.; Takács, I.; Winkler, S. 2013. Guidelines for Using Activated Sludge Models; IWA Scientific and Technical Report No. 22; IWA Publish-

ing: London.

19. Roeleveld, P.J.; van Loosdrecht, M.C.M. 2002. Experience with Guidelines for Wastewater Characterization in The Netherlands. Water Sci. Technol., 45 (6), 77–87.

20. Simson G. 2008. Pierwsze doświadczenia – test technologiczny z zastosowaniem preparatu BRENNTA-PLUS VP1 jako zewnętrzne źródło węgla organicznego do intensyfikacji procesu denitryfikacji w Białostockiej Oczyszczalni Ścieków. Forum Eksploatatora 44 (6), 21–23.

21. Sochacki A. , Knodel J., Geissen S.-U., Zambarda V., Bertanza G., Plonka L. 2009. Modelling and simulation of a municipal WWTP with limited operational data. Proceedings of a Polish-Swedish-Ukrainian Seminar, 2009; available online at: https://www.kth.se/polopoly_fs/1.650929!/JP-SU16p47.pdf (accesed January, 2017)

22. US EPA 1987. QUAL2E – The Enhanced Stream Water Quality Model EPA/823/B-95/003. Environmental Research Laboratory, Athens, GA, USA.

23. Vanrolleghem, P. A.; Insel, G.; Petersen, B.; Sin, G.; De Pauw, D.; Nopens, I.; Weijers, S.; Gernaey, K. 2003. A Comprehensive Model Calibration Procedure for Activated Sludge Models. Proceedings of the 76th Annual Water Environment Federation Technical Exhibition and Conference [CD-ROM]; Los Angeles, California, Oct 11–15; Water Environment Federation: Alexandria, Virginia.

24. Water Environment Federation: Wastewater Treatment Process Modeling, MOP31, 2nd Edition 2014. McGraw-Hill Professional, 2014, AccessEngineering.

22

Effect of Electrical Current and the External Source of Carbon on the Characteristics of Sludge from the Sequencing Batch Biofilm Reactors

Izabella Kłodowska[1], Joanna Rodziewicz[1*], Wojciech Janczukowicz[1]

[1] University of Warmia and Mazury in Olsztyn, Faculty of Environmental Sciences, Department of Environment Engineering, Warszawska 117a, 10-719 Olsztyn, Poland
* Corresponding author's email: joanna.rodziewicz@uwm.edu.pl

ABSTRACT

This work presents the results of an experiment on the effect of electrical current density (53, 105, 158 and 210 mA/m^2), the type of an external source of carbon (citric acid, potassium bicarbonate) and C/N$_{NO3}$ ratio (0.5, 1.0 and 1.5) on the quantity and quality of formed sludge. The experiment was conducted in sequencing batch biofilm reactors (SBBRs), under anaerobic conditions, with and without the passage of electrical current, under controlled pH of 7.5–8.0. The study demonstrated that in the reactors with electrical current passage and external source of carbon, the volume of sludge increased along with the current density increase from 53 to 158 mA/m^2. At its highest density (210 mA/m^2), the concentration of sludge was insignificantly lower. For all densities of electrical current and C/N$_{NO3}$ values, the concentrations of sludge formed in the reactors with potassium bicarbonate (1.00 to 1.26 g d. m./L) were lower than in the reactors with citric acid (1.26 to 1.30 g d. m./L). The concentration of organic matter was higher in the sludge from the reactors with electrical current passage and potassium bicarbonate, compared to the sludge from the reactors with citric acid. In the reactors with electrical current passage and external source of carbon, the total nitrogen content in the sludge decreased along with the C/N$_{NO3}$ ratio increase for current densities of 53 and 105 mA/m^2. For a higher electrical current density, the nitrogen content in the sludge was similar. Irrespectively of the current density, the nitrogen content in the sludge from the reactors with citric acid was higher than in the sludge from the reactors with potassium bicarbonate. For higher current densities (158 and 210 mA/m^2) the increase in the C/N$_{NO3}$ value caused an increase in the P content in the sludge. The electrical current density increase contributed to increasing the content of phosphorus in the sludge. The phosphorus content in the sludge from the reactors with citric acid was lower than in the sludge from the reactors with potassium bicarbonate. The CST values prove that the sludge formed during the wastewater treatment in electrobiological SBBR was characterized by very high dewaterability. The capillary suction time decreased along with increasing the electrical current density but was not significantly affected by the type of carbon source.

Keywords: bio-electrochemical reactor, denitrification, electrocoagulation, citric acid, potassium bicarbonate, sludge

INTRODUCTION

Poland, being poor in natural resources, has to use highly efficient technologies for wastewater treatment to prevent the effects of eutrophication [Janczukowicz and Rodziewicz, 2013; Attour et al., 2014]. It pertains to large, small and household wastewater treatment plants. Pursuant to the regulations binding since the 1st January 2016 stipulated in the regulation of the Minister of Natural Environment of 2014 [Regulations, 2014], the treated wastewater from households or farms discharged to the waters on the area of a municipal agglomeration should meet requirements set for PE agglomeration. This means that the household wastewater treatment plants based on the simplest solutions like a septic tank and pipe draining and located on the area of an agglomeration will have to be modernized or liqui-dated, whereas others – e.g. with activated sludge tanks and bed – will have to be modernized.

The solutions employing aerobic biofilm are often used in household and small wastewater treatment plants. However, such installations do not ensure sufficient nitrogen and phosphorus compounds removal [Klaczyński, 2013] that would be consistent with the regulations, as effluents from these plants still contain significant concentrations of nitrate nitrogen and orthophosphates. For this reason, they need to be modernized by, e.g. "coupling" them to the existing installations of a bio-electrochemical module which merges the biological and physicochemical processes of treatment in a single reactor. An example of such a solution is the sequencing batch biofilm reactor (SBBR) with the carrier in the form of disks, operating based on autotrophic (hydrogenotrophic) denitrification and electrocoagulation, having the potential of a facility constituting the third stage of wastewater treatment [Kłodowska et al., 2013].

Ample literature works [Karanasios et al., 2010; Feng et al., 2013; Shalaby et al., 2014; Kuokkanen et al., 2015] addressing the removal of nitrates and orthophosphates from wastewater subjected to the mechanical-biological pre-treatment, mainly described the results of the experiments conducted in the systems with separate reactors for the processes of denitrification and electrocoagulation. In turn, few works are available on the quantity and quality of sludge formed in bio-electrochemical reactors [Kuokkanen, 2016].

The authors of this manuscript have earlier demonstrated the effect of technological parameters and feeding external substrates of carbon to the reactor on the concentrations of nitrogen and phosphorus in the effluent from the bio-electrochemical sequencing batch biofilm reactor (SBBR) [Kłodowska et al., 2014; Kłodowska et al., 2016]. However, they have never analyzed the quantity nor the quality of the sludge formed during wastewater treatment in the reactor of this type.

Sludge stabilization is the key process which affects its properties in the perspective of its further management. The most common methods applied to this end in wastewater treatment plants include the aerobic and anaerobic stabilization [Bień et al., 1995; Baran and Turski, 1999]. During the process of stabilization, the organic matter of sludge is mineralized via biochemical transformations. It reduces the sludge demand for oxygen, minimizes the quantity of substances emitting noxious odor, reduces numbers of pathogens, and significantly decreases dry matter content. The American Agency for Environment Protection [Oleszkiewicz, 1999] divides sludge stabilization processes into: (1) those which cause stabilization and partial hygienization of sludge which is then suitable for being injected into the soil or for deposition on a landfill; and (2) those which cause advanced stabilization and hygienization of sludge which may further be used in agriculture (application on soil surface). Oleszkiewicz [1999] postulated adopting the dry matter content decrease by at least 38% as the criterion of a stabilized sludge.

Besides the dry matter content, an equally important parameter of sludge is its susceptibility to dewatering which is determined based on the capillary suction time (CST). The future use of sludge in the natural environment is also largely determined by the presence of biogenes.

Considering the characteristics of wastewater subjected to a two-stage pre-treatment and then treated in a bio-electrochemical reactor as well as the specificity of processes of hydrogenotrophic denitrification and electrocoagulation, the sludge formed in the SBBR as a result of wastewater treatment is expected to have low concentrations of organic compounds, high concentrations of nitrogen and phosphorus, and a short CST. What remains unknown is how the properties of sludge will change as a result of modifying the technological parameters of the process. The available research works provide no data in this respect.

The objective of this study was to determine the effect of the electrical current density, type of the source of organic and inorganic carbon and C/N_{NO3} ratio on the quantity and quality of the sludge formed in the bio-electrochemical SBBR reactor.

METHODS

The experiments were conducted simultaneously in vertical sequencing batch biofilm reactors (SBBR) with the volume of 3.0 L each (active volume – 2.0 L), under anaerobic conditions (Fig. 1). A set of 12 disks made of stainless steel with diameter of 0.10 m and total surface of 0.19 m^2 was installed in each reactor. The distance between disks was 5 mm. They were mounted coaxially on a vertical shaft rotating with the speed of 10 rpm; their submersion rate was 100%.

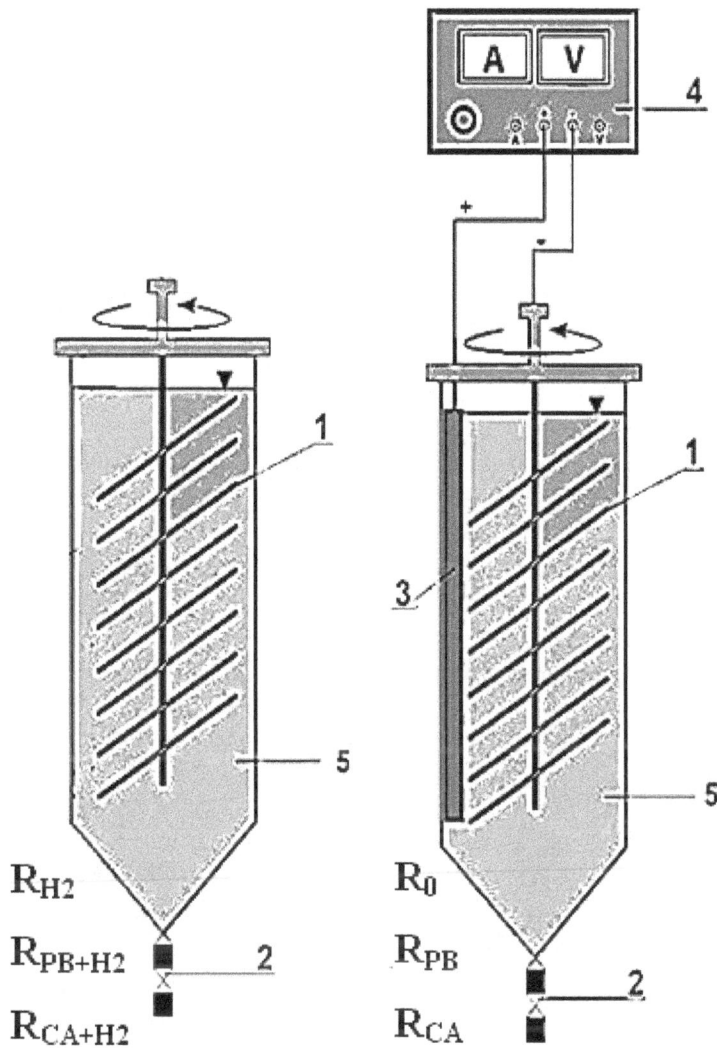

Fig. 1. Scheme of the experimental model: (1) cathode – discs with attached biofilm (stainless steel), (2) outlet, (3) anode (aluminum), (4) electric current source, (5) reactor

The experiment was conducted under the following conditions: without the passage of electrical current (reactors: R_0, R_{CA} and R_{PB}), and with the passage of electrical current (reactors: R_{H2}, R_{CA+H2} and R_{PB+H2}). In the control reactor (R_0), without the electrical current flow and without an external source of carbon, the synthetic wastewater was subjected to bio-treatment. In another two reactors (R_{CA} and R_{PB}) without the electrical current flow, citric acid and potassium bicarbonate, respectively, were used in the concentrations ensuring C/N_{NO3} ratios of 0.5, 1.0 and 1.5. In the reactors with the passage of electrical current, no external source of carbon was introduced in reactor R_{H2}, likewise in R_0, whereas citric acid and potassium bicarbonate were fed as carbon sources to reactors R_{CA+H2} and R_{PB+H2}, respectively. The wastewater retention time was 24 hours in each reactor.

The reactors were adapted for 3 months to achieve the appropriate structure of the biofilm and stable concentration of nitrogen compounds in the effluent using the activated sludge from denitrification tanks of the Municipal Wastewater Treatment Tank "Łyna" in Olsztyn as the inoculum. The analytical control of the treatment process was begun after the adaptation period.

In the reactors with the electrical current passage, provided by laboratory feeders – Programmable DC Power Supply – HANTEK PPS 2116A – (0–5A) (0–32V) and MANSON DC Power Supply – DPD 3030 (0–3A, 0–30V), disks with immobilized biofilm served as the cathode, whereas an aluminum plate with a total surface area of 0.033 m^2 served as the anode. The cathode and the anode were connected to the laboratory feeder to ensure the desired density of the electrical current, i.e. 53, 105, 158 and 210 mA/m^2 (the current intensity was 10, 20, 30 and 40 mA, the current voltage ranged from 3.0 to 5.0V). The experiment was conducted for 16 weeks under the conditions of controlled pH (pH 7.5–8.0).

The experiments were conducted with the synthetic wastewater characterized by high concentrations of nitrate nitrogen and total phosphorus, and by a low concentration of organic matter (COD). The parameters of wastewater corresponded to the parameters of municipal sewage subjected to bio-treatment in a municipal wastewater treatment plant with a high efficiency of organic compounds removal and enabling nitrification. The mean composition of wastewater flowing into the reactors was as follows: 50.68 ($\pm$1.61) mgN_{NO3}/L, 0.0 mgN_{NO2}/L, 0.0 mgN_{NH4}/L, 5.16 ($\pm$0.20) mgP/L, 70.30 ($\pm$10) mgO_2/L, TOC – 29.09 mgC/L, InOC – 47.04 mgC/L. The electrolytic conductance of wastewater reached 1.74($\pm$0) mS/cm. The analyses were conducted at a temperature of 25.3($\pm$0)°C.

The analyses of the sludge discharged from reactors included determinations of:

- concentration of sludge with the gravimetric method acc. to PN-EN 12880:2004,
- concentration of total nitrogen in the sludge with the spectrophotometric method based on the procedure developed by Lange LCK company,
- concentration of total phosphorus in the sludge with the spectrophotometric method based on the procedure developed by Lange LCK company,
- COD concentration in the sludge with the spectrophotometric method based on the procedure developed by Lange LCK company,
- capillary suction time (CST) with an electronic CST Meter by ProLabTech, 1112041 CE.

RESULTS AND DISCUSSION

In this experiment, we analyzed the effect of electrical current density, organic and inorganic substrate, and C/N_{NO3} ratio on the quantity and quality of the sludge formed in sequencing batch biofilm reactors (SBBRs), under anaerobic conditions with and without the passage of electrical current.

Our earlier investigations demonstrated high efficiencies of the nitrogen and phosphorus compounds removal in the reactors of this type, i.e. SBBR reactors. The efficiency of nitrogen removal increased along with the intensity of electrical

current and quantity of carbon fed to the reactor. Higher efficiencies were observed in the reactors fed with citric acid. The highest efficiency was recorded at the current density of 210 mA/m^2 in the reactor with citric acid. At the C/N_{NO3} of 0.5, this efficiency accounted for 83.05($\pm$1.16)%, while at C/N_{NO3} of 1.5 it reached 87.61($\pm$1.6)%. Similar tendencies were observed in the case of phosphorus. The highest efficiency of its removal reaching 97.69($\pm$2.1)% was determined in the reactor with citric acid used as the source of carbon and at the current density of 210 mA/m^2 and C/N_{NO3} ratio of 1.5. The respective value achieved in the reactor with potassium bicarbonate was slightly lower and reached 96.68%.

Literature works emphasize that the volume of the sludge formed in electrochemical reactors depends on the material the electrodes are made of. A study conducted by Lacasa et al. [2011] demonstrated lower production of sludge upon the use of aluminum electrode compared to the Fe electrode. During the analyses carried out at electrical current densities from 10 A/m^2 to 50 A/m^2 and with the Al electrode, the sludge formation ranged from 18.8 g/m^3 to 67.4 g/m^3, whereas with the use of Fe electrode – from 27.4 g/m^3 to 586.4 g/m^3. Significantly higher results were reported by Akyol [2012], who – applying the current density of 35 A/m^2 – determined the sludge concentration of 9.63 kg/m^3 using the Fe electrode and 7.73 kg/m^3 using the Al electrode. According to Gharibi et al. [2013], the concomitant processes of electrolysis and electrocoagulation contribute to the improved sludge dewaterability.

In our experiment, the quantity of sludge (Fig. 2) formed in the reactors without the electrical current passage was insignificantly lower than in the control reactor (R_0, 1.2 g d.m./L).

Simultaneously, higher quantities of formed sludge were observed in the R_{CA} reactors with an external source of organic carbon (citric acid), compared to the R_{PB} reactors fed with the inorganic carbon (potassium bicarbonate), i.e. from 1.1($\pm$0.09) to 1.2($\pm$0.09) g d.m./L and from 1.008($\pm$0.06) to 1.11($\pm$0.06) g d.m./L, respectively. The lower sludge quantity resulted from the predominance of the processes of heterotrophic denitrification in the reactors with citric acid and the processes of autotrophic denitrification in the reactors with potassium bicarbonate. The mass of autotrophic denitrifiers and, consequently, the quantity of sludge formed as a result of biofilm exfoliation is lower than in the case of heterotrophic denitrifiers [Grady et al., 1999].

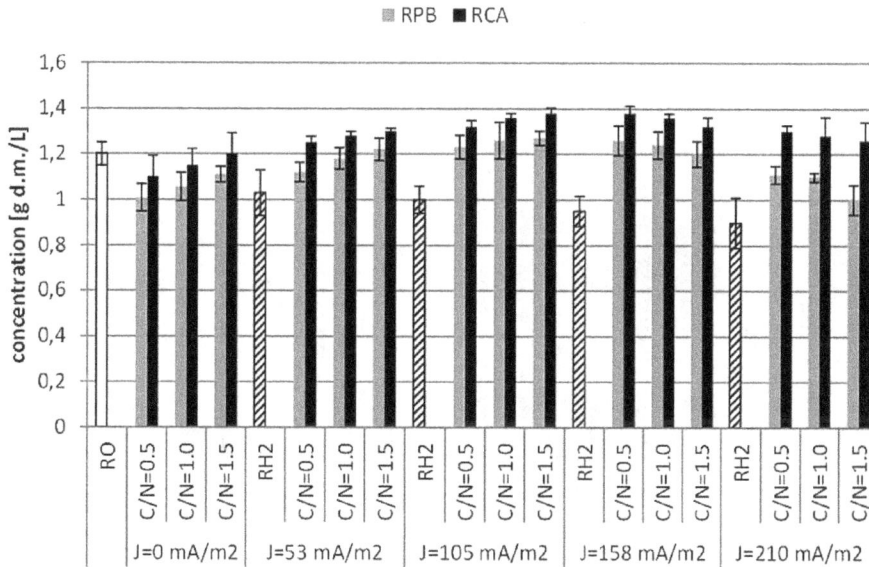

Fig. 2. The quantity of sludge formed in the reactors depending on electrical current density and C/N_{NO3} ratio

In the reactors with the passage of electrical current and with an external source of carbon (R_{CA+H2} and R_{PB+H2}), a tendency could be observed for increasing sludge formation along with the current density increase from 53 to 158 mA/m². The highest concentration of sludge (1.38(±0.03) g d.m./L) was determined in the R_{CA+H2} reactor at C/N_{NO3} of 0.5. At higher current density (210 mA/m²), the sludge concentrations were slightly lower and ranged from 1.26(±0.08) to 1.30(±0.03) g d.m./L. In the case of the reactors with potassium bicarbonate (R_{PB+H2}), the sludge concentrations were lower than in the reactors with citric acid – at all current densities and C/N_{NO3} values and fitted within the range from 1.00 (±0.07) to 1.26 (±0.06) g d.m./L. The concentration of sludge in the reactors with the electrical current passage and an external source of carbon was higher than in the R_{H2} reactor in which electrochemical processes (electrocoagulation and electrochemical reduction of nitrates) and autotrophic denitrification predominated when there was no external source of carbon [Rodziewicz, 2017]. The decreasing concentration of sludge along with the increasing current density observed in R_{H2} (from 1.03(±0.1) to 0.9(±0.11) g d.m./L) prove that the process of organic compounds oxidation to carbon dioxide was more intense at higher densities of the electrical current and at consequently higher temperatures of wastewater. Autotrophic denitrification was expected to predominate in this reactor [Rodziewicz, 2017].

The quantities of the formed sludge (below 1.3 kg/m³) determined in our experiment (fig. 2) were similar to those achieved by Akyol [2012], but many times lower than these reported by this author in the reactor with Al electrode (7.73 kg sludge/m³). They were also significantly lower than the values determined by Rodziewicz [2017] in her study on the treatment of wastewater originating from soilless crop cultivation in an electrobiological disk contactor (3.8 -5.25 kg/m³). This is mainly due to the differences in the quality of wastewater being treated and in the type of carbon source applied.

The lowest COD values (Fig. 3) were determined for the sludge from the control reactor R_0 (13.71(±1.2) mg/g d.m.). This was due to the lowest concentration of organic matter in the wastewater being treated.

The organic compounds present in the analyzed wastewater were consumed mainly by the biomass-forming organisms during the biofilm growth. Part of them was utilized by denitrifying heterotrophs. Feeding an additional source of organic carbon ($C_6H_8O_7$) and inorganic carbon ($KHCO_3$) to the reactor increased the content of organic compounds in the sludge. In the sludge from reactors R_{CA} and R_{PB}, the value of COD ranged from 22.17(±1.6) to 23.57(±1.56) mg/g d.m. and from 22.43(±0.1) to 29.41(±0.92) mg/g d.m., respectively, as a result of greater availability of organic compounds in the treated wastewater. In the R_{H2} reactor with the electrical current passage, the content of organic compounds in the sludge was affected by the electrical current den-

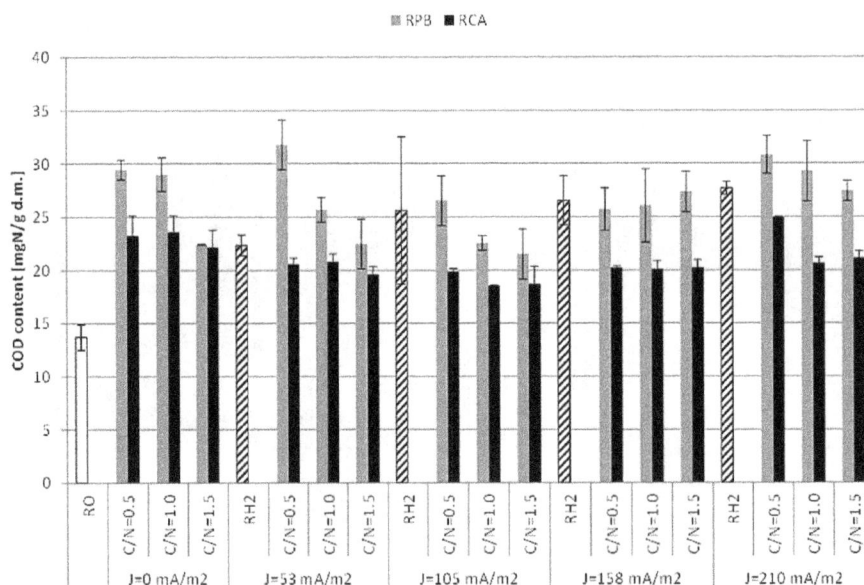

Fig. 3. Sludge COD values depending on electrical current density and C/N_{NO3} ratio

sity (22.33($\pm$0.96)-27.67($\pm$0.61) mg/g d.m.) and higher than in the control reactor R_0.

In the sludge formed in the reactors with electrical current passage and an external source of carbon (R_{CA+H2} and R_{PB+H2}), the COD values were higher in the reactors with potassium bicarbonate (R_{PB+H2}) than in the reactors with citric acid (R_{CA+H2}). This is an effect of a greater availability of carbon which was oxidized by microorganisms, consumed in the processes of heterotrophic or hydrogenotrophic denitrification or precipitated to sludge in the process of electrocoagulation depending on its form (organic, inorganic). This is also an effect of lower sludge concentrations in the reactors with potassium bicarbonate (R_{PB+H2}) than in the reactors with citric acid. The COD values determined in the experiment (Fig. 3) were much lower than these reported for municipal sewage sludge [Malej, 2000] and /or sludge from fish culture [Sikora, 2008].

The highest content of total nitrogen was determined in the sludge from the control reactor (Fig. 4). Considering the low concentration of organic matter in the wastewater being treated, nitrogen removal occurred mainly as a result of biomass growth. The total nitrogen content in the sludge from R_0 reached 12.1($\pm$0.7) mg N/g d.m..

In the reactors without the electrical current passage and external source of carbon, a decrease was noted in the nitrogen content along with the increasing C/N_{NO3} value, for both types of carbon sources. The nitrogen content was higher (range: 9.5($\pm$0.72) -12.0($\pm$0.22) mg N/g d.m.) in the sludge from the reactors with citric acid (R_{CA})

than in the sludge from the reactors with potassium bicarbonate (R_{PB}) where it decreased from 10.0($\pm$0.51) to 7.7($\pm$0.45) mg N/g d.m.

The lower contents of nitrogen in this sludge compared to the sludge from R_0 confirm that the autotrophic and heterotrophic denitrification did proceed in these reactors though with various intensity [Kłodowska et al., 2014].

In the reactor with the electrical current passage and an external source of carbon (R_{CA+H2} and R_{PB+H2}), the total nitrogen content in the sludge decreased along with the increasing C/N_{NO3} value at the current densities of 53 and 105 mA/m². At the higher densities (158 and 210 mA/m²), the total nitrogen content in the sludge was similar. Regardless of the current density and C/NO_3 value, the nitrogen content in the sludge from the reactors with citric acid was higher than in that from the reactors with potassium bicarbonate. In the case of the reactors with citric acid, the highest nitrogen content accounted for 8.2($\pm$0.53) mg N/g d.m. and the lowest one for 5.3($\pm$0.44) mg N/g d.m., whereas in the case of reactors with potassium bicarbonate, the respective values were 7.7($\pm$0.57) mg N/g d.m. and 3.6($\pm$0.31) mg N/g d.m. (fig. 4). These are significantly lower (two, three- and four-fold) from the values typical of the sludge from municipal wastewater treatment plants [Malej, 2000; Determining, 2004], the sludge from the treatment of wastewater from soilless crop cultivation [Rodziewicz, 2017], but two times higher than for the sludge from fish culture [Sikora, 2008]. This is an effect of highly efficient denitrification which proceeded in the

RPB RCA

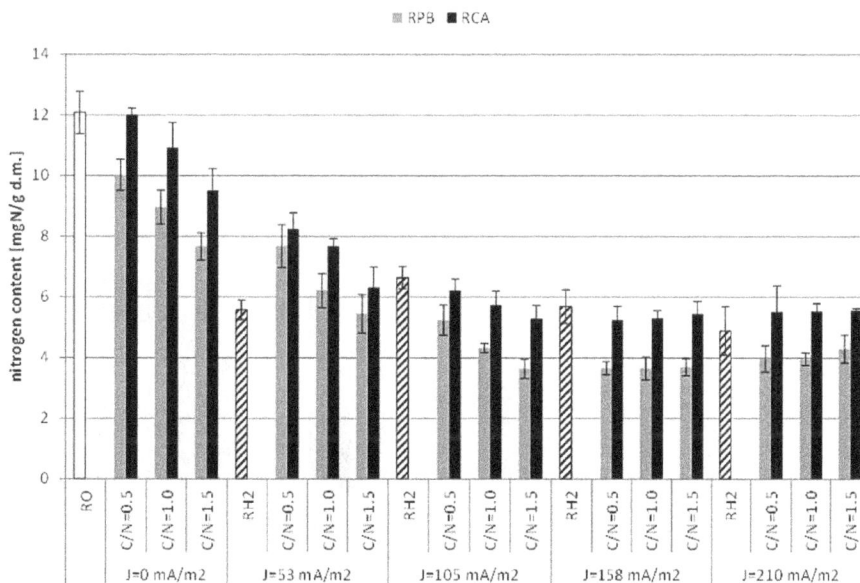

Fig. 4. Sludge nitrogen content depending on the electrical current density and the C/N_{NO3} ratio

reactors with the electrical current passage and of the specific character of the treated wastewater which contained only nitrate nitrogen.

In the case of phosphorus, its low content not exceeding 0.3($\pm$0.02) mg P/g d.m. (Fig. 5) was determined in the sludge from the control reactor.

The phosphorus removal in this reactor resulted exclusively from its incorporation into the biomass [Grady et al., 1999]. Introduction of carbon sources to the reactors contributed to a slight increase in the phosphorus concentration (0.4($\pm$0.05) mg P/g d.m.), as a result of the development of biomass with a different characteristics. In contrast, no significant differences were observed in the phosphorus concentrations in the sludge at different C/N_{NO3} values and different sources of carbon. A more intense phosphorus removal from the wastewater in the electrocoagulation process was already observed as a result of using aluminum electrode in the reactor with electrical current passage. The phosphorus content was higher in the sludge from the reactors with an inorganic source of carbon (R_{PB+H2}) than in the reactors with citric acid (R_{CA+H2}), probably because of the lower quality of sludge. For the current densities 158 and 210 mA/m^2 a tendency was observed for the P content increase in the sludge along with the C/N_{NO3} value increasing. Additionallz, the electrical current density increase contributed to an increasing content of phosphorus in the sludge formed during the wastewater treatment. This is a commonly known phenomenon described by many scientists. The effectiveness of the electrocoagulation process in

removing the phosphorus compounds from wastewater was reported in many works [Tchamango et al., 2010; Behbahani et al., 2013; Shalaby et al., 2014; Kuokkanen et al., 2016; Rodziewicz, 2017]. Kuokkanen et al. [2015], who used Al anode and Fe cathode, achieved a 79% efficiency of total phosphorus removal from the wastewater originating from the mining industry upon the use of electrical current with density of 100 A/m^2. Other authors obtained a 98.9% efficiency of synthetic wastewater treatment within 40 minutes [Ðuričić et al., 2016].

The phosphorus content determined in the sludge at the highest density of electrical current and the highest C/N_{NO3} value (1.67($\pm$0.06) mg P/g d.m. (0.17%), Fig. 5) in the reactor with potassium bicarbonate (R_{PB+H2}) is six-fold lower than the values reported by Rodziewicz (2017). In her study, the highest percentage content of phosphorus in sludge dry matter (reaching 0.84%) was obtained at the current density of 10.0 A/m^2 and HRT of 24 h. It is also six times lower than in the sludge from phosphorus precipitation with aluminum compounds and lower than the values noted for sludge formed during precipitation with lime [Oleszkiewicz, 1998]. The above-mentioned value is also lower than that determined in the sludge from trout culture [Sikora, 2008].

The sludge formed during the wastewater treatment in the electrobiological sequencing batch biofilm reactors was characterized by very high dewaterability. This was indicated by the determined values of capillary suction time (CST),

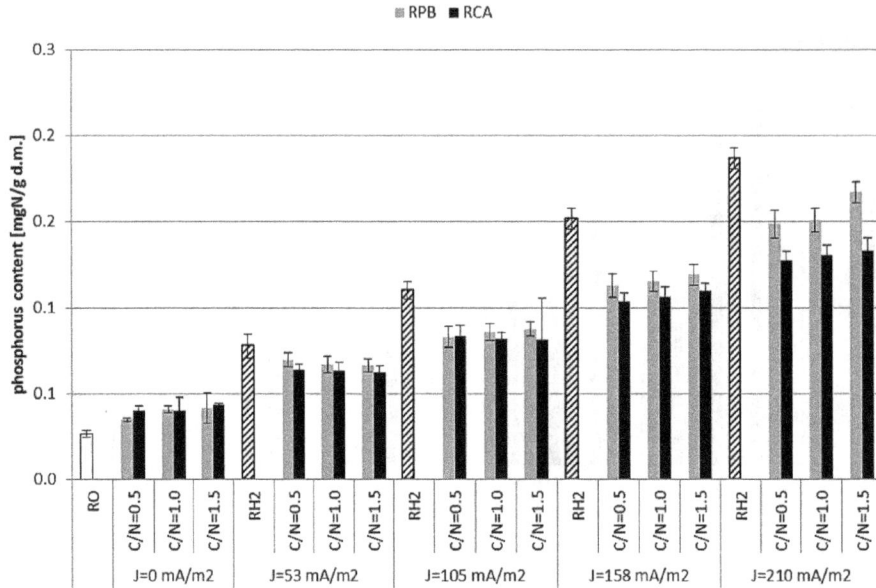

Fig. 5. Sludge phosphorus content depending on the electrical current density and the C/N_{NO3} ratio

which did not exceed 10 s for any of the analyzed sludge types (fig. 6). The dehydration rate of the sludge from the reactors with the electrical current passage and an external source of carbon (R_{CA+H2} and R_{PB+H2}) was higher than that of the sludge from the reactors with the additional carbon source (R_{CA} and R_{PB}). The longest dewatering time was reported for the sludge from the control reactor R_0 (9.89($\pm$0.54) s).

Worse dewaterability was observed in the case of the sludge from the reactors with citric acid being the source of organic carbon (with the highest CST value reaching 9.96($\pm$0.72) s), whereas a better one for sludge from the reactors with potassium bicarbonate being the source

of inorganic carbon (with the lowest CST value reaching 7.76($\pm$0.18) s).

Thus, the use of an inorganic substrate caused a shorter capillary suction time. Similar CST values, accounting for 8.7–9.2 s, were determined for the sludge from the chemical industry conditioned with the Fenton's reagent [Barbusuński and Filipek, 2000].

The susceptibility of the analyzed sewage sludge to dewatering was also demonstrated by other authors according to whom the CST values for crude municipal sludge were over 300 s [Dębowski et al., 2008]. These authors also showed the feasibility of shortening CST to ca. 50 s by sludge conditioning with the Fenton's

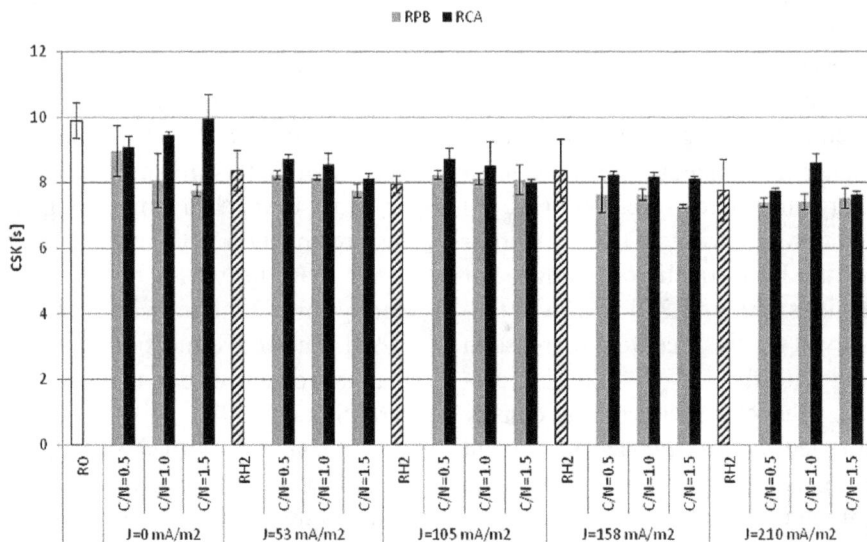

Fig. 6. Capillary suction time (CST) values depending on electrical current density and C/N_{NO3} ratio

method. Another work [Piotrowska-Cyplik and Czarnecki, 2005] reported the doses of polyelectrolytes which allowed for shortening the CST. With one of these polyelectrolytes, the capillary suction time was reduced to 14 s.

The results of our experiment prove that the sludge from the SBBR type bio-electrochemical reactor will not require conditioning and will dehydrate easily. Being rich in nutrients, it could be used for environmental purposes. It will pose no hazard to the soil nor to the aquatic environment due to its aluminum compounds, as evidenced earlier by Rodziewicz [2017] who demonstrated that at the electrical current density of 0.63 A/m^2 (i.e. higher than in our study), the concentration of aluminum (at HRT= 24 h) was at 0.3 mg/g d. m. sludge.

CONCLUSIONS

1. In the reactors with the electrical current passage and an external source of carbon, the volume of the produced sludge increased along with the current density increase from 53 to 158 mA/m^2. At its highest density (210 mA/m^2), the concentration of sludge was insignificantly lower.
2. For all densities of the electrical current and values of the C/N_{NO3} ratio, the concentrations of sludge formed in the reactors with potassium bicarbonate were lower than in the reactors with citric acid.
3. The content of organic matter (expressed as COD) was higher in the sludge from the reactors with the electrical current passage and potassium bicarbonate used as the external source of carbon, compared to the sludge from the reactors with citric acid.
4. In the reactors with the electrical current passage and an external source of carbon, the total nitrogen content in the sludge decreased along with the C/N_{NO3} ratio increase for the current densities of 53 and 105 mA/m^2. For higher current densities (158 and 210 mA/m^2), the total nitrogen content in the sludge was similar.
5. Irrespectively of the electrical current density, the nitrogen content in the sludge from the reactors with citric acid was higher than in the sludge from the reactors with potassium bicarbonate.
6. Irrespectively of the electrical current density, the phosphorus content in the sludge from

the reactors with citric acid was lower than in the sludge from the reactors with potassium bicarbonate.
7. For the current densities 158 and 210 mA/m^2, the increase in the C/N_{NO3} ratio caused an increase in the P content in the sludge.
8. The electrical current density increase contributed to increasing the content of phosphorus in the sludge formed during wastewater treatment.
9. The values of capillary suction time (CST) indicate that the sludge formed during the wastewater treatment in the electrobiological sequencing batch reactors with biofilm was characterized by high dewaterability.

Acknowledgments

This study was financed under Project No. 18.610.008–300 of the University of Warmia and Mazury in Olsztyn, Poland. The project was also funded by the National Science Centre, Poland (the decision nr DEC-2013/09/N/ST8/04163).

REFERENCES

1. Akyol A. 2012. Treatment of paint manufacturing wastewater by electrocoagulation. Desalination, 285, 91–99.
2. Attour A., Touati M., Tlili M., Ben Amor M., Lapicque F., Leclerc J.-P. 2014. Influence of operating parameters on phosphate removal from water by electrocoagulation using aluminum electrodes. Sep. Purif. Technol., 123, 124–129.
3. Barbusiński K., Filipek K., 2000. Aerobic Sludge Digestion in the Presence of Chemical Oxidizing Agents Part II. Fenton's Reagent. Pol. J. Environ. Stud., 9(3), 139–143.
4. Baran S., Turski R. 1999. Selected issues in the utilization and disposal of waste (in Polish). Wyd. Akademia Rolnicza, Lublin
5. Behbahani M., Moghaddam M.R.A., Arami M. 2013. Phosphate removal by electrocoagulation process: optimization by response surface methodology method. Environ. Eng. Manag. J., 12(12), 2397–2405.
6. Bień J., Stępniak L., Wolny L. 1995. Ultrasounds in water disinfection and preparation of sewage sludge before dehydration (in Polish). Seria Monografie Nr. 37, Częstochowa.
7. Determining the criteria for the use of sewage sludge outside agriculture (in Polish), 2004. Politechnika Częstochowska, Instytut Inżynierii

Środowiska, Częstochowa 2004

8. Dębowski M., Zieliński M., Krzemieniewski M. 2008. Efficiency of sewage sludge conditioning with the Fenton's method (in Polish). Ochr. Sr., 30(2), 43–47.

9. Đuričić T., Malinović B.N., Bijelić D. 2016. The phosphate removal efficiency electrocoagulation wastewater using iron and aluminum electrodes. Bulletin of the chemists and Technologists of Bosnia and Herzegovina. 47, 33–38.

10. Feng H., Huang B., Zou Y., Li N., Wang M., Yin J., Cong Y., Shen D. 2013. The effect of carbon sources on nitrogen removal performance in bio-electrochemical systems. Bioresource Technol., 128, 565–570.

11. Gharibi H., Sowlat M.H., Mahvi A., Keshavarz M., Safari M.H., Lotfi S., Abadi M.B., Alijanzadeh A. 2013. Performance evaluation of a bipolar electrolysis/electrocoagulation (EL/EC) reactor to enhance the sludge dewaterability. Chemosphere, 90(4), 1487–1494.

12. Grady C.P.L, Daigger G.T, Lim H.C. 1999. Biological Wastewater Treatment, Second Edition, Marcel Dekker, Inc. New York, Basel.

13. Janczukowicz W., Rodziewicz J. 2013. Carbon sources in the processes of biological removal of nitrogen and phosphorus compounds (in Polish). 114. Monografie Komitetu Inżynierii Środowiska PAN. Lublin.

14. Karanasios K.A., Vasiliadou I.A., Pavlou S., Vayenas D.V. 2010. Hydrogenotrophic denitrification of potable water: a review. J. Hazard. Mater., 180(1–3), 20–37.

15. Klaczyński E. 2013. Sewage treatment plant – chemical removal of phosphorus (in Polish). Wodociągi i Kanalizacja, 2(108), 26–28.

16. Kłodowska I., Rodziewicz J., Janczukowicz W. 2014. Removal of nitrogen compounds in the process of autotrophic denitrification in a Sequencing Batch Biofilm Reactor (SBBR). Pol. J. Nat. Sci., 29(4), 359–369.

17. Kłodowska I., Rodziewicz J., Janczukowicz W., Cydzik-Kwiatkowska A., Parszuto K. 2016. Effect of citric acid on the efficiency of the removal of nitrogen and phosphorus compounds during simultaneous heterotrophic-autotrophic denitrification (HAD) and electrocoagulation. Ecol. Eng., 95, 30–35.

18. Kuokkanen V., Kuokkanen T., Rämö J., Lassi U., Roininen J. 2015. Removal of phosphate from wastewaters for further utilization using electrocoagulation with hybrid electrodes – Techno-economic studies. J. Water Process Eng., 8, 50–57.

19. Kuokkanen V. 2016. Utilization of electrocoagulation for water and wastewater treatment and nutrient recovery, Acta Universitatis Ouluensis C, Technica, 562.

20. Malej J. 2000. Properties of sewage sludge and selected methods of their neutralisation, processing and utilization (in Polish), Rocznik Ochrona Środowiska, 2, 69–101.

21. Lacasa E., Caňizares P., Sáez C., Fernández F.J., Rodrigo M.A. 2011. Electrochemical phosphates removal using iron and aluminium electrodes. Chem. Eng. J., 172, 137–143.

22. Oleszkiewicz J. 1998. Sewage sludge management. Decider's guide. Kraków.

23. Piotrowska-Cyplik A., Czarnecki Z. 2005. Determination of the capillary suction time (CST) as a method for estimation of optimal dose of flocculants dewatering of municipal sewage sludge. J. Res. Appl. Agr. Eng., 50(1), 21–23.

24. Rodziewicz J., Krzemieniewski M. 2015. Patent application P.411116 – Sequential batch reactor with rotating biological contactor for wastewater treatment.

25. Rodziewicz J. 2017. Removal of nitrogen and phosphorus compounds from wastewater originating from soilless cultivation of plants in a rotating electrobiological contactor (in Polish). Rozprawy i monografie. 202. Wydawnictwo Uniwersytetu Warmińsko–Mazurskiego w Olsztynie.

26. Regulations of the Minister of Environment from 18th of November 2014 on conditions to be met for disposal of treated sewage into water and soil and concerning substances harmful to the environment (Dz.U. 2014. no. 1800), (in Polish).

27. Shalaby A., Nassef E., Mubark A., Hussein M. 2014. Phosphate removal from wastewater by electrocoagulation using aluminium electrodes. Am. J. Environ. Eng. Sci., 1(5), 90–98.

28. Sikora J. 2008. Analysis of the efficiency of conditioning and stabilization of sludge generated in fish farming under the influence of ultrasonic waves and Fenton reactions, Praca doktorska, UWM Olsztyn.

29. Tchamango S., Nanseu-Njiki C.P., Ngameni E., Hadjiev D., Darchen A. 2010. Treatment of dairy effluents by electrocoagulation using aluminium electrodes. Sci. Total Environ., 408(4), 947–952.

23

Ecological Evaluation of Sustainable Development in the Studied Farms of Przysucha County

Niewęgłowski Marek[1], Gugała Marek[1], Włodarczyk Bogusław[2], Anna Sikorska[3*]

[1] Siedlce University of Natural Sciences and Humanities, Faculty of Natural Science, ul. Prusa 14, 08-110 Siedlce, Poland

[2] Mazovian Agricultural Advisory Centre, ul. Czereśniowa 98, 02-456 Warszawa, Poland

[3] Department of Agriculture, The State Higher School of Vocational Education in Ciechanów, Narutowicza 9, 06-400 Ciechanów, Poland

* Corresponding autor's e-mail: anna.sikorska@pwszciechanow.edu.pl

ABSTRACT

The subject of the research involved the agricultural farms from the Przysucha county (Masovian Voivodeship, Poland). The assessment of ecological results from farms was the purpose of the thesis. Evaluation was made by using selected indicators: minerals balance, soil's organic substances balance and vegetation cover of soil's index. The research was carried out among 100 chosen agricultural farms, situated on light soil, i.e. rye soil. The ecological assessment of the examined farms showed that all of minerals balances (N, P, K) and soil's organic substances balances were positive. In the case of nitrogen, balances exceeded the limit value 30 kg N·ha[-1]. Vegetation cover of soil's index, as regards arable land, did not reach the recommended value, i.e. at least 60%. However, the cover of utilised agricultural area soil was similar to the recommended level (>70%). That was because of the large orchards and permanent crops share in horticultural farms, as well as large permanent grassland share in bovine and mixed farms.

Keywords: ecological assessment, result, agricultural farm

INTRODUCTION

The general concept of sustainable development covers all areas of human activity. The area of agriculture, which is featured by many connections with the natural environment [Fotyma, 2000] is particularly important. The rough definition of FAO, which is closer to the area of agriculture, recognizes sustainable development as management of natural resources, their protection and such a direction of technological and institutional changes which meets the needs of people now and in the future [Faber, 2007].

In practical terms, sustainable agriculture should simultaneously and harmoniously fulfill four main goals: production, economic, environmental and social ones [Fotyma, 2000]. The production goal is to produce the right amount of agricultural products (raw materials) with the qualities required by the consumer or the processing industry. The economic goal is to generate agricultural income that ensures a decent standard of living for the farmer and his family as well as enables the development of the farm. The ecological goal is to ensure the long-term balance of the agrosystem and to prevent the degradation of the natural environment. On the other hand, the social goal is defined quite generally as being reduced to the acceptance of non-agricultural part of the society for the actions of agricultural producers [Fotyma, 2000]. Thus, the essence of sustainable farming in agriculture can be defined as the aspiration to obtain both stable as well as economically and socially acceptable production, in a manner that does not threaten the natural environment.

The goals, tasks and ways of implementing sustainable development in Poland were defined in strategy programs [Strategy 1999; Strategy

2012], and a set of environment-friendly agricultural practices, the application of which will ensure sustainable development in the field of agricultural production, is included in the code of good agricultural practice [Duer ed al. 2002].

The issue of sustainable development of agriculture at the level of the region and the country as well as farms is the subject of investigation of research institutes and universities. In the studies carried out by research institutes at the farm level, the ecological (environmental) and economic criteria are mainly applied. The IAFE-PIB research uses the GUS (The Polish Central Statistical Office) data and the data from the FADN accounting system. Within the given criterion, individual research units apply slightly different sets of analytical indicators. The strength of the surveyed farms is very different, often depending on the implementation capacity and the sources of financing (FADN system, long-term programs, research projects, statutory topics). In this respect, IAFE – PIB is featured by the most comprehensive scope of research, using data from the Central Statistical Office (over 2,300,000 farms) and the FADN system (over 12,000 farms) [Wrzaszcz, 2012, Zegar, 2013]. In other scientific units studies, the number of farms is much smaller.

In the absence of a consistent approach to measuring sustainability, various criteria for assessing sustainable development are adopted. Three criteria (dimensions, ranges) of the assessment are most often taken into account: economic, ecological and social [Kotosz, 2012; Häni, 2004; Faber, 2007; Majewski, 2008; Faber et al.2010; Baum, 2011; Sadowski, 2012; Harasim 2014]. According to Majewski [2008, 2009], a sustainable and permanent development is featured by actions that should be economically viable, ecologically safe and socially acceptable. The ecological criterion is also defined as environmental or agri-environmental [Faber, 2007, Toczyński et al. 2009, Wrzaszcz, 2011, 2012].

Moreover, there are proposals for a wider scope of assessment by adding other dimensions: institutional [Piontek, 2002, Adamowicz and Dresler, 2006, Florczak, 2008], spatial [Borys, 1998, Piontek, 2002, Adamowic and Dresler, 2006], and cultural [Bombik and Marciniuk-Kluska, 2010], moral [Piontek, 2002, Adamowicz and Dresler, 2006] and ethical [Runowski, 2007, Siemiński, 2011]. There are proposals of assessing the sustainable and permanent development taking into account six criteria, i.e. economic, ecological, spatial, technical and technological, socio-cultural and ethical [Siemiński, 2011], or in the assessment of administrative units (communes) divided into five ranges: economic, ecological, social, institutional and spatial [Adamowicz and Dresler, 2006]. According to Majewski [2008, 2009], five criteria of assessment: economic, ecological, social, organization of production and management, and quality of the production space, can be adopted as the basis for constructing the synthetic indicator of durability of the farm. In the opinion of Zegar [2007], the diversity of agriculture, i.e. natural conditions, economic entities, production technology and other circumstances, creates a difficulty in establishing uniform criteria for assessing the sustainability of agricultural holdings.

The basic elements of the ecological assessment of farms provide the balance of nutrients and soil organic matter, as well as the degree of soil cover with vegetation [Harasim and Włodarczyk, 2016]. In the case of crop production, the management of mineral components is very important, because its proper process determines the satisfactory crop yield, positively affects the change of soil fertility and reduces the soil and groundwater contamination with biogenes. The management of mineral components (nitrogen, phosphorus and potassium) and organic matter should be based on balance sheets, which take into account the revenues of components from all the sources and their outflows along with crops harvested from the field.

The nitrogen balance is generally upset and difficult to maintain at a certain level, because on farms there may be losses that are difficult to predict. Its gaseous forms can oxidize and infiltrate into the atmosphere, and nitrate forms can be washed out to deeper layers of soil and groundwater [Sainju, 2017]. According to the Code of Good Agricultural Practice [Duer et al., 2002], the environmentally safe balance of nitrogen should not exceed 30 kg N/ha UAA. In the case of phosphorus and potassium, their balances should be sought for (revenue = use). On the soils with very low phosphorus and potassium content, it is recommended to use higher doses of fertilizers (by about 50%) in relation to the absorption of these components in the yield of plants [Jadczyszyn, 2005]. However, on soils with very high abundance of these nutrients, their doses in fertilizers can be reduced by 50% in relation to the absorption along with the crops yield.

RESEARCH METHODOLOGY

The source material consists of the research carried out in Przysucha province, located in the southern part of Masovia Province in Poland. One hundred farms located on light soils, i.e. rye complexes were examined. The information and the source data from farms were obtained during a direct interview with the use of a questionnaire. A purposeful selection of research objects from farms cooperating with the Mazovian Agricultural Advisory Center was used. The division of the farms under examination into groups was carried out within particular evaluation criteria, such as the direction of production, the size of the farm area, the quality of the soils of agricultural land and the intensity of production.

When developing the material, a tabular-descriptive method with elements of horizontal analysis was used.

RESEARCH RESULTS

The assessment of sustainable development of farms can be made on the basis of various criteria, e.g.: production, economic, ecological. For the purpose of this study, the assessment of ecological conformity of agricultural practices to the principles of sustainable development of farms was made on the basis of quantitative indicators. The balance of minerals at the field level was evaluated (balance N, P, K in $kg \cdot ha^{-1}$ UAA) and organic matter in the soil ($t \cdot ha^{-1}$ AL) and soil cover with vegetation (% AF and UAA) (Table 1).

In the ecological assessment, the mineral balance at the field level was calculated taking into account as the incomings of the components used in mineral and natural fertilizers including straw for plowing, and in the case of nitrogen the extra amount of this component from atmospheric precipitation (10 kg N·year). On the outflows side, there were certain quantities of components extracted from the soil with main and side yields. The content of minerals (NPK) in the main and side yields as well as in the manure and straw (plowed) was adopted as the standard ones from the literature [Krusze, 1984, Fotyma and Mercik, 1985, Maćkowiak and Żebrowski, 2000, Gorlach and Mazur, 2001]. The soil balance of the organic matter was calculated with the use of the reproduction and degradation rates of humus [Duer et al., 2002]. The indices of arable land cover

with vegetation in winter and cover of agricultural land with vegetation during the year were calculated in accordance with the methodology included in Harasim's study [2004]. The study also assessed the intensity of the organization of agricultural production according to the Kopec's method [1987]. The above-mentioned indicators, due to the availability of source data, are most often included in the works pertaining to the problem of sustainable development of farms [Fotyma and Kuś 2000; Faber 2001; Kuś and Krasowicz, 2001; Kopiński, 2002; Krasowicz, 2005, 2006; Kuś, 2006; Harasim and Włodarczyk, 2007].

In the analysis of selected ecological indicators, the amount of the soil balance of the organic substance (t d.m·ha AL) and the index of soil cover with vegetation (%) were very important. The balance of organic matter in soil determines the fertility and productivity of soils. If the balance was negative in the long term, the soil could degrade. The index of soil covering with vegetation significantly affects the impact of an agricultural holding on the environment [Kuś and Krasowicz, 2001]. Higher values of this index indicate a lower risk of nitrate leaching and better soil protection against erosion. The essence of the correct crop structure on arable land is to run such an economy to have as large area of "green fields" in the winter time as possible. From the point of view of the principles of good agricultural practice in flat areas, the cover of soil with vegetation should reach at least 60% of arable land [Duer et al., 2002].

Table 2 presents the data connected to the balances of the basic minerals and soil organic substance as well as the soil coverage index. In regard to the nitrogen balance, it can be concluded that all types of farms exceeded the safe value of the balance of this component (30 kg N·ha⁻¹). The largest exceeding of the recommended bal-

Table 1. Indicators for assessing the ecological conformity of agricultural practices with the principles of sustainable development

Quantitative indicators	Hazard values
Balance:	
- nitrogen (N)	< 30 kg N·ha⁻¹ UAA > 0
- phosphorus (P_2O_5)	balance ≥ 0
- potassium (K_2O)	balance ≥ 0
- organic matter in the soil	balance ≥ 0
Soil cover by plants:	
- arable land	> 60% area AL
- utilised agricultural area	> 70% area UAA

Table 2. Ecological indicators featuring particular types of farms.

Specyfication	Farms in total	Direction of production			
		orchards	vegetables	mixed	cattle
Balances of mineral components (inflow-outflow), (kg/ha of UAA):					
- N	49.6	62.4	53.5	46.9	39.0
- P_2O_5	49.6	35.1	54.8	48.7	56.7
- K_2O	66.7	78.8	88.0	57.2	51.2
Amount of soil organic substance balance (t d.m.·ha⁻¹ AL)	0.34	0.47	0.33	0.15	0.37
Index of soil cover with vegetation (%):					
- arable land	42.6	36.3	33.7	51.4	48.5
- utilised agricultural area	67.5	79.7	53.8	69.7	67.3

ance by 108% was recorded in the fruit farms, followed by the vegetable farms with a surplus of 78% and next the mixed (56%) and the cattle ones (30%). Nitrogen surplus may affect the pollution of deeper layers of soil and groundwater with nitrates. In cattle farms, exceeding the recommended nitrogen balance was the lowest. In this group of farms, fertilization was based largely on its own natural fertilizers and supplemented in mineral form. In contrast, the other types of farms, especially those with horticultural production, were mostly based on mineral fertilizers and did not conduct a rational management of this component. In the case of phosphorus and potassium, it is recommended that their balances on soils with medium abundance in these components should be balanced (income = expenses) [Harasim and Włodarczyk 2016].

The lowest phosphorus balance occurred in the fruit farms (Table 2). In the other farms, too high surpluses of this component have no rational justification. In the case of potassium balance, they were higher than with phosphorus. The largest balance of potassium was found in the vegetable farms (88.0 kg K_2O·ha⁻¹) and the orchards (78.8 kg K_2O·ha⁻¹). In the other types of farms, i.e. the mixed and cattle ones, the potassium balance was significantly lower (Table 2). Such significant balances of all components prove that the farmers did not examine the soil and did not use fertilization on the basis of analysis results and recommendations of the advisory services, and the fertilizer doses resulted rather from their own habit of using them with individual crops (higher fertilization in the horticultural production, and lower in the other crops). In this case, the principle of rational fertilization was not respected, and the use of excessive fertilization may have had a negative impact on the natural environment and negatively affected the efficiency of crop production.

The balance of organic matter determines the fertility and productivity of soils (Duer et al. 2002). Out of the 4 types of farms considered, the fruit farms were marked by the largest amount of the soil organic substance balance (Table 2), which results from the use of a large amount of natural fertilizers on the fields for new plantings of fruit trees. The lowest balance of organic matter was found in the mixed farms, featured by a high share of cereals in the crop structure (82.6%), which significantly contribute to the degradation of organic matter in the soil.

The index of soil cover with vegetation in winter, so-called "green fields" is also a very important indicator of ecological assessment [Harasim, 2004]. As regards AL, the highest ratio was found in the mixed farms (51.4%) and the lowest in the vegetable ones (33.7%) with the lowest share of cereals in crops (63.3%). However, no type of farms had a 60% level of covering arable land with green plants during the winter, which could expose soil to erosion and nitrate leaching to groundwater. With regard to the area of arable land, only the group of the vegetable farms with the index of 53.8% did not exceed the safe level of 60% of the level of soil cover with green plants (Table 2). However, the largest index (79.7%) characterized fruit farms. The level of this index in the mixed and cattle farms was affected by the share of permanent grassland in the structure of utilized agricultural area. This share for the mixed farms was 25.2%, and for the cattle ones – 34.8%.

The ecological indicators did not show a clear dependence on the area of farms (Table 3). In the group of farms with the area of 7–15 ha UAA, the highest values were achieved by the balance of nitrogen and phosphorus. Only the phosphorus balance increased along with the growing area of the farm. The amount of the soil organic matter balance was the highest in the large farms with the area > 15 ha of UAA, which is related to their livestock production. They had their own manure, but their crop structure was more varied.

The index of soil cover with plants on arable land reached the highest values in small (<7 ha) and large (> 15 ha) farms (Table 3). In small farms, the higher index comparing to the medium farms can be explained by the cultivation of green plants for plowing in the spring to enrich the fields with organic matter for new planting of fruit trees or by spring vegetables production, which partially replaces the manure fertilization. In the farms over 15 ha, the higher index indicates a high share of winter cereals in the crop structure. However, no type of farms achieved a safe 60% for the rate of soil cover with vegetation. The correlation of this index on the utilized agricultural area was similar. Farms with the area of <7 ha were marked by the highest index (71.2%), which is connected with fruit production. In medium and large farms (7–15 and> 15 ha of UAA), permanent grassland appeared, being the main feed base for the cattle kept on the cattle and mixed farms.

In the assessment of the ecological indicators, depending on the quality of utilized agricultural area, the largest balances of basic nutrients (nitrogen, phosphorus and potassium) were recorded in farms with very poor soils (Table 4). The use of higher fertilization on very poor soils has a logical explanation, because farmers wanting to obtain satisfactory yields use a higher level of fertilization. However, it should be remembered that very poor (light) soils are particularly susceptible to leaching the excess components into ground and surface waters. Therefore, mineral fertilization on these soils should be under special control, so as not to expose the natural environment to pollution with these components [Harasim and Włodarczyk, 2007].

The amount of soil organic substance balance was higher in the farms with very poor and medium soils. The index of soil cover with vegetation on arable land was the highest in the farms with very poor soils (52.6%), which is associated with a large share of winter cereals in the crop structure. The lowest index of soil coverage (28.7%) occurred on medium soils which were used more often for fruit and vegetable cultures. The index of coverage of the utilized agricultural area by vegetation did not show a clear dependence on its quality.

Table 3. Ecological indicators depending on the area of the farm

Specification	Farms in total	Size of a farm (area of UAA in ha)		
		<7	7–15	>15
Balances of mineral components (inflow-outflow), (kg/ha of UAA)				
- N	49.6	49.1	54.4	40.5
- P_2O_5	49.6	44.6	47.5	52.9
- K_2O	66.7	66.6	76.5	54.8
Amount of soil organic substance balance (t d.m·ha^{-1} AL)	0.34	0.37	0.27	0.54
Index of soil cover with vegetation (%):				
- arable land	42.6	46.0	37.1	47.0
- utilised agricultural area	67.5	71.2	65.2	66.2

Table 4. Ecological indicators depending on the quality of utilized agricultural area

Specification	Farms in total	Soil quality (UAA valuation indicator)		
		very poor (<0.5 points)	poor (0.5–0.7 points)	average (>0.7 points)
Balances of mineral components (inflow-outflow), (kg/ha of UAA)				
- N	49.6	56.2	46.0	50.8
- P_2O_5	49.6	58.3	48.2	43.0
- K_2O	66.7	76.8	63.0	68.9
Amount of soil organic substance balance (t d.m.·ha^{-1} AL)	0.34	0.44	0.31	0.42
Index of soil cover with vegetation (%):				
- arable land	42.6	52.6	43.9	28.7
- utilised agricultural area	67.5	66.7	67.4	68.8

Table 5. Ecological indicators depending on the intensity of production (direct costs)

Specification	Farms in total	Intensity of production (direct costs in PLN/ha UAA)		
		extensive (<600 PLN)	mid-intensive (600–1200 PLN)	intensive (>1200 PLN)
Balances of mineral components (inflow-outflow), (kg/ha of UAA)				
- N	49.6	45.2	42.5	60.0
- P_2O_5	49.6	51.3	44.1	57.8
- K_2O	66.7	58.4	64.3	79.1
Amount of soil organic substance balance (t d.m.·ha⁻¹ AL)	0.34	0.31	0.30	0.47
Index of soil cover with vegetation (%):				
- arable land	42.6	49.6	47.7	31.8
- utilised agricultural area	67.5	69.0	68.9	65.2

Table 5 presents the ecological indicators depending on the intensity of production measured by the level of incurred direct costs. The highest balances of minerals (NPK) occurred under the conditions of intensive production, which features the profile of fruit and vegetable farms. Lower balances of nitrogen and phosphorus appeared in the medium-intensive farms, and in the case of potassium – in the extensive ones.

The amount of soil organic substance balance was the highest in the intensive farms (Table 5). The index of soil cover with vegetation on arable land was more dependent on the intensity of production than the index of utilized agricultural area cover. In the first case, the least favorable soil cover with vegetation was found in farms with intensive production with a small share of winter cereals in crops. This concerns mainly the farms specializing in the horticultural and vegetable production. The level of the index referring to utilized agricultural area was quite similar in the groups of farms with varying intensity of production.

CONCLUSIONS

1. To sum up the ecological assessment, it can be concluded that the balances of all mineral components (NPK) and soil organic matter were positive. In the case of nitrogen, the balance exceeded the limit value of 30 kg N·ha⁻¹. Under the conditions of very light soils and with a higher level of production intensity, the balances of mineral components reached the highest values.
2. The principle of rational fertilization in the surveyed farms was not respected, and the use of excessive fertilization may have had a negative impact on the natural environment and negatively affected the efficiency of crop production.
3. The index of soil cover with vegetation in relation to arable land did not reach the recom-

mended value, i.e. at least 60%. In the case of utilized agricultural area the index of soil cover with vegetation was close to the recommended level (> 70%), which was affected by a large share of orchards and permanent plantations in the fruit farms and permanent grassland in the cattle and mixed farms.

Acknowledgements

The results of the research carried out under the research theme No. 363/S/13 were financed from the science grant granted by the Ministry of Science and Higher Education.

REFERENCES

1. Adamowicz M., Dresler E. 2006. Sustainable development of rural areas based on selected municipalities of the Lublin province. Zesz. Nauk. AR Wroc., Roln., 540(87), 17–24. (in Polish)
2. Baum R. 2011. Evaluation of sustainable development in agriculture (methodical study). Rozpr. Nauk. UP Poznań, 434. (in Polish)
3. Bombik A., Marciniuk-Kluska A. 2010. Indicators in Sustainable Rural Areas Development Modelling. Acta Sci. Pol., Oecon., 9(1), 29–37. (in Polish)
4. Borys T. 1998. Theoretical aspects of constructing eco-development indicators (W:) Ecodevelopment control.Red. B. Poskrobko. Wyd. Politechniki Białostockiej, Białystok. (in Polish)
5. Duer I., Fotyma M., Madej A. (red.) 2002, Kodeks dobrej praktyki rolniczej. MRiRW – MŚ – FAPA Warszawa, 56. (in Polish)
6. Faber A. 2001. Indicators proposed for research on the balance of agricultural development. Fragm. Agron., 1, 31–44. (in Polish)
7. Faber A. 2007. Review of agri-environmental indicators recommended for use in the assessment of sustainable farming in agriculture. Studia i Raporty IUNG-PIB, 5, 9–24. (in Polish)

8. Faber A. 2001. Indicators proposed for research on the balance of agricultural development. Fragm. Agron., 1, 31–44. (in Polish)

9. Faber A., Pudełko R., Filipiak K., Borzęcka-Walker M., Borek., Jadczyszyn J., Kozyra J., Mizak K., Świtaj Ł. 2010. Assessment of the degree of sustainability of agriculture in Poland in various spatial scales. Studia i Raporty IUNG-PIB, 2010, 20, 9–27. (in Polish)

10. Florczak W. 2008. Sustainable development indicators. Wiad. Statyst., 3, 14–34. (in Polish)

11. Fotyma M. 2000. Problems of sustainable agriculture in the light of IUNG conference in Puławy in June 2000. Biul. Inf. IUNG, 14, 3–8. (in Polish)

12. Fotyma M., Kuś J. 2000. Sustainable development of a farm. Pam. Puł., 120/l, 101–116. (in Polish)

13. Fotyma M., Mercik S. 1995. Agricultural chemistry. Wyd. Nauk. PWN Warszawa. (in Polish)

14. Gorlach E., Mazur S. 2001. Agricultural chemistry. Wyd. Nauk. PWN Warszawa. (in Polish)

15. Harasim A. 2014. Guide to assessing the sustainability of agriculture at different levels of management. IUNG-PIB Puławy. (in Polish)

16. Harasim A. 2004. Indicators of the soil-protecting action of plants. Post. Nauk Rol., 4, 33–43. (in Polish)

17. Harasim A., Włodarczyk B. 2016. Assessment of the sustainability of different types of farms on light soils. Roczniki Naukowe SERiA, t. XVIII, 2, 112 (in Polish)

18. Harasim A., Włodarczyk B. 2007. Possibilities of sustainable development of farms with different production directions on light soils. Rocz. Nauk. SERiA, 9(1), 167–171. (in Polish)

19. Häni F. 2004. Holistic sustainability assessment at the farm level. http://old.shl.bfh.ch/fed/docs/Subotica.pdf

20. Jadczyszyn T. 2005. Determining the doses of fertilizers. Wieś Jutra, 6, 28–29. (in Polish)

21. Kopiński J. 2002. Comparison of indicators for the development of sustainable farms with different intensity of agricultural production. Rocz. Nauk Rol., G, 89(2), 66–72. (in Polish)

22. Kotosz B. 2012. Measuring sustainable development at macro level. (W:) Global Commodity Markets: New Challenges and the Role of Policy. International Scientific Days 2012, Slovak University of Agriculture in Nitra, 707–712. https://spu.fem.uniag.sk/mvd2014/proceedings/ISD_2012.pdf

23. Krasowicz S. 2005. Evaluation of the possibility of sustainable development of farms with different production directions. Rocz. Nauk. SERiA, 7(1), 144–149. (in Polish)

24. Krasowicz S. 2006. Ways of implementing the idea of sustainable development in an agricultural holding. Zesz. Nauk. AR Wroc. Roln., 540(87), 255–261. (in Polish)

25. Krusze N. (Eds.). 1984. Gardening in tables. PWRiL Warszawa. (in Polish)

26. Kuś J. 2006. Possibilities of sustainable development of specialized farms. Probl. Inż. Rol., 2, 5–14. (in Polish)

27. Kuś J., Krasowicz S. 2001. Natural and organizational conditions for the sustainable development of farms. Pam. Puł., 124, 273–288. (in Polish)

28. Maćkowiak Cz., Żebrowski J. 2000. Chemical composition of manure in Poland. Nawozy i Nawożenie, 4, 119–130. (in Polish)

29. Majewski E. 2009. Economic and ecological sustainability of a farm. Rocz. Nauk. Rol., G, 96(3), 140–151. (in Polish)

30. Majewski E. 2008. Sustainable development and sustainable agriculture – theory and the practice of farms. SGGW Warszawa. (in Polish)

31. Piontek B. 2002. The concept of sustainable and durable Poland development. Wyd. Nauk. PWN Warszawa. (in Polish)

32. Runowski H. 2007. Searching for economic, ecological and ethical balance in milk production. Rocz. Nauk Rol., G, 93(2), 13–26. (in Polish)

33. Sadowski A. 2012. Sustainable development of farms taking into account the impact of the Common Agricultural Policy of the European Union. Rozpr. Nauk. UP Poznań, 447. (in Polish)

34. Sainju U.M. 2017. Determination of nitrogen balance in agroecosystems. Elsevier, MethodsX 4, 199–208. https://doi.org/10.1016/j.mex.2017.06.001

35. Siemiński J.L. 2011. The concept of "Sustainable development" – development of sustainable sustainable rural areas in Poland (opportunities and opportunities).IUNG-PIB Puławy, 1–17. (in Polish)

36. Strategy for the sustainable development of Poland until 2025. 1999. Ministry of Environmental Protection. http://www.mos.gov.pl/1materiałyinformacyjne/raporty_opracowania/strategia/index1.html (in Polish)

37. Strategy for sustainable development of rural areas, agriculture and fisheries for 2012–2020 (2012). Uchwała Rady Ministrów nr 163 z 25 kwietnia 2012 r. MP z 9 listopada 2012 r., poz. 839. (in Polish)

38. Toczyński T., Wrzaszcz W., Zegar J. S. 2009. From research on socially sustainable agriculture (8). Sustainability of Polish agriculture in the light of public statistics data. IERiGŻ-PIB Warszawa, 161. (in Polish)

39. Wrzaszcz W. 2011. Level of environmental sustainability of individual farms in Poland (based on FADN data). Rocz. Nauk. SERiA, 13(5), 70–75. (in Polish)

40. Wrzaszcz W. 2012. Level of environmental sustainability of individual farms in Poland (based on FADN data). Studia i Monogr., IERiGŻ-PIB Warszawa, 155. (in Polish)

41. Zegar J.S. 2007. Social aspects of sustainable agricultural development. Fragm. Agron., 4, 282–298. (in Polish)

42. Zegar J.S. (Eds.). 2013. Sustainability of Polish agriculture. Powszechny Spis Rolny 2010. GUS Warszawa. (in Polish)

24

Analysis of Heat Loss of a Biogas Anaerobic Digester in Weather Conditions in Poland

Tomasz Janusz Teleszewski[1*], Mirosław Żukowski[1]

[1] Department of HVAC Engineering, Faculty of Civil and Environmental Engineering, Bialystok University of Technology, Wiejska 45E, 15-351 Bialystok, Poland
[*] Corresponding author's e-mail: t.teleszewski@pb.edu.pl

ABSTRACT

Currently in Poland, the construction of biogas plants as alternative energy sources is increasing. Often, the technical solutions for design and building of biogas plants are transferred to Poland from the countries in which these technologies are developed without taking into account the specific climatic conditions prevailing in Poland. It does occur that newly built biogas plants have a problem maintaining a sufficiently high temperature in the winter, which is caused by the insufficient insulation of the biogas anaerobic digester envelope. This paper presents an analysis of heat loss, depending on the climatic conditions prevailing in Poland and the working conditions of a biogas plant, based on an existing facility located in Ryboly (Poland). The work is supplemented with the results of tests using a thermal imaging camera. It should be noted that currently there are no requirements in the literature regarding the design of a thermal insulating layer in biogas installations in Poland.

Keywords: biogas anaerobic digester, heat demand, heat losses

INTRODUCTION

Biogas plants are one of the most extensively developing and promising renewable energy sources in Poland. This green technology can also limit the environmental nuisance resulting from agricultural activities (Białowiec et al. 2015, Sadecka 2016) and wastewater treatment (Lebiocka 2010, Kogut et al. 2012, Montusiewicz 2014, Pilarska 2018). The use of the waste as a substrate makes a biogas plant a pro-ecological alternative energy source that contributes to the improvement of the quality of the environment, and this technology can be considered as an integral part of the environmental protection.

Biogas plants can also reduce the environmental nuisance resulting from the agricultural activities. The main advantages of this technical solution include:
- The use of farmyard manure, farm slurry, agricultural and municipal waste, sewage, crops, and many kinds of plant materials as a fuel.
- Heat and electricity cogeneration with high efficiency.
- Improvement of the sanitary conditions of the environment.
- Increase of the energy security at the local level by reducing the consumption of fossil fuels.
- Independence of energy production from the atmospheric conditions and seasons.
- Reduction of the greenhouse gas emissions.
- Aid in recycling of phosphorus and nitrogen.
- Reduction of unemployment in rural areas.

The biogas plants also have certain disadvantages:
- Relatively high investment costs (a medium-sized biogas plant costs about 3 million euros).
- Introduction of monocultures in crops.
- Competition for food production.
- Common problems with finding customers for the generated heat energy.
- Lack of clear regulation and a large bureaucracy in Poland.

According to the Polish Energy Policy until 2030, an average of one biogas plant in each municipality will be constructed by 2020. Therefore, the issues related to the production of biogas will become increasingly widespread. These types of energy sources operate in two temperature ranges: 25–37°C (mesophilic fermentation) and 45–55°C (thermophilic fermentation). In both cases, the temperature inside the bioreactor chamber (anaerobic digester) is above the ambient temperature. Thus, there is always a heat loss to the outside environment.

The thermal performance of the tubular digester located on the Agronomy Campus in Cusco (Peru) was developed by Perrigault et al. (2012). They created a one-dimensional heat transfer (radiative, convective, and conductive) model that included the following parameters: the geometry of the modelled object, solar radiation, ambient temperature, and wind velocity, while the results of the calculations were the temperature of slurry, biogas, walls, cover, holding membrane, and the greenhouse air. The correctness of the model was verified positively by comparison with the measurement results. According to the authors, their model may be useful in predicting the influence of the geometry and materials on the thermal performance of the anaerobic digester.

Biogas plants have enjoyed great popularity in China. In 2010, there were over 30 000 average-sized and large installations. A typical stirred bioreactor treating pig manure was tested by Guo et al. (2013). The optimum fermentation temperature and organic loading rates were the main parameters that could be determined using a mathematical model created by the above-mentioned team of scientists. The results obtained from the calculations agreed with the experimental data for an ambient temperature only greater than 20°C. The authors suggested that the maximum energy production can be achieved when the digesters run at organic dry matter of 4.6–5.4 kg/m³ and the temperature of mesophilic fermentation is around 26°C for an external temperature around 10–20°C.

Shaheen and Nene (2014) carried out an analysis of the heat transfer from the slurry and the gas dome to the external environment. The biogas plant was located at the Maharashtra Institute of Technology (India). The model of this object was created using Matlab software. The results of the thermal simulations can be used for optimal control of the biogas plant operation.

The energy simulation of an anaerobic digester buried in the ground was done by Terradas-III et al. (2014) in order to obtain the production of biogas. A one-dimensional heat transfer model was applied to obtain the slurry and biogas temperature. As shown by the results of the calculations, the proposed model allowed accurate prediction of the amount of biogas production and the temperature inside the digester chamber.

Hreiz et al. (2017) developed a computer tool that can be useful for engineers in the design of semi-buried anaerobic digesters. The model proposed by the authors was based on the heat transfer balance and allowed calculation of the temperature in the fermenter as a function of the following parameters: ambient temperature, solar irradiation, wind speed, and intensity of rainfall. As it turned out, the results of the operating temperature simulations were in good agreement with the experimental measurements. A significant achievement of this work was showing the major sources of heat losses from the digester chamber.

As it turned out, there is no information on the heat balance of biogas plants under the Polish climatic conditions in the literature. In most cases, such investments implemented in Poland are based on ready-made projects of biogas plants drawn up for conditions prevailing in Germany. Therefore, the authors of the current paper decided to pursue this subject, i.e. the estimation of the heat losses of the digester under the Polish climate conditions.

DESCRIPTION OF THE RESEARCH OBJECT UNDER INVESTIGATION

The biogas plant subjected to analysis, shown in Figure 1a, is located near the village of Ryboly (Poland). The research object consists of two cylinder-shaped anaerobic digesters. They have the same dimensions, i.e. diameter of 30 m and height of 6 m. The fermentation chambers are buried 1 m into the ground. The cylindrical outer wall and the bottom are made of 25 cm-thick reinforced concrete. They were insulated using extruded polystyrene with a thickness of 10 cm. The chambers are covered with a flexible dome in a shape similar to a cone. The four heating loops connected in parallel, made of steel pipes with a diameter of 60 mm, are attached to the walls of the tank in its lower part. Due to the fact that both

a)

b)

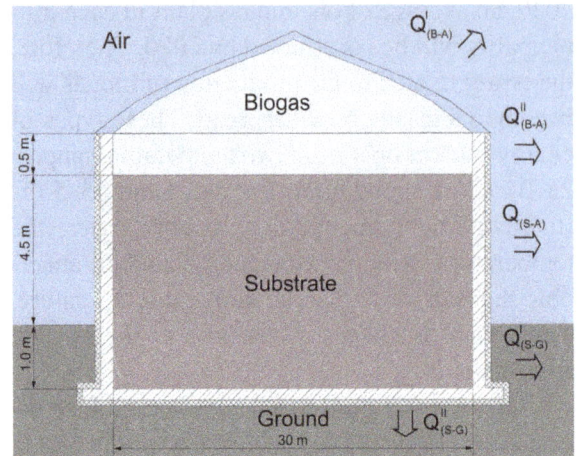

Figure 1. Ryboly Biogas Plant: a) fermentation chambers and the storage tank, b) physical model of biogas digester (not to scale)

fermentation chambers have identical dimensions, further computational analysis was carried out for a single digester.

ANALYSIS OF HEAT LOSS FROM A BIOGAS ANAEROBIC DIGESTER

The computational analysis presented in this article considers the anaerobic digester shown in Figure 1b. The total heat losses Q_T through the digester presents the following sum of the partial heat losses (Fig. 1b):

$$Q_T = Q^I_{(B-A)} + Q^{II}_{(B-A)} + Q_{(S-A)} + \\ + Q^I_{(S-G)} + Q^{II}_{(S-G)}, \; [\text{W}] \quad (1)$$

where: $Q^I_{(B-A)}$ are the heat losses due to heat transfer through the pneumatic cover of the reactor (gas products – air),

$Q^{II}_{(B-A)}$ are the heat losses due to heat transfer through the vertical wall (gas products – air),

$Q_{(S-A)}$ are the heat losses due to heat transfer through the vertical wall (substrate – air),

$Q^I_{(S-G)}$ are the heat losses due to heat transfer through the vertical wall (substrate – ground)

$Q^{II}_{(S-G)}$ are the heat losses due to heat transfer through the bottom of the tank (substrate – ground).

The fermentation chamber has the shape of a cylinder (inner diameter of 30 m) with a cone-shaped cover with an angle of inclination forming

a cone to the base plane of 30 degrees. In many practical cases, when the thickness of the cylindrical wall is small compared to the diameter, the equations relating to heat transfer through flat walls can be used for thermal calculations. The heat losses through the reactor flat walls are determined according to the following relationship:

$$Q_n = U \cdot A_n \cdot (T_i - T_e), \; [\text{W}] \quad (2)$$

where: U is the heat transfer coefficient [W/(m²K)],

A is the area of the partition [m²],

T_i is the temperature in the fermentation chamber

T_e is the ambient temperature.

Figure 2 presents the distribution of average monthly temperature inside the fermentation chamber T_i and the average monthly external temperature T_e measured in 2016. Inside the fermentation chamber, the average annual temperature $T_i=40°C \pm 0.5°C$ was assumed. Under the foundation slab, a constant temperature of 8°C was adopted, while for the walls in contact with the ground, the average temperature was determined according to the dependence presented in the literature (Biernacka, 2010).

In order to determine the heat transfer coefficient of the biogas walls, the PN-EN ISO 6946 standard was used:

$$U = \cfrac{1}{R_{si} + \sum_{i=1}^{n} \cfrac{d_i}{k_i} + R_{se}}, \; \left[\frac{\text{W}}{\text{m}^2\text{K}}\right] \quad (3)$$

where: k is the thermal conductivity coefficient,

d_i is the thickness of the material layer in the component,

R_{si} and R_{se} are the resistances of heat transfer on the inner and outer surface of the walls of the digester, respectively.

These values were assumed in accordance with the EN ISO 6946 (2017) standard. The convective heat transfer coefficient from the substrate was determined based on Dewil et al. (2007) and was equal 500 W/(m²K). The fermentation chamber is covered with a polymer mesh and a pneumatic roof made of EPDM rubber. Tab. 1 presents the results obtained from the calculations of the heat transfer coefficient from the formula (3) for the partitions of the digester. The highest heat transfer coefficient is for the covering of the fermentation chamber, while the smallest value is for the foundation plate.

In order to simulate the heat losses in the digester for various climate zones, a computer program was written in the Fortran language, in which formulas (1–3) were used. The simulations of heat transfer through the components of the digester were made for the outdoor temperatures from -25°C to 20°C. Figure 3

presents the results of calculations of the heat loss through the partitions of the digester, depending on the outside temperature, assuming a constant temperature of the substrate inside the reactor. The temperature inside the biogas plant, determined by its designer (T_i=40°C), was set in the calculations. In the case of insufficient thermal insulation of the fermentation chamber, this temperature may change (Fig. 2). With the decrease of the outside temperature, the heat losses through the walls of the digester increase. However, the greatest increase in heat losses takes place through walls in contact with the external air. In the winter period, the impact of the heat loss on the temperature inside the fermentation chamber is visible by lowering its value below 40°C (Fig. 2), which also affects the efficiency of the digester.

In order to estimate the accuracy of determination of heat losses through the digester, the surface temperature of the walls (substrate – air) and covers of the digester (gas products – air) estimated from the calculations were compared to the average temperature measured with a thermal imaging camera. The average temperature of the selected surfaces was obtained from the thermal image according to the following relationship:

$$T_m = \frac{1}{A} \iint_A T dA, \; \left[{}^\circ C \right] \tag{4}$$

where: A is the surface area of the wall.

As it turned out, the maximum relative differences in the average temperature was less than 9.5%, which proves the high accuracy of computer simulations.

Table 1. Heat transfer coefficients for the partitions of the digester

Description	U (W/m²K)
Covering of the anaerobic digester (gas products – air)	2.08
Wall (gas products – air)	0.32
Wall (substrate – air)	0.34
Wall (substrate – ground)	0.30
Foundation plate (substrate – ground)	0.15

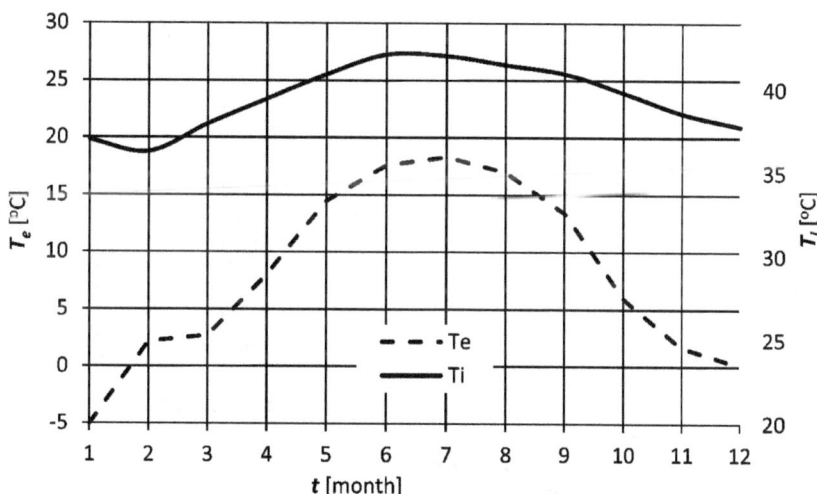

Figure 2. Average monthly temperature T_i and T_e measured in 2016

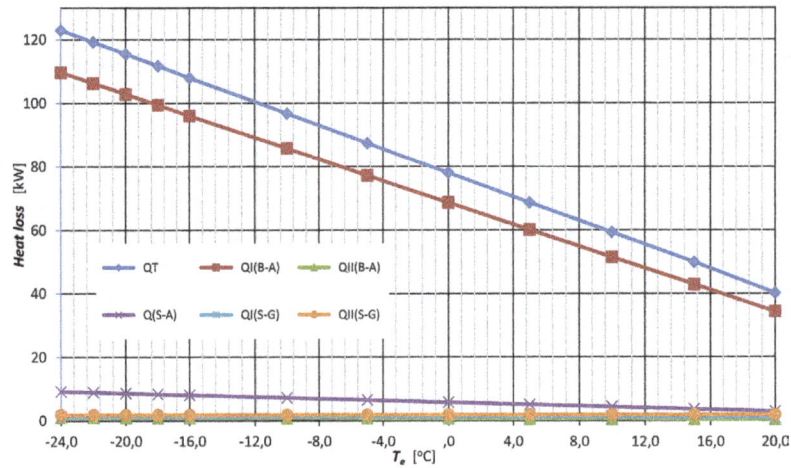

Figure 3. Computational heat losses through the walls of the digester depending on the outside temperature

RESULTS AND DISCUSSION

The basic parameter necessary to determine the heat loss is the design external temperature. The influence of the external temperature in various geographic locations on the operation of biogas plants is described extensively in the literature (El-Mashad et al. 2006, Gebremedhin et al. 2005, Wu and Bibeau 2006, Merlin et al. 2012). Marlin et al. (2012) determined the temperature in the unheated digester versus various ambient air and wastewater temperatures.

The design external temperature for Poland is found in the PN-EN-12831 (2017) standard. According to this standard, Poland is divided into five climatic zones, where the outdoor temperatures are: -16°C, -18°C, -20°C, -22°C, -24°C for zones I, II, III, IV, V, respectively. The heat losses determined on the basis of the design external temperature for the presented biogas plant

are shown in Figure 4. The largest total heat losses are generated by the biogas plants in the fifth climatic zone, and these losses are higher by 13.88%, 10.06%, 6.49%, 3.14% for I, II, III, and IV zones, respectively.

Figure 5 shows the percentage share of heat losses through building partitions in relation to the total heat losses of the digester in the fourth climatic zone. The largest heat losses are through the covering of the digester, which amounts to as much as 90% of the total heat loss. The smallest heat losses occur through the walls and the foundation slab. It should be noted that this condition may be influenced by the depression in the ground of the digester and the shape of the fermentation chamber. Gebremedhin et al. (2005) described the effect of these two parameters on the heat loss of the digester. According to this study, the heat loss by a covering the cylindrical digester at a depth of 1 m in the ground

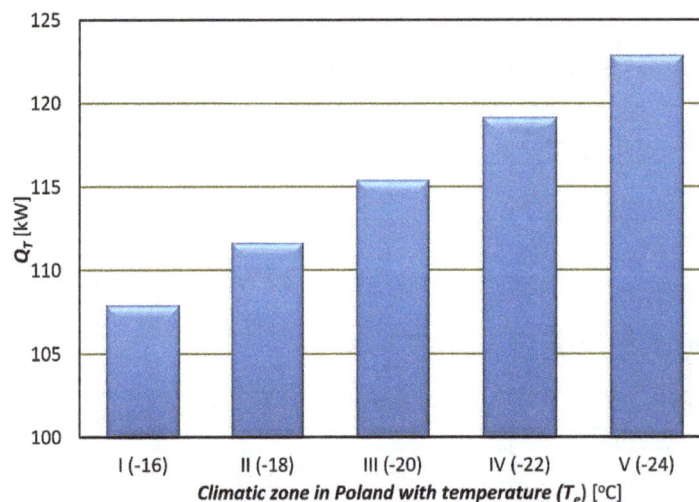

Figure 4. Total heat losses from the one digester chamber in five climatic zones in Poland

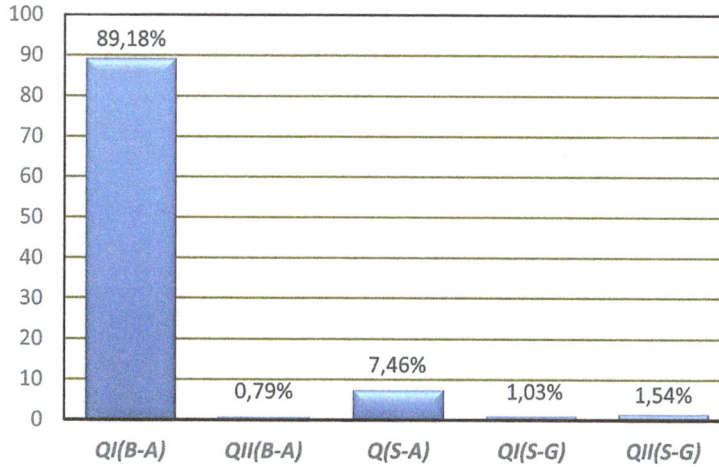

Figure 5. Share of heat losses through partitions in relation to the total heat loss of the digester

a)

b)

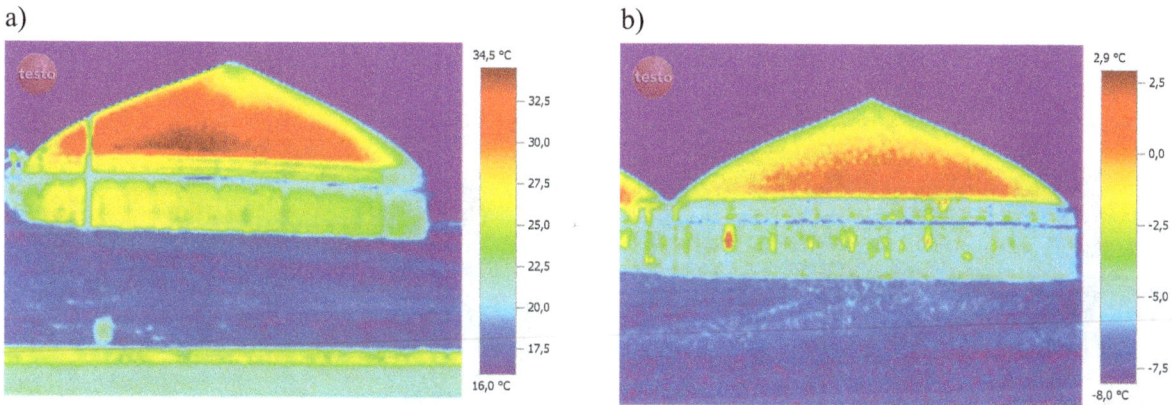

Figure 6. Thermal image of the digester in two different seasons: a) summer period (T_e=20°C), b) winter period (T_e=-6°C)

is about 74% of total heat losses. Gebremedhin et al. (2005) shows that the cylindrical digester with a flat top represents the best geometry for minimizing heat loss. In the presented work, the covering of the fermentation chamber is made in the shape of a cone, which additionally increases the heat loss. Figure 6 shows a view of two thermal images of the fermentation chamber in two different seasons: in summer, where the outside temperature was 20°C (Fig. 7a), and in winter, where the outside temperature was -6°C (Fig. 7b). These two thermal images were taken on a cloudy day. In both cases, i.e. in summer and winter, the highest temperature can be found on the cover of the digester. It should be noted that the presented work does not include the heat gains from beam solar radiation. On a cloudy day, only a diffuse radiation occurs.

Figure 7 presents the dependence of the energy measured in 2016 necessary for heating fermentation chambers depending on the month. The highest consumption of energy was in autumn and winter, while the lowest was in the summer. For example, in December 2016, the energy consumption for heating digesters was up to eight times higher than in July 2016. A similar work trend of the heat exchanger was calculated by Hreiz et al. (2017) for a semi-buried agricultural anaerobic digester located in north-eastern France, near the city of Nancy in the Lorraine region.

Thermal bridges also cause a certain amount of heat loss. The main reason for the occurrence of thermal bridges involves the design or assembly errors. The most common thermal bridges are caused by: breaking the continuity of the insulation layer, insufficient thickness of the thermal insulation layer and inhomogeneity of the partition structure, i.e. occurrence of elements that conduct heat better in the construction of the partition. Figure 8 shows an example of a linear thermal bridge, which is caused by too thin a layer of insulation between the wall of the digester and the ground. Figure 9 shows the thermal bridge of a manifolds box of the heating chamber distributor. The next figure (Fig. 10) shows the thermal bridge of a poorly insulated pipe supplying the

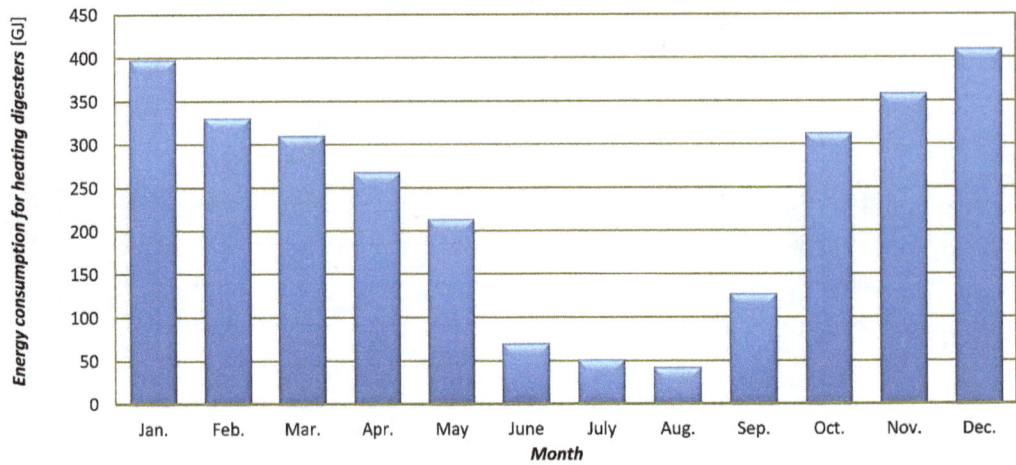

Figure 7. Energy consumption for heating of one digester measured in 2016

a)

b)

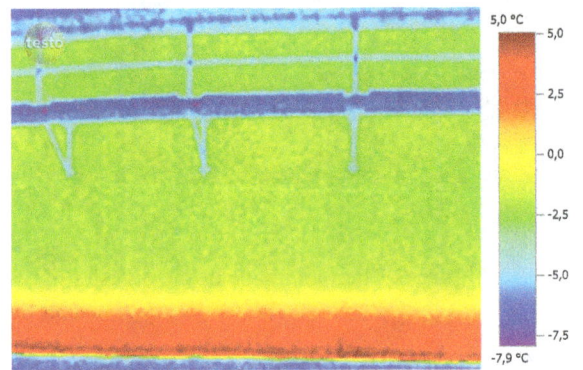

Figure 8. An example of a linear thermal bridge of the vertical wall of the digester:
a) view of the wall b) thermal image

a)

b)

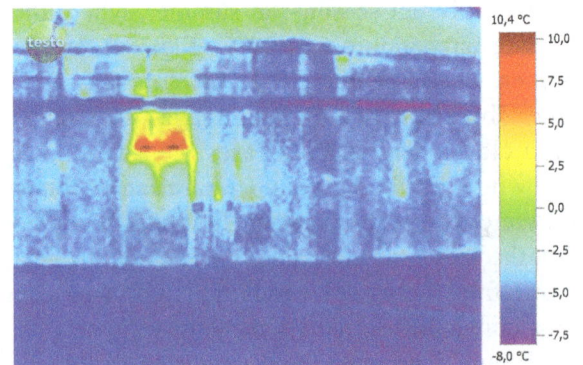

Figure 9. An example of a thermal bridge located in the heat distributor box on the wall
of the fermenting chamber: a) general view, b) thermal image

a) b)

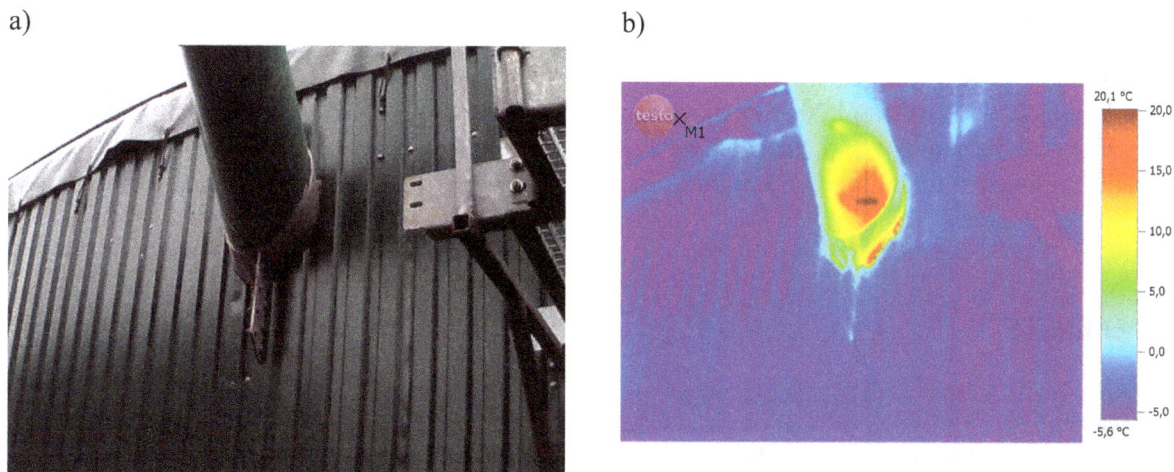

Figure 10. An example of a thermal bridge caused by the lack of proper insulation of the pipe supplying the substrate to the digester: a) general view, b) thermal image

a) b)

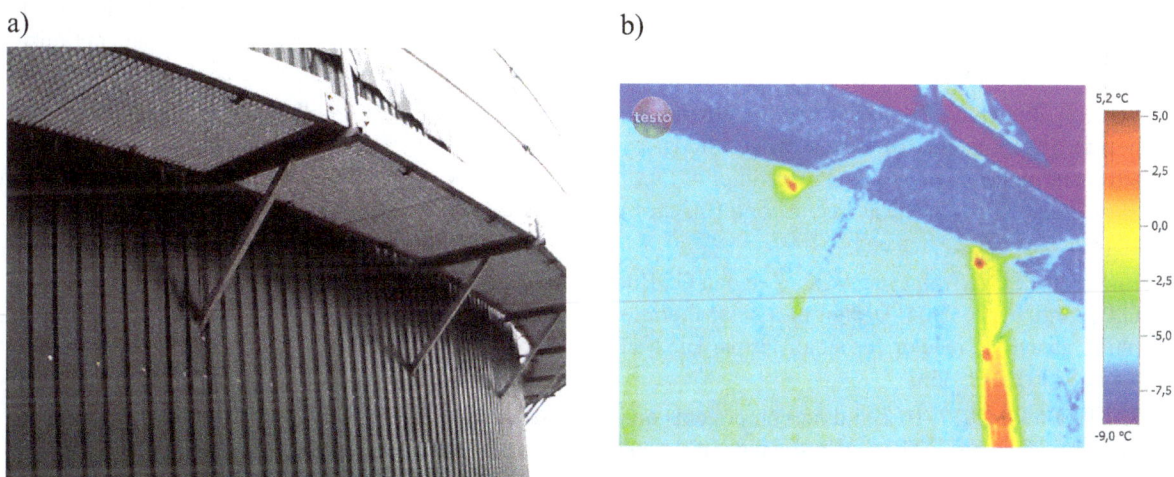

Figure 11. An example of point thermal bridges caused by the use of thermally non-insulated pins for attaching the platform to the wall of the digester: a) general view b) thermal image

substrate to the digester. On the other hand, Figure 11 shows the point thermal bridges caused by incorrect pins connecting the platform with the wall of the fermentation chamber.

CONCLUSION

The paper presents a heat loss analysis for a typical biogas plant under the Polish climatic conditions. The simulations were made for an existing object. It should be noted that properly made thermal insulation of the biogas plant contributes to the stabilization of the temperature inside the fermentation chamber, and especially prevents the temperature drops inside the digester in winter. On the basis of the conducted research, the following conclusions were drawn:

1. Heat losses through the cover account for about 90% of the digester heat losses. In order to reduce this effect, the authors propose additional insulation of the cover or heating of the air space in the cover with warm air in the winter. The heating of the fermentation chamber cover can be made using the waste heat from a biogas plant.

2. The smallest heat losses occur at the interface between the walls of the digester and the soil. Therefore, in order to improve the energy balance, maximum thermal insulation foundation of the digester in the ground is recommended.

3. In this work, thermal bridges were not assumed in the calculations. However, it should be remembered that bridges can have a relatively high impact on the energy losses from the digester. Many thermal bridges were detected using a thermal imaging camera. In order to reduce the heat loss, it is recommended to eliminate thermal bridges both at the design stage and at the construction of the digester.

Acknowledgments

This work was performed within the framework of Grant No. S/WBIIS/4/2014 of Bialystok University of Technology and financed by the Ministry of Science and Higher Education of the Republic of Poland.

REFERENCES

1. Białowiec A., Wiśniewski D., Pulka J., Siudak M., Jakubowski B., Myślak B. 2015. Biodrying of the Digestate from Agricultural Biogas Plants. Annual Set The Environment Protection, 17(2), 1554–1568 (in Polish).

2. Biernacka B. 2010. Semi-empirical formula for the natural ground temperature distribution in Bialystok city region. Civil and Environmental Engineering, 1, 5–9 (in Polish).

3. Dewil R., Appels L., Baeyens J. 2007. Improving the heat transfer properties of waste activated sludge by advanced oxidation processes. Proceedings of European Congress of Chemical Engineering (ECCE-6) Copenhagen, 16–20 September 2007.

4. El-Mashad H.M., Van Loon W.K., Zeeman G., Bot G.P., Lettinga G. 2004. Design of a solar thermophilic anaerobic reactor for small farms. Biosyst. Eng., 87 (3), 345–353.

5. EN ISO 6946:2017–10 Building components and building elements – Thermal resistance and thermal transmittance – Calculation method.

6. Gebremedhin K.G., Wu B., Gooch C., Wright P., Inglis S. 2005. Heat transfer model for plug-flow anaerobic digesters. Trans. ASAE, 48 (2), 777–785.

7. Guo J., Dong R., Clemens J., Wang W. 2013. Thermal modelling of the completely stirred anaerobic reactor treating pig manure at low range of mesophilic conditions. Journal of Environmental Management, 127, 18–22.

8. Hreiz R., Adouani N., Jannot Y., Pons M.-N. 2017. Modeling and simulation of heat transfer phenomena in a semi-buried anaerobic digester. Chemical Engineering Research and Design, 119, 101–116.

9. Kogut P., Piekarski J., Dąbrowski T., Kaczmarek F. 2012. Biogas production plants as a method of utilisation of sewage sludge in relation to the polish legislation. Annual Set The Environment Protection, 14, 299–313 (in Polish).

10. Lebiocka M., Montusiewicz A., Zdeb M. 2010. Anaerobic co-digestion of sewage sludge and old landfill leachate. Polish Journal of Environmental Studies, Series of Monographs, 2, 141–145.

11. Merlin G., Kohler F., Bouvier M., Lissolo T., Boileau H. 2012. Importance of heat transfer in an anaerobic digestion plant in a continental climate context. Bioresour. Technol. ,124, 59–67.

12. Montusiewicz A. 2014. Co-digestion of sewage sludge and mature landfill leachate in pre-bioaugmented system. Journal of Ecological Engineering, 15(4), 98–104.

13. Perrigault T., Weatherford V., Martí-Herrero J., Poggio D. 2012. Towards thermal design optimization of tubular digesters in cold climates: A heat transfer model. Bioresource Technology, 124, 259–268.

14. Pilarska A. A. 2018. Anaerobic co-digestion of waste wafers from confectionery production with sewage sludge. Polish Journal of Environmental Studies, 27(1), 237–245.

15. PN-EN 12831–1:2017–08 Heating systems in buildings – Method for calculation of the design heat load.

16. Sadecka Z., Suchowska-Kisielewicz M. 2016. Cofermentation of Chicken Manure. Annual Set The Environment Protection, 18, 609–625 (in Polish).

17. Shaheen M., Nene A. A. 2014. Thermal simulation of biogas plants using Matlab. International Journal of Engineering Research and Applications, 4, 24–28.

18. Terradas-III G., Triolo J. M., Pham C. H., Martí-Herrero J., Sommer S. G. 2014. Thermic model to predict biogas production in unheated fixed dome digesters buried in the ground. Environmental Science and Technology, 48, 3253–3262.

19. Wu B., Bibeau E.L. 2006. Development of 3-D anaerobic digester heat transfer model for cold weather applications. Trans. Am.Soc. Agric. Eng., 49 (3), 7749–7757.

25

Assessment of the aquatic environment quality of high Andean lagoons using multivariate statistical methods in two contrasting climatic periods

María Custodio[1*], Richard Peñaloza[2], Fernán Chanamé[1], Raúl Yaranga[1], Rafael Pantoja[1]

[1] Universidad Nacional del Centro del Perú, Facultad de Zootecnia, Instituto de Investigación en Alta Montaña, Av. Mariscal Castilla No. 3989-4089, Huancayo, Perú

[2] Universidad Nacional Agraria La Molina, Av. La Molina s/n La Molina, Lima, Casilla Lima 12, Perú

[*] Corresponding author's e-mail: mcustodio@uncp.edu.pe

ABSTRACT

The quality of the aquatic environment of high Andean lagoons was evaluated by means of multivariate statistical methods in two contrasting climatic periods. The water samples and benthic macroinvertebrates were collected in 22 sampling sites during the rainy and dry seasons. In each lagoon DO, DTS, EC, temperature and pH were determined in situ. The results revealed that the physicochemical parameters comply with the environmental quality standards for water, except COD and BOD_5. In the PCA of the physicochemical parameters, the first two axes explained 73% of the total variation. The gradual analysis in pairs showed significant differences. The SIMPER analysis determined an average of four families of benthic macroinvertebrates per lagoon that showed more than 70% contribution. The ANOSIM revealed that Tragadero lagoon differs significantly from the others. The DistLM showed a value of 0.46 of coefficient of determination. Therefore, the quality of the environment of high Andean lagoons evaluated by means of multivariate statistical methods presents important differences or dissimilarities not only in the physicochemical characteristics of the water, but also in the composition of the benthic macroinvertebrate communities.

Keywords: aquatic environment, high Andean lagoons, benthic macroinvertebrates, water quality.

INTRODUCTION

Water is the renewable natural resource that exercises the most limiting action in human development and in all forms of life [Bharti 2011]. The availability and quality of water is a concern for humanity, due to the increasing deterioration that continental aquatic ecosystems are experiencing as a result of natural and anthropogenic pressures. Therefore, the control of water quality has become very important to maintain the sustainability of water resources [Wen et al. 2012]. In developed countries, Strategic Framework Directives have been implemented, such as the Water Framework Directive, basin and marine water quality in 2000 (WFD, 2000/60/EC), 2006 (QBWD, 2006/7/EC) and 2008 (MSFD, 2008/56 / EC), respectively [Petus et al. 2014]. On the other hand, in developing countries, the supply and treatment of water are the most important issues and they allocate most of the investments in water management. However, in practice, less than 20% of the total wastewater effluents are treated before they are discharged into water bodies [Swiech et al. 2012, Seiler et al. 2015].

In Peru, water pollution is one of the biggest environmental problems, not only due to the rapid growth of the country's urban centers, but also due to the large volumes of wastewater that it generates. Currently, it is estimated that wastewater is treated only in 29.1% through 143 treatment plants, the rest being discharged without any treatment to the continental surface water bodies and the sea [MINAM 2011]. Degradation of water quality directly affects all types of water applications [Article 2015]; for example, the population and recreational use, coastal and continental marine extraction and cultivation activities, the irrigation of vegetables and animal drink, as well as the use for the conservation of the aquatic environment.

The focus of water quality monitoring in Peru is based on the comparison of physical, chemical and bacteriological parameters with environmental quality standards for water established by the Ministry of the Environment. However, the use of this methodology allows identifying sources of pressure, but does not enable to better understand the processes of alteration of water quality. The complex data matrices that are generated in each monitoring need to be systematized and analyzed with various statistical methods. The application of multivariate statistical methods for data analysis offers a better understanding of the quality of the aquatic environment [Muangthong and Shrestha 2015].

In this context and given the reduced importance that continental aquatic ecosystems receive, which become increasingly fragile in the face of natural and anthropogenic pressures, the need to apply tools that allow monitoring the quality of these ecosystems arises in order to achieve a sustainable management of these resources. The objective of the study was to evaluate the quality of the aquatic environment of high Andean lagoons by means of multivariate statistical methods in two contrasting climatic periods.

MATERIAL AND METHODS

Description of the study area

The Pomacocha, Tragadero, Cuncancocha, Incacocha and Ñahuinpuquio lagoons considered in the study are located in the Mantaro river basin. Which is located in the Central Andes of Peru, between latitudes: 10 ° 34' S – 13 ° 35' S and longitudes: 73 ° 55' W – 76 ° 40' W, with altitudes ranging from 500 masl to 5350 masl (Figure 1). Associated with these lagoons are bofedales with typical vegetation cover (*Distichia muscoides and Oxychloe andina*).

Sampling and measurement procedures

The water samples and benthic macroinvertebrates were collected in five lagoons, in 22 sampling sites. In each lagoon, dissolved oxygen (mg/L), total dissolved solids (mg/L), conductivity (µS/cm), temperature (°C) and pH, were determined in situ by means of Hanna Instruments portable equipment. Additionally, the water samples were collected in sterile glass bottles for bacteriological analysis and in disposable plastic bottles of two liters for the analysis of nitrates,

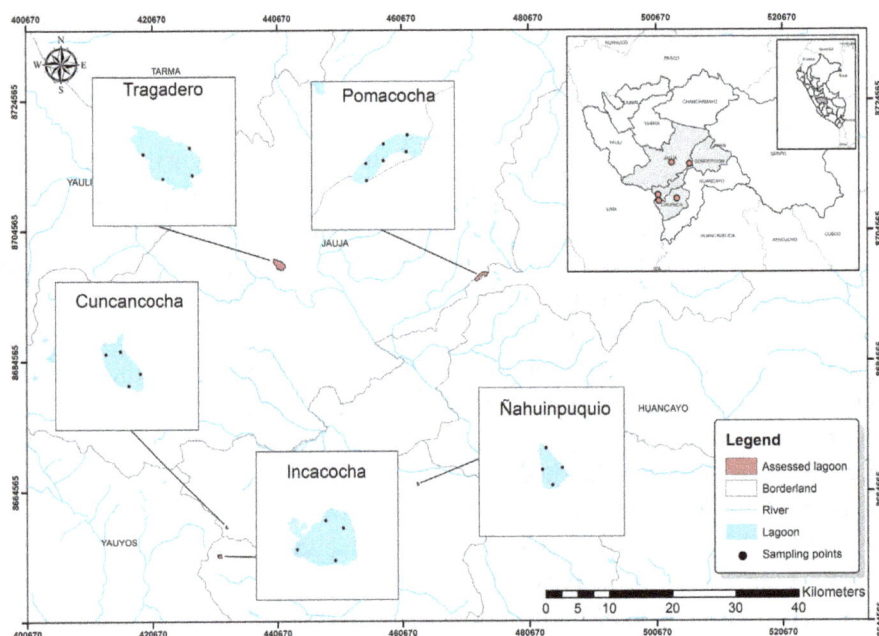

Figure 1. Geographical location of the study area and water sampling points and benthic macroinvertebrates in high Andean lagoons

total phosphorus, COD, BOD$_5$ and heavy metals. Previously, the bottles were labeled, treated with a 10% nitric acid solution for 24 hours and rinsed with bidistilled water. Then, one liter of water from each of the samples was added 1.5 mL of concentrated nitric acid, for conservation.

The preparation of the sample consisted of placing 250 ml of water in a beaker, which was boiled, until obtaining 100 ml. Immediately, 5 ml of nitric acid and 5 ml of concentrated hydrochloric acid were added for the destruction of the organic matter and again it was boiled (until the water was consumed and a pasty consistency was obtained). It was allowed to cool and then 10 ml of distilled water was added, filtered and stuck in a 100 ml fiola, with 1% nitric acid [Clesceri et al. 2012]. The quantitative determination of heavy metals (copper, zinc, iron and lead) was carried out with the flame atomic absorption spectrophotometry method, according to the methodology recommended by the FAO (1983), using an AA-6800 Atomic Absorption Spectrophotometer equipment, Shimadzu brand, for which the standard copper, zinc, iron and lead solutions were previously prepared and read in the increasing order of concentration, with which the calibration curve was prepared and then the samples were read. The collection of benthic macroinvertebrates was carried out using the Ekman-Birge dredger from Hydro-Bios. The samples were fixed with 70% alcohol to carry out the taxonomic determination later.

Statistical data analysis

The data on the lagoon factors and sampling season (rain and low water) were analyzed using the Primer-E v7 statistical package (Massey University, New Zealand) and PAST v3.19. Initially, the normality of the data was evaluated and then a Principal Component Analysis (PCA) was performed to analyze the physicochemical parameters of the water in the 22 sampling points. Next, we proceeded to evaluate the effects of the lagoon factors and sampling period with the PERMANOVA model of two factors with Euclidean distance for physicochemical parameters as a measure of similarity [Clarke and Gorley 2015].

The analysis of the general patterns of composition of benthic macroinvertebrate communities was carried out using the PCO method (analysis of principal coordinates, similarity of Bray-Curtis for Cluster analysis) in order to obtain the percep-

tual map. Next, the PERMANOVA analysis was carried out to determine significant differences between the sites and distribution patterns of species and abundances [Clarke and Gorley 2006]. In order to determine the most influential taxa in the composition of the communities of benthic macroinvertebrates in each of the lagoons, an analysis of the similarity percentages (SIMPER) was conducted from a similarity matrix. Likewise, the ANOSIM similarity analysis was performed for the pair-wise biological comparison of the biological variables between lagoons and to find significant differences of the relative abundances of the macroinvertebrates between lagoons and seasons. The individual and group results of the macroinvertebrate communities were compared using MDS 2nd stage [Ceschia et al. 2007], applying the Spearman range to determine the coefficient of similarity of the distribution in relation to the matrix of the physicochemical parameters.

The BioEnv method was used to identify the physicochemical parameters that have a strong correlation with macroinvertebrate communities and find the best explanation of spatial patterns [Ignatiades et al. 2009]. DistLM (linear model based on distance) was applied to obtain the value of permutational regression R^2, of greater importance, between the matrices of biological variables and physicochemical parameters of water individually and in groups, in order to determine the contribution relative of these variables on the structure of benthic macroinvertebrate communities [Primo et al. 2012]. The analysis of BEST with R^2 as a selection criterion was used to determine which of the physicochemical parameters of the water better explain the composition pattern of the macroinvertebrate community.

RESULTS AND DISCUSSION

Analysis of water quality

Table 1 shows the mean values and standard deviation of the physicochemical parameters of water quality of high Andean lagoons, considering the sampling period. The pH values of the water presented the variations in each of the lagoons evaluated, with the values ranging from $6,790 \pm 0.582$ in the Cuncancocha lagoon during the dry season to 8.611 ± 0.190 in the Incacocha lagoon, during the rainy season. These variations would be related to the edaphic conditions of the

Table 1. Mean and standard deviation of water quality parameters at different sampling points of high Andean lagoons measured during the rain and dry periods

Indicator	Pomacocha		Tragadero		Cuncancocha		Incacocha		Ñahuinpuquio	
	Rain	Dry	Rain	Dry	Rain	Dry	Rain	Dry	Rain	Dry
pH	6.841±0.126	6.796±0.083	7.563±0.415	7.531±0.616	7.159±0.248	6.790±0.582	8.611±0.190	8.860±0.044	7.634±0.058	7.745±0.080
EC (µS/cm)	254.244±3.033	248.171±5.847	269.813±7.503	264.343±14.426	260.352±4.390	253.192±1.677	262.534±4.024	250.750±4.226	258.771±41.41	252.837±48.342
COD (mg/L)	31.092±0.923	30.581±1.025	59.637±0.234	56.492±0.690	46.875±1.250	45.388±0.216	38.833±0.312	37.167±0.136	34.521±0.401	33.813±1.055
BOD$_5$ (mg/L)	18.068±0.398	18.079±0.466	28.806±0.678	28.430±0.754	18.679±0.869	18.442±2.081	18.513±0.374	18.413±0.285	20.800±0.400	20.704±0.537
DO (mg/L)	6.680±0.105	6.628±0.159	6.122±0.068	6.016±0.209	6.496±0.164	6.384±0.114	6.432±0.046	6.376±0.017	7.118±0.083	7.324±0.140
Temperature (°C)	11.739±0.775	11.552±1.342	17.704±0.160	19.475±0.952	11.377±0.353	10.521±0.077	10.509±0.405	9.283±0.304	10.609±0.371	9.283±0.304
DTS (mg/L)	3.096±0.069	2.862±0.081	9.200±1.397	9.238±1.704	2.577±0.066	2.363±0.308	2.625±0.0.199	2.513±0.160	2.752±0.061	2.354±0.219
Total phosphorus (mg/L)	0.042±0.006	0.041±0.007	1.519±0.036	1.498±0.012	0.802±0.019	0.792±0.001	0.500±0.038	0.503±0.046	0.740±0.008	0.760±0.003
Nitrates (mg/L)	0.396±0.023	0.409±0.039	0.768±0.029	0.762±0.034	0.466±0.003	0.466±0.003	1.506±0.023	1.314±0.510	1.555±0.037	1.480±0.011
Total iron (mg/L)	0.034±0.001	0.034±0.001	0.044±0.000	0.054±0.000	0.033±0.000	0.033±0.000	0.017±0.000	0.017±0.000	0.038±0.000	0.038±0.000
Total copper (mg/L)	0.002±0.000	0.002±0.000	0.005±0.001	0.006±0.001	0.002±0.000	0.002±0.000	0.002±0.000	0.002±0.000	0.002±0.001	0.002±0.001
Total lead (mg/L)	0.000±0.000	0.001±0.000	0.000±0.000	0.000±0.000	0.000±0.000	0.000±0.000	0.001±0.000	0.000±0.000	0.000±0.000	0.000±0.000
Total zinc (mg/L)	0.000±0.000	0.001±0.001	0.000±0.000	0.000±0.000	0.000±0.000	0.000±0.000	0.000±0.000	0.000±0.000	0.000±0.000	0.000±0.000

body of water [Rim et al. 2017] and to wastewater discharges [Christia et al. 2014]. However, these variations are within the natural ranges for aquatic life, according to the environmental quality standards for water in Peru [MINAM 2015], as well as within the ranges established by the WHO (2011) and the Canadian Council of Ministers of the Environment [CMME 2007]. Similar behavior was exhibited by the electrical conductivity, registering the highest values in the Tragadero lagoon in both sampling seasons, this result would be related to the pollutant load of organic matter contributed by the wastewater of the anthropogenic activities that develop in its surroundings [Alam et al. 2016].

The highest average values of the chemical oxygen demand – COD – were registered in the Tragadero lagoon, in both seasons, and were higher than the RCTs (40 mg/L). On the other hand, the BOD$_5$ recorded in all the study gaps far exceeded the environmental quality standards for water (10 mg/L), revealing the deterioration of quality that these bodies of water are experiencing. Moreover, the average concentrations of dissolved oxygen-DO were lowest in the Tragadero lagoon, during the dry season, indicating the deterioration of the water quality in this lagoon and the effect on the chemical, photosynthetic and respiratory processes that occur in the water [Kuzmanovic et al. 2017]. The mean values of the highest temperature were recorded in the lagoon of Ñahuinpuquio (19.475 ± 0.952) during the dry season. The mean values of total dissolved solids-TDS ranged from 2.363 ± 0.308 to

9.238 ± 1.704 in the Cuncancocha and Tragadero lagoons, respectively.

The mean values of total phosphorus were higher than the environmental quality standards for water (0.035 mg/L), in all the lagoons evaluated, mainly in the Tragadero lagoon that presents values that triplicate to the phosphorus concentrations of the rest of the lagoons, confirming the problems of eutrophication that this body of water is experiencing [Copetti et al. 2015]. Additionally, the highest average values of nitrates were registered in this lagoon. The obtained results of these two nutrients reveal the pressure that these bodies of water are supporting, such as the discharge of domestic and industrial wastewater; as well as the drainage of the cultivated fields installed in its surroundings. Regarding the average concentration of Cu, Pb and Zn, the results found in the lagoons evaluated did not exceed the environmental quality standards of Peru, unlike the Fe that exceeded the values of the referred quality standards.

The PCA result of the physicochemical parameters and the sampling points in the five lagoons reveals the variability explained by each of the axes of variation (Figure 2). The first two axes explained 73% of the total variation and corresponded to the sampling points of the Tragadero lagoon. The first axis was closely related to high concentrations of iron and total dissolved solids. The second axis of variation was characterized by high concentrations of total phosphorus, COD and BOD$_5$. The rest of the physicochemical indicators explain little of the total variability.

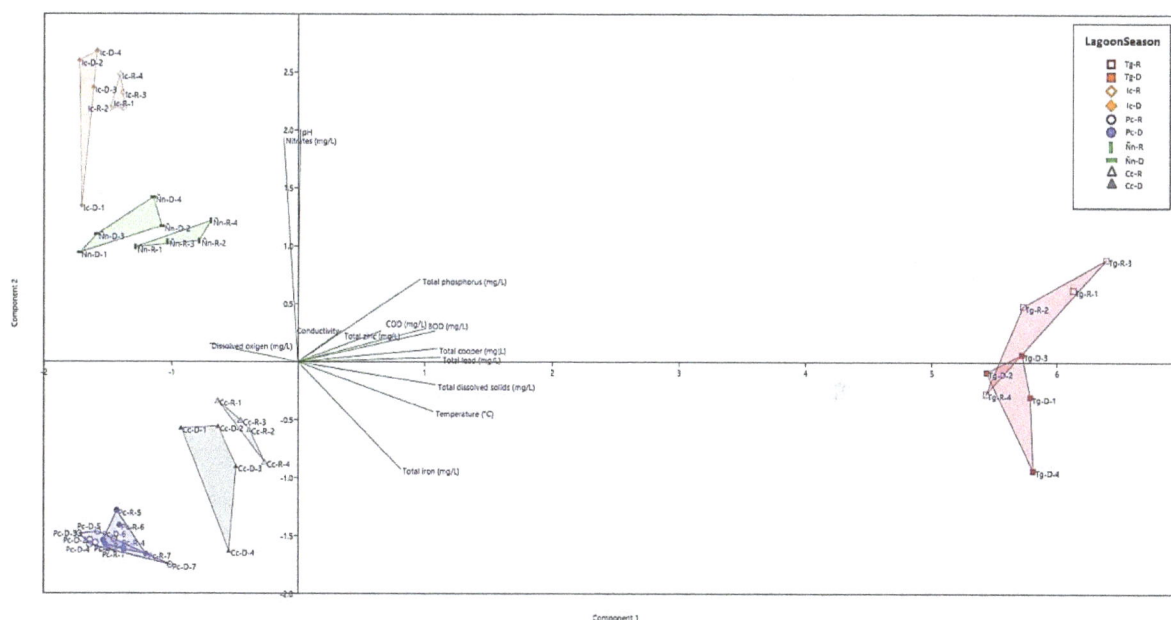

Figure 2. Analysis of the main components (PCA) of the physicochemical parameters of the water of high Andean lagoons, according to the time of sampling. Sampling included the lagoons: Incacocha in rainy season (Ic-R) and dry season (Ic-D); Ñahuinpuquio (Ñn-R and Ñn-D), Cuncancocha (Cc-R and Cc-D), Pomacocha (Pc-R and Pc-D) and Tragadero (Tg-R and Tg-D).

The PERMANOVA analysis for the interaction of the lagoon factors by sampling period, at a level of significance of 0.01, reveals that at least one of the observations resulting from this interaction, with respect to the physicochemical parameters of the water, is different from the others. In addition, the gradual analysis in pairs for the rainy season exhibits that the lagoons evaluated present significant differences, except the Pomacocha lagoon. The Tragadero and Incacocha lagoons show significant differences in relation due to the effect of the sampling period factor, that is, the values of the physicochemical parameters of the water are different both in the rainy season and in the dry season. On the other hand, for the Cuncancocha and Ñahuinpuquio lagoons, the sampling period factor does not influence the physicochemical characteristics of the water and it is possible to affirm that they are statistically similar to a level of significance of 0.05.

Analysis of benthic macroinvertebrate communities

A total of 4905 individuals of benthic macroinvertebrates were captured during the sampling periods in five high-Andean lagoons of the Junín region, which corresponded to nine orders and 14 families. The largest number of individuals of benthic macroinvertebrates was presented by the order Diptera with 3992 individuals, followed by the order Ephemeroptera with 429 individuals. The highest density of benthic macroinvertebrates was recorded in the Tragadero lagoon in both sampling seasons, corresponding to the order Diptera. However, the composition of the taxa by taxonomic orders showed slight variations during the sampling periods, Diptera, Ephemeroptera, Coleoptera and Trichoptera were the orders with the highest taxa richness.

The SIMPER analysis carried out to analyze the composition of the benthic macroinvertebrate communities of the five high Andean lagoons determined an average of four families per lagoon that showed more than 70% contribution in the analysis. The families that contributed the most in the differentiation of these communities were: Chironomidae (41.30%), Ceratopogonidae (11.97%), Psychodidae (11.14%) and Hidrophilidae (9.11%). The contribution of the Chironomidae family in the composition of the macroinvertebrate communities varied according to time. The largest contributions of the Chironomidae family were made in the Tragadero lagoons, with 69.90% and Incacocha with 53.80%, during the dry season and rain, respectively. These results coincide with those recorded in other studies in the aquatic environments with low oxygen levels, dominance of the Chironomidae and decrease in the density of other macroinvertebrates [Graeber

Table 2. Average abundance and percentage of contribution (in parentheses) of families of benthic macroinvertebrates according to the lagoon factor in the SIMPER analysis

Taxa	Pool all groups Contr. %	Rainy season										Dry season									
		Pc		Tg		Cc		Ic		Ñn		Pc		Tg		Cc		Ic		Ñn	
		x̄	Contr. %	x̄	Contr. %	x̄	Contr. %	x̄	Contr. %	x̄	Contr. %	x̄	Contr. %	x̄	Contr. %	x̄	Contr. %	x̄	Contr. %	x̄	Contr. %
Chironomidae	41.3	19.3	37.7	78.5	67.1	15.3	49.3	21.0	53.8	20.8	41.4	64.3	55.4	84.0	69.9	20.8	34.9	23.3	44.1	23.8	49.3
Ceratopogonidae	12.0	9.3	18.2	9.3	7.9	8.5	27.4	10.5	26.9	9.5	18.9	7.9	6.8	4.5	3.7	11.8	19.8	11.8	22.3	6.8	14.0
Psychodidae	11.1	1.6	3.1	15.5	13.2	2.5	8.1	1.0	2.6	9.3	18.6	7.9	6.8	19.0	15.8	7.5	12.6	3.8	7.1	9.0	18.6
Hydrophilidae	9.1	5.6	10.9	0.0	0.0	1.3	4.0	2.3	5.8	7.5	14.7	7.1	6.2	1.5	1.2	7.3	12.2	5.3	9.9	5.5	11.4
Baetidade	8.5	6.4	12.6	2.8	2.4	1.3	4.0	1.5	3.8	2.8	5.5	10.3	8.9	2.0	1.7	5.5	9.2	5.3	9.9	0.0	0.0
Leptophlebiidae	4.9	4.6	8.9	0.0	0.0	0.0	0.0	0.0	0.0	0.0	0.0	8.4	7.3	1.0	0.8	0.0	0.0	0.0	0.0	0.0	0.0
Tubificidae	3.2	0.9	1.7	4.3	3.6	0.8	2.4	1.5	3.8	0.5	1.0	1.3	1.1	2.3	1.9	1.5	2.5	1.0	1.9	1.3	2.6
Limnephilidae	3.1	1.9	3.6	0.0	0.0	1.5	4.8	1.3	3.2	0.0	0.0	2.4	2.1	0.0	0.0	1.8	2.9	2.0	3.8	1.3	2.6
Lymnaeidae	2.6	0.0	0.0	5.8	4.9	0.0	0.0	0.0	0.0	0.0	0.0	1.0	0.9	3.5	2.9	1.0	1.7	0.0	0.0	0.8	1.6
Psychomyiidae	1.6	1.0	2.0	0.0	0.0	0.0	0.0	0.0	0.0	0.0	0.0	3.3	2.8	0.0	0.0	0.0	0.0	0.0	0.0	0.0	0.0
Planariidae	1.2	0.4	0.8	1.0	0.9	0.0	0.0	0.0	0.0	0.0	0.0	1.1	1.0	0.0	0.0	1.3	2.1	0.5	0.9	0.0	0.0
Glossiphoniidae	0.8	0.0	0.0	0.0	0.0	0.0	0.0	0.0	0.0	0.0	0.0	0.3	0.2	2.5	2.1	0.5	0.8	0.0	0.0	0.0	0.0
Hyalellidae	0.6	0.3	0.6	0.0	0.0	0.0	0.0	0.0	0.0	0.0	0.0	0.7	0.6	0.0	0.0	0.8	1.3	0.0	0.0	0.0	0.0

et al. 2017]. The Pomacocha and Cuncancocha lagoons presented the lowest contributions of the Chironomidae family (Table 2).

The analysis of the differences in the relative abundances of the benthic macroinvertebrates in the five lagoons evaluated through the ANOSIM shows that the Tragadero lagoon differs significantly with respect to the others, except for the Pomacocha lagoon during the dry season. In addition, it is observed that the Cuncancocha, Incacocha and Ñahuinpuquio lagoons have significant similarities (Table 3). These results reveal that the great majority of benthic macroinvertebrate species have low tolerance to contamination [Johnson and Ringler 2014].

The non-metric multidimensional scaling analysis shows an average value of stress level of 0.14, which according to the range given by Kruskal indicates a regular interpretation in the perceptual map. Most of the sampling points of the lagoons evaluated are grouped, regardless of the sampling period. It is observed that the sampling points in the Tragadero lagoon tend to be clearly separated from the rest of the sampling points of the other lagoons. A similar behavior is shown by one of the sampling points in the Pomacocha lagoon. This is explained by the marked differences in the abundances of Chironomidae species that these lagoons have. The abundance of these species in the rest of the lagoons is low.

The PERMANOVA analysis for the interaction of the lagoon factors by sampling period, at a level of significance of 0.05, reveals that at least one of the observations resulting from this interaction influences the biological values, that is, at least one of the observations in a certain lagoon and sampling period is different from the others in relation to the number of species and abundances. Moreover, when performing the test as a pair (Pair-Wise) it was found that for the rainy

Table 3. ANOSIM pair-wise test results for the multivariate diversity test

	PcR	TgR	CcR	IcR	ÑnR	PcD	TgD	CcD	IcD	ÑnD
PcR										
TgR	0.0025									
CcR	**0.175**	0.0292								
IcR	**0.5226**	0.0307	**0.7173**							
ÑnR	**0.0536**	0.0259	**0.084**	**0.1109**						
PcD	0.0005	**0.1502**	0.0022	0.0055	0.0057					
TgD	0.0033	**0.8046**	0.027	0.0275	0.0287	**0.0615**				
CcD	0.0462	0.0298	**0.1395**	**0.1933**	**0.9141**	0.0054	0.029			
IcD	**0.1893**	0.0309	**0.1113**	**0.6279**	**0.8039**	0.0063	0.0286	**0.9727**		
ÑnD	0.018	0.0287	**0.0898**	**0.057**	**0.7706**	0.009	0.0296	**0.4315**	**0.2281**	

season the Incacocha and Cuncancocha lagoons are highly similar for the biological indexes. For the Pomacocha and Incacocha lagoons, the distribution of observations is greatly dispersed. In contrast, the Laguna Tragadero maintains a significant difference between all the lagoons. The Cuncancocha, Ñahuinpuquio and Incacocha lagoons are statistically similar, making these lagoons grouped in relation to their biological indices, forming three similarity groups, the Pomacocha lagoon that differs from Tragadero, and this differs from the Cuncancocha, Ñahuinpuquio and Incacocha lagoons (Figure 3).

The stepped analysis shows that in the Pomacocha lagoon, the effect of the sampling period factor influences the biological indexes, showing that they differ and change in number of species and abundances due to the effect of the sampling period. In the case of the Tragadero, Cuncancocha, Incacocha and Ñahuinpuquio lagoons, the sampling period factor does not influence diversity indices.

Correlation of biological and environmental variables

The correlation of the biological variables (abundance of benthic macroinvertebrates) and environmental variables (physicochemical parameters of the water), for the rainy season, through the Spearman correlation coefficient showed a Rho value of 0.60. This result reveals that the distribution of the dissimilarities of the environmental variables is similar to the biological variables (Figure 4). While performing matrix resemblance pairing analysis, it was found that the Rho is 0.32 (Spearman rank), using the two factors (lagoon and sampling time). This result indicates that the biological variables show the same dispersion pattern as the environmental variables at 32% similarity. Although this value is very low, it is possible that the dispersion of the biological data is influencing the decrease of this value, especially in the dry season, where the

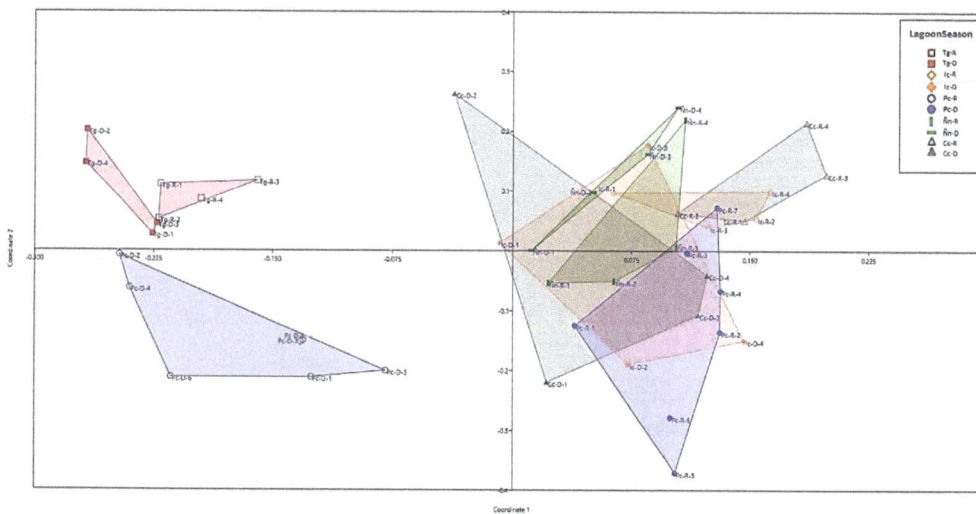

Figure 3. Analysis of principal coordinates (PCO) of the composition of the communities of benthic macro invertebrates, using similarity of Bray – Curtis for each high Andean lagoon, during the rainy and dry periods.

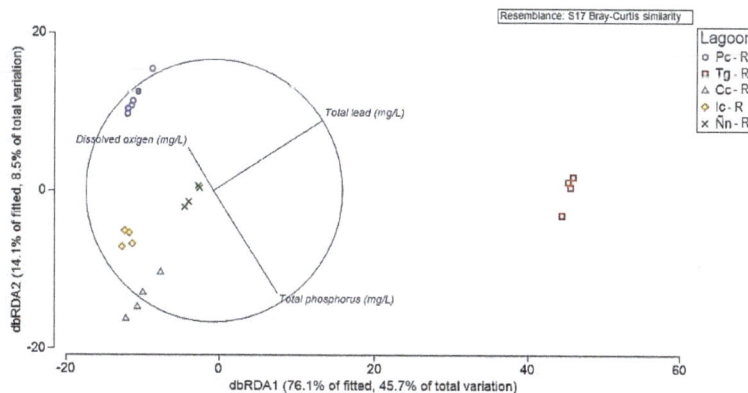

Figure 4. Analysis of linear models based on the comparison distance according to the lagoon factor evaluated for the rainy period.

biological indices are very different. However, the distribution given by the main environmental vectors would be revealing that the abundance of the species of the families Leptophlebiidae and Psychodidae in the Pomacocha lagoon would be determined by the low values of the environmental variables in general. The species of the Lymnaeidae and Glossiphoniidae families are species that are present in areas with high values of total phosphorus, COD, BOD_5, zinc, copper, lead, total dissolved solids, temperature and iron; as in the Tragadero lagoon.

It is normal that as the number of variables increases, the value of the relationship decreases; but this large decrease indicates that in the dry season there are phenomena that condition the increase in the number of species and abundance in the Pomacocha lagoon especially, causing the biological matrix to have a poor relationship with the distribution of the environmental variables. According to the individual analysis using the lagoon factor for the dry season, the correlation value is 0.30, showing the Pomacocha lagoon ex-

periences a great increase in its biological values, simultaneously increasing the presence of species that are similar to the characteristics of the Tragadero lagoon, the most contaminated, with similarities in the number of species and 60% abundance. According to the analysis of linear models based on distance, this behavior in the Tragadero and Pomacocha lagoons is explained by the increase in temperature in the water and the high levels of lead, especially in the Tragadero lagoon (Figure 5).

The sequential analysis of linear models based on distance (DistLM) using the two factors (lagoon and period) shows that the predominant predictors presented a value of 0.46 of coefficient of determination out of a total of 0.58, where all the physicochemical variables participated, being observed that the values of total lead, phosphorus and conductivity are the most important in relation to the distribution of biological data. However, the highest total lead values are presented by the Tragadero lagoon and phosphorus lagoons, the second important variable, the Tra-

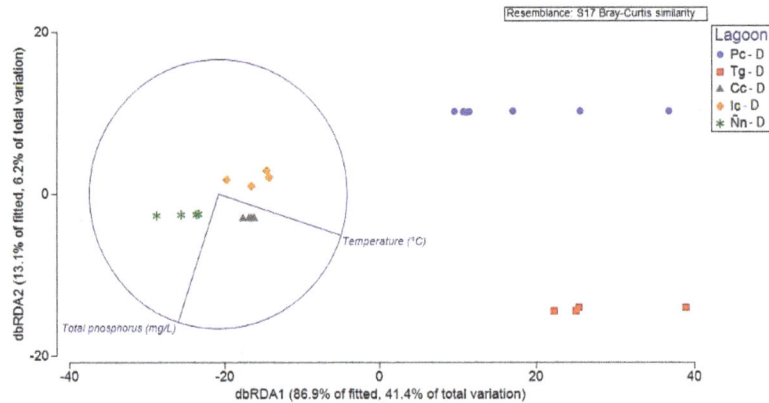

Figure 5. Analysis of linear models based on the comparison distance according to the lagoon factor evaluated for the dry period.

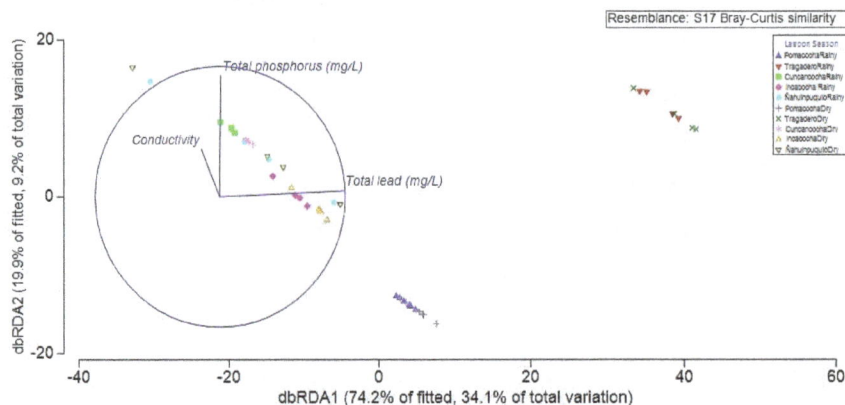

Figure 6. Analysis of linear models based on comparison distance between the interaction of sampling time and lagoon.

gadero and Pomacocha lagoons; which explains the greater abundance of benthic macroinvertebrates in these bodies of water, mainly of the Chironomidae family (Figure 6), considered as the resistant and resilient taxon against anthropogenic pressures [Ruaro 2014].

The BEST (Biota and enviromental matching) analysis reveals that using all the observations divided by the two evaluation factors (lagoon and period), the biological matrices (with a Rho = 0.462) are explained by the environmental variables of temperature, phosphorus and lead total.

CONCLUSIONS

The quality of the aquatic environment of high Andean lagoons evaluated by multivariate statistical methods presents important differences not only in the physicochemical characteristics of the water, but also in the composition of the benthic macroinvertebrate communities. The anthropogenic activities developed in and around the lagoons are affecting the quality of the water and as a consequence to the biological communities, altering their composition; such as it is observed in the Tragadero lagoon. In addition, the sampling period factor influences the number and abundance of benthic macroinvertebrate species. In the rainy season, the correlation of the physicochemical variables has a good degree of adjustment (around 60%), indicating that there is a good correspondence and explanation of the behavior exhibited by the biological communities. Thus, the Tragadero lagoon has higher indices of abundance and dominance of the Chironomidae family, in response to the strong pressures exerted by anthropogenic activities. Therefore, the most important predictors of the environmental quality of these bodies of water are total lead, phosphorus and conductivity.

Acknowledgements

The authors express their gratitude to the General Research Institute of the National University of Central Peru for the financing of the study, to the Water Research Laboratory for allowing us to make use of the equipment and materials for this study, to M. Sc. Samuel Pizarro for his assistance in field sampling and to Eng. Cirilo Sedano Fuentes for his valuable collaboration in the translation of the manuscript.

REFERENCES

1. Alam, M.Z., Carpenter-Boggs, L., Rahman, A., Haque, M.M., Miah, M.R.U., Moniruzzaman, M., Qayum, M.A., Abdullah, H.M., 2016. Water quality and resident perceptions of declining ecosystem services at Shitalakka wetland in Narayanganj city. Sustainability of Water Quality and Ecology. doi: 10.1016/j.swaqe.2017.03.002

2. Article, O., 2015. Development and evaluation of the Lake Multi-biotic Integrity Index for Dongting Lake, China. J. Limnol 74, 594–605. doi: 10.4081/jlimnol.2015.1186

3. Bharti, N., 2011. Water quality indices used for surface water vulnerability assessment. International Journal of Environmental Sciences 2, 154–173.

4. Canadian Council of Ministers of the Environment (CMME), 2007. Canadian water quality guidelines for the protection of aquatic life, Canadian water quality guidelines.

5. Ceschia, C., Falace, A., Warwick, R., 2007. Biodiversity evaluation of the macroalgal flora of the Gulf of Trieste (Northern Adriatic Sea) using taxonomic distinctness indices. Hydrobiologia 580, 43–56. doi: 10.1007/s10750–006–0466–8

6. Christia, C., Giordani, G., Papastergiadou, E., 2014. Assessment of ecological quality of coastal lagoons with a combination of phytobenthic and water quality indices. Marine Pollution Bulletin 86, 411–423. doi:10.1016/j.marpolbul.2014.06.038

7. Clarke, K., Gorley, R., 2015. Primer v7: User Manual/Tutorial, in: Plymouth, Uk. pp. 1–296. doi: 10.1111/j.1442–9993.1993.tb00438.x

8. Clarke, K.R., Gorley, R.N., 2006. PRIMER v6: User Manual/Tutorial. PRIMER-E, Plymouth UK 192 p. doi: 10.1111/j.1442–9993.1993.tb00438.x

9. Clesceri, L.S., Greenberg, A.E., Eaton, A.D., 2012. Standard Methods for the Examination of Water and Wastewater, in: Standard Methods for the Examination of Water and Wastewater. p. 733.

10. Copetti, D., Finsterle, K., Marziali, L., Stefani, F., Tartari, G., Douglas, G., Reitzel, K., Spears, B.M., Winfield, I.J., Crosa, G., D'Haese, P., Yasseri, S., Lürling, M., 2015. Eutrophication management in surface waters using lanthanum modified bentonite: A review. Water Research 97, 162–174. doi: 10.1016/j.watres.2015.11.056

11. Graeber, D., Jensen, T.M., Rasmussen, J.J., Riis, T., Wiberg-Larsen, P., Baattrup-Pedersen, A., 2017. Multiple stress response of lowland stream benthic macroinvertebrates depends on habitat type. Science of the Total Environment 599–600, 1517–1523. doi: 10.1016/j.scitotenv.2017.05.102

12. Ignatiades, L., Gotsis-Skretas, O., Pagou, K., Krasakopoulou, E., 2009. Diversification of phytoplankton community structure and related pa-

rameters along a large-scale longitudinal east-west transect of the Mediterranean Sea. Journal of Plankton Research 31, 411–428. doi: 10.1093/plankt/fbn124

13. Johnson, S.L., Ringler, N.H., 2014. The response of fish and macroinvertebrate assemblages to multiple stressors: A comparative analysis of aquatic communities in a perturbed watershed (Onondaga Lake, NY). Ecological Indicators 41, 198–208. doi: 10.1016/j.ecolind.2014.02.006

14. Kuzmanovic, M., Dolédec, S., de Castro-Catala, N., Ginebreda, A., Sabater, S., Muñoz, I., Barceló, D., 2017. Environmental stressors as a driver of the trait composition of benthic macroinvertebrate assemblages in polluted Iberian rivers. Environmental Research 156, 485–493. doi: 10.1016/j.envres.2017.03.054

15. NME, 2015. Supreme Decree N° 015–2015-NME – National Environmental Quality Standards for Water. Official Newspaper El Peruano 3.

16. NME, 2011. National Environmental Action Plan 2011–2021. NME 2021.

17. Muangthong, S., Shrestha, S., 2015. Assessment of surface water quality using multivariate statistical techniques: case study of the Nampong River and Songkhram River, Thailand. Environmental Monitoring and Assessment 187. doi: 10.1007/s10661-015-4774-1

18. Petus, C., Marieu, V., Novoa, S., Chust, G., Bruneau, N., Froidefond, J.M., 2014. Monitoring spatio-temporal variability of the Adour River turbid plume (Bay of Biscay, France) with MODIS 250-m imagery. Continental Shelf Research 74, 35–49.

doi: 10.1016/j.csr.2013.11.011

19. Primo, A.L., Marques, S.C., Falcão, J., Crespo, D., Pardal, M.A., Azeiteiro, U.M., 2012. Environmental forcing on jellyfish communities in a small temperate estuary. Marine Environmental Research 79, 152–159. doi: 10.1016/j.marenvres.2012.06.009

20. Rim, A., Charff, A., Ayed, L., Khadhar, S., 2017. Effect of Water Quality on Heavy Metal Redistribution-Mobility in Agricultural Polluted Soils in Semi-Arid Region. Pedosphere 160. doi: 10.1016/S1002-0160(17)60367-9

21. Ruaro, R., 2014. A scientometric assessment of 30 years of the Index of Biotic Integrity in aquatic ecosystems : Applications and main flaws. Ecological Indicators 29, 105–110. doi: 10.1016/j.ecolind.2012.12.016

22. Seiler, L.M.N., Helena, E., Fernandes, L., Martins, F., Cesar, P., 2015. Evaluation of hydrologic influence on water quality variation in a coastal lagoon through numerical modeling. Ecological Modelling 314, 44–61. doi: 10.1016/j.ecolmodel.2015.07.021

23. Swiech, T., Ertsen, M.W., Pererya, C.M., 2012. Estimating the impacts of a reservoir for improved water use in irrigation in the Yarabamba region, Peru. Physics and Chemistry of the Earth 47–48, 64–75. doi: 10.1016/j.pce.2011.06.008

24. Wen, S., Shan, B., Zhang, H., 2012. Metals in sediment/pore water in Chaohu Lake: Distribution, trends and flux. Journal of Environmental Sciences (China) 24, 2041–2050. doi: 10.1016/S1001-0742(11)61065-6

25. WHO, 2011. WHO | Guidelines for drinking-water quality, fourth edition [WWW Document]. WHO.

PERMISSIONS

LIST OF CONTRIBUTORS

Łukasz Malinowski and Iwona Skoczko
Bialystok University of Technology, Faculty of Civil and Environmental Engineering, Department of Technology and Environmental Engineering Systems, Wiejska 45A, 15-351 Białystok, Poland

Lilianna Bartoszek, Piotr Koszelnik and Renata Gruca-Rokosz
Department of Environmental and Chemistry Engineering, Rzeszów University of Technology, al. Powstańców Warszawy 12, 35-959 Rzeszów, Poland

Justyna Zamorska and Monika Zdeb
Department of Water Purification and Protection, Rzeszów University of Technology, al. Powstańców Warszawy 12, 35-959 Rzeszów, Poland

Witold Niemiec
Department of Water Purification and Protection, Rzeszow University of Technology, Al. Powstańców Warszawy 6, 35-959 Rzeszów, Poland

Tomasz Trzepieciński
Department of Materials Forming and Processing, Rzeszow University of Technology, Al. Powstańców Warszawy 8, 35-959 Rzeszów, Poland

Magda Dudek, Paulina Rusanowska, Marcin Zieliński and Marcin Dębowski
University of Warmia and Mazury in Olsztyn, Department of Environmental Engineering, ul. Warszawska 117a, 10-720 Olsztyn, Poland

Anna Baryła, Agnieszka Karczmarczyk and Agnieszka Bus
Warsaw University of Life Sciences – SGGW, Faculty of Civil and Environmental Engineering, Department of Environmental Improvement, Nowoursynowska 166, 02-787 Warszawa, Poland

Beata Fortuna-Antoszkiewicz, Jan Łukaszkiewicz and Edyta Rosłon-Szeryńska
Department of Landscape Architecture, Warsaw University of Life Sciences – SGGW, Nowoursynowska 166 St., 02-787 Warsaw, Poland

Czesław Wysocki
Department of Environment Protection, Warsaw University of Life Sciences – SGGW, Nowoursynowska 166 St., 02-787 Warsaw, Poland

Piotr Wiśniewski
Department of Environmental Protection of Mokotów District, City of Warsaw, Poland

Piotr Zawadzki and Edyta Kudlek
Silesian University of Technology, Faculty of Energy and Environmental Engineering, Institute of Water and Wastewater Engineering, Konarskiego 22B, 44-100 Gliwice, Poland

Mariusz Dudziak
Silesian University of Technology, Faculty of Energy and Environmental Engineering, Institute of Water and Wastewater Engineering, Division of Water Supply and Sewage Systems, Konarskiego 18, 44-100 Gliwice, Poland

Joanna Kostecka, Mariola Garczyńska, Agnieszka Podolak and Grzegorz Pączka
Department of Natural Theories of Agriculture and Environmental Education, Faculty of Biology and Agriculture, University of Rzeszów, 35-601 Rzeszów, Ćwiklińskiej 1A Str., Poland

Janina Kaniuczak
Department of Soil Science, Environmental Chemistry and Hydrology, Faculty of Biology and Agriculture, University of Rzeszów, 35-601 Rzeszów, Ćwiklińskiej 1A Str., Poland

Tobiasz Gabryś and Beata Fryczkowska
Institute of Textile Engineering and Polymer Materials, University of Bielsko-Biala, ul. Willowa 2, 43-309 Bielsko-Biala, Poland

Anna Kwarciak-Kozłowska
Institute of Environmental Engineering, Czestochowa University of Technology, Czestochowa, Poland

Henryk Grzywna, Paweł B. Dąbek and Beata Olszewska
Wrocław University of Environmental and Life Sciences, Institute of Environmental Protection and Development, pl. Grunwaldzki 24, 50-363 Wrocław, Poland

Mateusz Rybak
Department of Agroecology, Faculty of Biology and Agriculture, University of Rzeszów, Ćwiklińskiej 1A, 35–601 Rzeszów, Poland

Teresa Noga
Department of Soil Studies, Environmental Chemistry and Hydrology, Faculty of Biology and Agriculture, University of Rzeszów, Zelwerowicza 8B, 35–601 Rzeszów, Poland

Robert Zubel
Department of Botany and Mycology, Maria Curie-Skłodowska University, ul. Akademicka 19, 20-033 Lublin, Poland

Martin Juriga, Vladimír Šimanský and Nora Polláková
Department of Soil Science, Faculty of Agrobiology and Food Resources, Slovak University of Agriculture, Tr. A. Hlinku 2, 949 76 Nitra, Slovakia

Ján Horák, Elena Kondrlová and Dušan Igaz
Department of Biometeorology and Hydrology, Horticulture and Landscape Engineering Faculty, Slovak University of Agriculture, Hospodárska 7, 949 01 Nitra, Slovakia

Natalya Buchkina and Eugene Balashov
Agrophysical Research Institute, 14 Grazhdansky prospekt, St. Petersburg, 195220, Russia

Milena Rusin and Janina Gospodarek
Department of Agricultural Environment Protection, University of Agriculture, al. Mickiewicza 21, 31-120 Krakow, Poland

Tomasz Jóźwiak, Urszula Filipkowska, Paula Bugajska and Tomasz Kalkowski
Department of Environmental Engineering, Faculty of Environmental Sciences, University of Warmia and Mazury in Olsztyn, ul. Warszawska 117, 10–720 Olsztyn, Poland

Jacek Antonkiewicz
Department of Agricultural and Environmental Chemistry, University of Agriculture in Krakow, Poland

Barbara Kołodziej
Department of Industrial and Medicinal Plants, University of Life Sciences in Lublin, Poland

Elżbieta Jolanta Bielińska
Institute of Soil Science, Environment Engineering and Management, University of Life Sciences in Lublin, Lublin, Poland

Katarzyna Gleń-Karolczyk
Department of Agricultural Environment Protection, University of Agriculture in Krakow, Poland

Gusti Irya Ichriani
Postgraduate Programme, Faculty of Agriculture, Brawijaya University, Jl. Veteran, Malang 65145, Indonesia
Faculty of Agriculture, University of Palangka Raya, Jl. Yos Sudarso, Kota Palangka Raya 74874, Central Kalimantan, Indonesia

Syehfani and Yulia Nuraini
Department of Soil Science, Faculty of Agriculture, Brawijaya University, Jl. Veteran, Malang 65145, Indonesia

Eko Handayanto
Research Centre for Management of Degraded and Mining Lands, Brawijaya University, Jl. Veteran, Malang 65145, Indonesia

Lidiia Davybida, Dmytro Kasiyanchuk, Liudmyla Shtohryn, Eduard Kuzmenko and Mariia Tymkiv
Department of Geotechnogenic Safety and Geoinformatics, Ivano-Frankivsk National Technical University of Oil and Gas, 15 Karpatska Street, Ivano-Frankivsk, 76019, Ukraine

Rahayu Siwi Dwi Astuti
Master Program of Environmental Studies, School of Postgraduate Studies Diponegoro University, Semarang, Indonesia

Hady Hadiyanto
Master Program of Environmental Studies, School of Postgraduate Studies Diponegoro University, Semarang, Indonesia
Chemical Engineering Departement, Faculty of Engineering Diponegoro University, Semarang, Indonesia

Alicja Machnicka
Institute of Environmental Protection and Engineering, University of Bielsko-Biała, ul. Willowa 2, 43-309 Bielsko-Biała, Poland

Dariusz Andraka, Jacek Dawidowicz and Wojciech Kruszyński
Białystok University of Technology; Faculty of Civil and Environmental Engineering; Wiejska 45E, 15-351 Bialystok; Poland

Iwona Kinga Piszczatowska
Wodociągi Białostockie (Bialystok Water Supply) Sp. z o.o., Młynowa 52/1, 15-590 Białystok, Poland

Izabella Kłodowska, Joanna Rodziewicz and Wojciech Janczukowicz
University of Warmia and Mazury in Olsztyn, Faculty of Environmental Sciences, Department of Environment Engineering, Warszawska 117a, 10-719 Olsztyn, Poland

Niewęgłowski Marek and Gugała Marek
Siedlce University of Natural Sciences and Humanities, Faculty of Natural Science, ul. Prusa 14, 08-110 Siedlce, Poland

Włodarczyk Bogusław
Mazovian Agricultural Advisory Centre, ul. Czereśniowa 98, 02-456 Warszawa, Poland

Anna Sikorska
Department of Agriculture, The State Higher School of Vocational Education in Ciechanów, Narutowicza 9, 06-400 Ciechanów, Poland

Tomasz Janusz Teleszewski and Mirosław Żukowski
Department of HVAC Engineering, Faculty of Civil and Environmental Engineering, Bialystok University of Technology, Wiejska 45E, 15-351 Bialystok, Poland

María Custodio, Fernán Chanamé, Raúl Yaranga and Rafael Pantoja
Universidad Nacional del Centro del Perú, Facultad de Zootecnia, Instituto de Investigación en Alta Montaña, Av. Mariscal Castilla No. 3989-4089, Huancayo, Perú

Richard Peñaloza
Universidad Nacional Agraria La Molina, Av. La Molina s/n La Molina, Lima, Casilla Lima 12, Perú

Index